NOW IT CAN BE TOLD

THE STORY OF THE MANHATTAN PROJECT

BY

LESLIE R. GROVES

NEW INTRODUCTION BY
EDWARD TELLER

DA CAPO PRESS

Library of Congress Cataloging in Publication Data

Groves, Leslie R., 1896–1970.
    Now it can be told.

    (A Da Capo paperback)
    Reprint. Originally published: New York: Harper, 1962.
    1. United States. Army. Corps of Engineers. Manhattan District—History. 2.
Atomic bomb—United States—History. I. Title. II. Title: Manhattan project.
QC773.A1G7 1983   623.4'5119'0973   82-23649
ISBN 0-306-80189-2

This Da Capo Press paperback edition of *Now It Can Be Told: The Story of
the Manhattan Project* is an unabridged republication of the first edition
published in New York in 1962, here supplemented with a new introduction by
Edward Teller. It is reprinted by arrangement with the author's estate.

Published by Da Capo Press, Inc.
A Member of the Perseus Books Group

Manufactured in the United States of America

INTRODUCTION

Recently, several television dramas have examined the personalities and events of the atomic bomb project in Los Alamos, and interest in this watershed project has grown markedly. The recent seven-hour BBC production is perhaps the best of the offerings, but even there, historical inaccuracies and skewed perspectives are abundantly present. The readers of General Groves's own account are to be complimented for choosing to learn directly from one of the major participants. History in some ways resembles the relativity principle in science. What is observed depends on the observer. Only when the perspective of the observer is known can proper corrections be made.

General Groves was, perhaps above all else, a most straightforward man. He said what he thought with little consideration for the effect his opinions might have on others. Deviousness was totally foreign to his nature. Vannevar Bush, the head of all scientific wartime projects, interviewed General Groves prior to his appointment to the Los Alamos project. Bush suggested to the office of the Secretary of State that Groves might lack sufficient tact for such a sensitive role. It is typical of Groves that he has reported this opinion in his book.

The general was a direct man of practical action. His strengths included not only enormous dedication and a great capacity for work, but also an unconcerned approach to complicated problems. He very often managed to ignore complexity and arrive at a result which, if not ideal, at least worked. Groves was also incredibly versatile, a quality required for such an unprecedented job as organizing the enterprise that led to the first atomic bomb. He had to worry both about the diffusion of uranium hexafluoride molecules and about the problems faced by the wives in Los Alamos. (As Groves mentions, contrary to local gossip, Los Alamos was not an establishment for the care of pregnant WACs). He organized the intelligence effort to determine the Nazi prospects of building an atomic weapon of their own, and he influenced, perhaps not in the right way, the decision to drop the bomb on Hiroshima.

For Groves, the Manhattan Project seemed a minor assignment, less significant than the construction of the Pentagon. He was deeply disappointed at being given the job of supervising the development of an atomic weapon since it deprived him of combat duty. He started with, and partially retained, thorough doubts about the feasibility of the project. Yet in convincing the leaders at DuPont that they should participate, he appeared totally confident in order to overcome the incredulity of those overly sane chemical engineers. I would ascribe his behavior in this case less to a lack of openness than to his unwavering sense of duty as an Army officer.

Groves, as history records, was eminently successful as the military director of the Los Alamos Lab. Perhaps his most outstanding decision was the choice of J. Robert Oppenheimer to administer the scientific work at the laboratory. In the fantastic story of the atomic bomb, this less obvious, less remembered but no less vital story of the highly effective cooperation between two individuals, poles apart in every way, remains unique.

Oppie clearly saw the importance of the project and envisioned the new age that would arise from the old. While Oppie resembled General Groves in the intensity of his effort and dedication, his gentle demeanor, social grace — even his physique — showed a marked contrast to the General's. But the greatest difference between the two men lay in the complexity of their personalities. Oppenheimer was so extraordinarily complicated and clever that he could mask these qualities with an appearance of simplicity. Groves's astuteness is most clearly demonstrated in that, despite opposition from the military security division whose value and function he supported and even overestimated, he selected Oppenheimer as scientific director. The contrast and cooperation of Groves and Oppenheimer is the most striking human story in the "Manhattan Project."

Much of my life has been spent in laboratories of similar size and complexity to the Los Alamos Laboratory. I have known many directors intimately. For a short time, I was even a director myself. I know of no one whose work begins to compare in excellence with that of

Oppenheimer's.

Oppie knew in detail the research going on in every part of the laboratory, and was as excellent at analyzing human problems as the countless technical ones. Of the more than 10,000 people who eventually came to work at Los Alamos, Oppie knew several hundred intimately, by which I mean that he understood their relationships with one another, and what made them tick. He knew how to lead without seeming to do so. His charismatic dedication had a profound effect on the successful and rapid completion of the atomic bomb.

Some of Oppenheimer's qualities come through in the recent BBC production. However, General Groves, in this television drama, is rather inadequately presented. (Even his girth was underestimated.) Obviously no one with so little intelligence, as the General Groves presented by the BBC, could have met the massive responsibilities of providing shelter, equipment, and materials with so little delay and impediment to the project. I must confess, however, that between 1943 and 1945 General Groves could have won almost any *un*popularity contest in which the scientific community at Los Alamos voted.

I remember a meeting early in 1943 at which Oppie announced a revision in security clearance procedures, made necessary by the fact that so many of us could not be cleared under the usual regulations. The new rules called for us to clear one another by vouching for our good intentions and backgrounds. Someone piped up, "Does that mean that clearance will be based on our *scientific* good names?" Oppie responded, "Grove doesn't believe we have any other." We laughed. General Groves, or "His Nibs" as we called him, could hardly have been completely unaware of Oppie's and our attitudes. He chose to disregard them.

No one should be surprised that a group of independent scientists found General Groves and his regulations irritating. Secrecy runs contrary to the deepest inclinations of every scientist; we were willing to make sacrifices, even when they seemed ludicrous, because of the war. But military regulations affected every detail of our lives and, worse, they were worded as if we lacked the common sense of a five-year old. I recall a directive issued in early 1945 during the only serious water shortage at Los Alamos. The order did not carry General Groves's signature, but we all attributed it to him. It detailed the ways in which the shortage would be met and concluded with the memorable sentence:

"Residents will not use showers except in case of emergency."

Adopting the attitude that General Groves, rather than we ourselves, lacked native intelligence helped to decrease our frustration. If we failed to appreciate him, his recognition of us was similarly lacking. Although Groves made the following statement almost a decade after the Los Alamos years, he conveyed these feelings fully during the time that we were together.

> We (the military leaders at Los Alamos) came up through kindergarten with them (the scientists). While they could put elaborate equations on the board, which we might not be able to follow in their entirety, when it came to what was so and what was probably so, we knew just about as much as they did. So when I say that *we were responsible for the scientific decisions,* I am not saying that we were extremely able nuclear physicists, because actually we were not. We were what might be termed 'thoroughly practical nuclear physicists.'[1]

Earlier in the same discussion, Groves referred to the three possible states of plutonium as "solid, gas or electric." Clearly we spoke different languages. Yet, in reading his book, I was struck by his comment that, since the weather forecasters at the Trinity test had failed to predict the rain that was falling, "I soon excused them. After that it was necessary for me to make my own weather predictions—a field in which I had nothing more than very general knowledge." As in other complicated issues, General Groves made the right decision.

My contact with General Groves during the Los Alamos period was very limited. I remember only one incident with clarity. One of my jobs at Los Alamos was to assure the safety conditions in the gas diffusion plant. The main hazard was that in advanced stages of separating U_{235} and U_{238}, contamination with water or some other substance might cause the diffusing gases to solidify, at which point an unwanted chain reaction might result. This part of my job took me from time to time to New York, and one morning (at 4:00 a.m. Los Alamos time) I woke to hear the General's voice at the other end of my telephone, instructing me to go to his Washington office immediately.

The emergency, I discovered, was a chemical explosion at a gas diffusion plant on the East Coast; Groves wanted to question me about

---

[1] *In the Matter of J. Robert Oppenheimer.* Transcript of Hearing before Personnel Security Board, U.S. Government Printing Office (Washington, D.C., 1954), p. 164 (emphasis added).

the possibility of serious malfunction in our separation process. After a preliminary discussion, Groves assembled a group of his staff at a long table. I sat on his right and was kept wide awake by a barrage of hypothetical questions while the General slouched, with eyes closed, seemingly half asleep. Periodically, he would open both eyes, look me square in the face and state, "But after all, Professor, this is only theory."

Toward the beginning of the third hour of this inquisition, a colonel at the end of the table asked if it were not possible that all the U_{235} atoms might assemble at one end of the apparatus by pure chance, and thereby cause a nuclear explosion. "Of course," I answered, "this is a possibility, but it is as probable as that all the air molecules in the room will assemble under the table, causing us all to suffocate."

Groves immediately sat up and said, "But Doctor, you did say this is possible." Conant intervened with, "What Dr. Teller intends to say is that such an assembly is really quite impossible." From this moment on, General Groves treated me with exquisite politeness. Apparently, I had passed his test as to whether or not I could be trusted.

Neither through contact nor through rumor did I ever learn of Grove's sense of humor. Yet in reading his book, I discovered not only that he was quite sufficiently endowed with one but that he could laugh at himself. He tells the story of a search on a wall-sized map for a small town in Germany, which was finally located two feet above the floor. Secretary Stimson, his aide, a Chief of Staff, and Groves then all got down almost on their hands and knees, "gazing at this point barely off the floor," at which time, continues his account, a photographer "might well have caught one of World War II's more interesting photographs." Groves would surely have stood out in the picture, a fact the reader might miss, but which Groves in telling the story clearly appreciated.

A measure of his resourcefulness can be illustrated by a story involving Ernest Lawrence. About 1943, General Groves, visiting the Berkeley Radiation Laboratory which was separating U_{235} by electromagnetic means, attempted to spur Lawrence on by saying to him, "Your reputation is at stake here." Later over a nice rum drink, Lawrence said to him, "You know, General, my reputation has been made, but yours is at stake here." Groves did not respond. However, a couple of years later, Groves in addressing a group at Los Alamos commented: "When all of this is over, you will go back to your universities, regardless of the outcome, but my reputation is at stake here."

I did not begin seriously to examine the validity of my wartime impression of Groves until four years later, and then only because of an unusual conversation. In September, 1949, I was in England discussing possible safety measures for the introduction of commercial nuclear-generated electricity with our British counterparts. Toward the end of my visit, Sir James Chadwick, who had headed the wartime British scientific delegation to Los Alamos, invited me to dinner at his home in Caius College. Sir James was well-known in the scientific community for his taciturn nature, but his wife was a charming conversationalist. She drew me out about our mutual friends and acquaintances from Los Alamos, and eventually inquired about General Groves. My response, I am afraid, reflected an unflattering opinion of him.

At that point, a miracle occurred. Sir James, who had spoken perhaps twenty words that evening, became talkative to the point of being almost uninterruptible. He told me most emphatically and repeatedly that the atomic bomb project would never have succeeded without General Groves. I pointed out how often Groves had made plain his dislike of the British. Sir James brushed aside my comment. That made no difference. What was important, Sir James went on, was that Groves understood the overriding importance of the project better than some of the leading American scientists. Without Groves, he said, the scientists could never have built the bomb.

I have rarely seen anyone — even an ordinarily effusive talker — so insistent on making his point. However, Sir James's tirade carried no trace of reproach for my inappropriate remark about General Groves. At the end of the evening, my host walked me back to my inn. On parting, he told me to remember what he had said as I might "have need of it."

Shortly after this evening, I was back in the United States and gained some new information. It then dawned on me that during our conversation Chadwick probably had known what I had just learned: the Soviets had exploded an atomic bomb. Chadwick knew that American scientists, who had less direct an experience with World War II than their British colleagues, many of whose homes and families were in peril, had not realized the urgency and importance of the atomic bomb project. General Groves, on the other hand, having considered military matters throughout his career, knew exactly what it meant to be inadequately defended. In the struggle against totalitarian military might in the ensuing years, the awareness of the overriding importance of defense has rested with men who, like Groves, understand that only strength can counter an adversary determined to enforce his goals by physical force.

General Groves's book deals with a period which precedes the memories of most of the present generation. In the intervening years, the world has changed so much it is hardly recognizable in terms of the past. Today, national security and technology have become inseparable. Yet the gulf between the military establishment and the scientific community is as great as ever. General Groves was one of the pioneers who, with difficulty but ultimate success, managed to throw a bridge across the abyss.

I do not see much hope for the survival of our democratic form of government if we cannot rebuild that bridge made by General Groves and J. Robert Oppenheimer. We must find ways to encourage mutual understanding and significant collaboration between those who defend their nation with their lives and those who can contribute the ideas to make that defense successful. Only by such cooperation can we hope that freedom will survive, that peace will be preserved.

—EDWARD TELLER
Stanford, California
December, 1982

TO THE MEN AND WOMEN OF THE MANHATTAN PROJECT,
AND TO ALL THOSE WHO AIDED THEM IN THEIR
YET UNPARALLELED ACCOMPLISHMENT

CONTENTS

FOREWORD *xiii*

PART I

1 THE BEGINNINGS OF THE MED *3*

2 FIRST STEPS *19*

3 THE URANIUM ORE SUPPLY *33*

4 THE PLUTONIUM PROJECT *38*

5 LOS ALAMOS: I *60*

6 HANFORD: I *68*

7 HANFORD: II *78*

8 OAK RIDGE *94*

9 NEGOTIATIONS WITH THE BRITISH *125*

10 SECURITY ARRANGEMENTS AND PRESS CENSORSHIP *138*

11 LOS ALAMOS: II *149*

12 THE COMBINED DEVELOPMENT TRUST *170*

13 MILITARY INTELLIGENCE: ALSOS I—ITALY *185*

14 A SERIOUS MILITARY PROBLEM *199*

15 MILITARY INTELLIGENCE: ALSOS II—FRANCE *207*

xi

16 THE PROBLEM OF THE FRENCH SCIENTISTS *224*

17 MILITARY INTELLIGENCE: ALSOS III—GERMANY *230*

PART II

18 TRAINING THE AIR UNIT *253*

19 CHOOSING THE TARGET *263*

20 TINIAN *277*

21 ALAMOGORDO *288*

22 OPERATIONAL PLANS *305*

23 HIROSHIMA *315*

24 THE GERMANS HEAR THE NEWS *333*

25 NAGASAKI *341*

PART III

26 THE MED AND CONGRESS *359*

27 THE DESTRUCTION OF THE JAPANESE CYCLOTRONS *367*

28 TRANSITION PERIOD *373*

29 THE AEC *389*

30 POSTWAR DEVELOPMENTS *401*

31 A FINAL WORD *413*

APPENDIXES *417*

INDEX *445*

A section of illustrations will be found following page 242

FOREWORD

Atomic physics* is not an occult science. It is true that those who have devoted their adult lives to its study know far more about this highly specialized field than I or the average layman can ever hope to learn. Yet the same might be said of almost any other group of specialists and, just as it is possible for many laymen to understand the laws of economics that govern our markets and the laws of mechanics by which our automobiles run, so it is possible for them to have a general comprehension of the basic laws of atomic energy.

Man's understanding of nature is usually a cumulative and gradual process. Certainly this has been the case throughout the growth of atomic physics. No single stroke of genius delivered up the finished product. Rather, its present state of development derives from the labors of many individuals from many countries, operating in many fields of endeavor, over a span of many years.

About the great milestones in this evolutionary process which led eventually to the Manhattan Project and to its end product, the atomic bomb, much has already been written, and to that story there is nothing I can add. I have recorded here only that which I am qualified to write about—my own experiences during the development of atomic energy between September 17, 1942, and December 31, 1946, the period during which I was in charge of the Manhattan Project. Insofar as practicable, my account deals with those matters with which I was

---

* Throughout this book, the terms "atomic energy" and "atomic physics" have usually been used instead of the currently accepted "nuclear energy" and "nuclear physics." The reason for so doing is that during the period with which this narrative deals the word "atomic" was the one which was generally employed, as being more understandable in all areas outside the highly scientific.

personally concerned. Matters outside my direct cognizance are discussed only to the extent that some knowledge of them is essential to the reader's understanding of our work and of the problems we faced.

Until quite recently such an undertaking would have been impossible, simply because the best interests of the United States did not allow a broad enough disclosure of the facts to permit an adequate discussion of our activities. However, the tremendous advances that American technology has made since the days of the Manhattan Project have served to remove the limits upon the dissemination of information about what is now becoming clearly a part of the past. Gradually more and more of the details of our work have been declassified and, with the issuance of an executive order in May, 1959, the curtain was drawn aside on the story of the project. Even though some of its details must continue to remain secret, enough can now be told to give the reader a good understanding of the project as a whole, and of the way in which it was conducted.

In writing this story I have tried first of all to fill in as many as possible of the gaps existing in the American public's understanding of the project. Altogether too many of these gaps have given rise to misinformed conjecture, and as a result many Americans tend to feel embarrassed or discomfited by their country's greatest single scientific success.

Secondly, I want to emphasize the cohesive entity that was the Manhattan Project, a factor in its success that has been largely overlooked.

Finally, I want to record the lessons that I learned while in charge of the project. Throughout that time we had few precedents to follow: the work and the problems that I and my associates faced were unique. We learned a great deal from our successes and from our failures. It is my hope that this knowledge, much of it gained the hard way, will be of value to others who must venture into uncharted fields, whether as agents of the government or as members of private organizations. While ours was the first large organization of its kind, it surely will not be the last. For this reason alone, the story of the Manhattan Project is worth telling.

Some of those who have already written about the development of atomic energy during the war years participated in the Manhattan Project in various capacities, but while in many cases their work was vitally important, their points of view were inevitably somewhat limited and their accounts of our activities focused on only one phase or part of the whole. Other writers, who had no direct connection with the project, have attempted to deal with it on a broader scale. Interesting and informative as these accounts have often been, they have suffered from their authors' lack of access to many important facts. Since my responsibilities in the project were personal, broad and all-encompassing, my point of view is quite different, in many respects, from those of the others who have written on this subject. To the same extent my narrative differs from their accounts.

The command channels of the Manhattan Engineer District (MED) —the name given to the atomic bomb project—had no precedent. They grew up with the project and were changed as conditions changed. Yet the basic concept—that of always keeping authority and responsibility together—never changed.

Although there have been numerous stories about how obscure and devious our channels were, nobody who was directly involved ever had any doubt about what he was supposed to do. We made certain that each member of the project thoroughly understood his part in our total effort—that, and nothing more. Even the Joint Chiefs of Staff, as an organization, were not involved in the approval of our plans or told of their purpose. The four individual Chiefs were kept informed only insofar as their specific duties required.

Dr. Vannevar Bush, Chairman of the Office of Scientific Research and Development (OSRD), and Dr. James B. Conant, Chairman of the National Defense Research Committee (NDRC), a subdivision of OSRD, were primarily responsible for President Roosevelt's decision to transform the atomic energy development program from a research project into a program aimed at producing a decisive military weapon. Once the military purpose of the project became governing, Lieutenant General Brehon Somervell, Chief of the Army Services of Supply, and Major General W. D. Styer, his Chief of Staff, entered the picture. Within a few months they brought me in to head up the project, subject, of course, to the personal approval of General Marshall, Secretary of War Stimson, and, finally, the President. After I was brought in Bush, Conant, Styer and Rear Admiral W. R. E. Purnell were made responsible for overseeing my performance, serving to provide constant reassurance to Secretary Stimson and to the President that the project was being properly run.

At first, I was responsible only for the engineering, construction and operation of the plants to produce bomb materials. Had our work been routine and clearly defined, my responsibility probably would have ended there. However, it soon became evident to Dr. Bush and to me that if serious delays to our work were to be avoided the MED should expand its research activities, and take over control of all the atomic research projects then under the management of the OSRD, thus uniting authority with responsibility. This transfer was effected without friction during the fall and winter of 1942 by the simple device of allowing the OSRD contracts to continue in force until they expired, at which time they were replaced by new ones in which the MED was the contracting agency. The transition was so smooth, indeed, that, as I have read accounts of that period by some of the people involved, I have been struck by the fact that they did not seem to be aware of just when the transfer of authority actually took place.

Gradually, I had to take over other unforeseen responsibilities, such as security and counterintelligence. I also became responsible for military intelligence on atomic developments throughout the world, as well as for insuring that the postwar position of the United States in the field of atomic energy would not be unfavorable.

The fact that I could not operate without becoming deeply involved in future planning soon projected me into matters of extremely high-level policy, including international relations. And since my routine duties required me to have an intimate knowledge of the details of our work, which others less closely associated with the project could not hope to have, I gradually came to be more and more responsible for the initial formulation of general policy and for the translation of policy into action.

Thus I became responsible, particularly to General Marshall, Secretary Stimson and President Truman, for the over-all success of the use of the bomb against Japan. This assignment included selecting the target cities, subject to the approval of the Chief of Staff and the Secretary of War; preparing the orders and instructions for the bombing operations; and arranging for Army and Navy units to provide the necessary support to our overseas effort. By the time these supporting echelons were brought into the picture, the situation was too complicated and changing too quickly to permit the decentralization of authority that we habitually sought; authority and responsibility for this operation were retained in Washington.

I repeat, there was never the slightest doubt in the mind of anyone in the command channels about his own responsibilities, or about to whom he should look for direction or assistance. There was never any breakdown in these channels. That some historians do not seem to have completely grasped our command system is unquestionably due to the fact that they were unfamiliar with our methods of operation. Unfortunately, in any secret operation, it is impossible to give full information to everyone who feels that he is entitled to it, and inevitably some hurt feelings result. Our case was no exception.

Despite the passage of nearly two decades, it is still too early to write a completely objective story of the development of the first atomic bombs. I have tried, but only the passage of a much longer period of time than is allotted to me will permit a final judgment of some of the more controversial features of the project. Yet time also operates to remove from the scene those who have firsthand knowledge of the facts and who hold firsthand opinions. Gradually, the bases of their actions and their opinions will fall more and more into the realm of historians' conjectures. So I am recording here a number of facts that I feel should be known, together with some of my opinions and my reasons for holding them. I do this in order that there can be no doubt about the ways in which I tried to carry out my responsibilities for the conduct of the project.

I have covered mainly those things which most required my personal attention. Naturally, these matters tended toward the extreme. The bulk of the project moved ahead by dint of the hard work and the feeling of urgency of everyone concerned and without requiring any personal supervision on my part.

Consideration for the reader and the limitations of space do not permit me to mention by name a great many of the people and organizations who contributed importantly to the success of the project. However, I want to emphasize as strongly as I can my gratitude to and admiration for the tens of thousands of devoted, hard-working men and women whose combined efforts made possible the greatest scientific and technical achievement of all time. The debt our country owes them is not measurable.

PART I

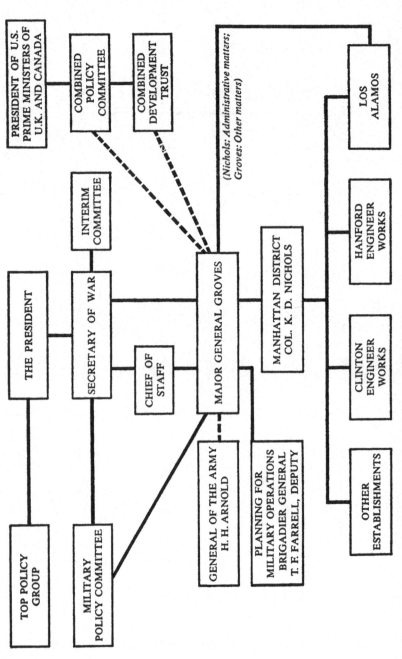

Organization of Atomic Project, May, 1945 (Simplified chart)

CHAPTER 1

THE BEGINNINGS OF THE MED

One day in mid-September, 1942, about a month and a half before the invasion of North Africa, I was offered an extremely attractive assignment overseas. At that time I had been on duty in Washington for over two years as Deputy Chief of Construction of the Army Corps of Engineers. Under the general supervision of my immediate superior, Major General T. M. Robins, I was in charge of all Army construction within the United States as well as in our off-shore bases. This included the building of camps, airfields, ordnance and chemical manufacturing plants, depots, port facilities and the like. But though the responsibility was great and the work essential, I was, like every other regular officer, extremely eager for service abroad as a commander of combat troops; and I now replied with deep pleasure that any duty in an active theater of operations appealed to me. However, I added, I would have to secure the consent of Lieutenant General Brehon Somervell, Commanding General of the Army Services of Supply, before I could definitely say yes. I promised to see him and to give my answer by noon the next day.

The following morning, a few minutes after I had finished testifying before a Congressional committee on a military housing bill, I met Somervell outside the hearing room, and asked him whether he had any objection to my being relieved from my Army construction duties. To my great surprise, he told me that I could not leave Washington. He went on to say: "The Secretary of War has selected you for a very important assignment, and the President has approved the selection."

"Where?"

"Washington."

"I don't want to stay in Washington."

"If you do the job right," Somervell said, "it will win the war."

My spirits fell as I realized what he had in mind. "Oh, that thing," I said. Somervell went on, "You can do it if it can be done. See Styer[1] and he will give you the details."

My initial reaction was one of extreme disappointment. I did not know the details of America's atomic development program at that time, but, because of the nature of my responsibilities, as I shall explain, I knew of its existence and its general purpose— through the use of uranium to produce an atomic bomb which it was hoped might be of unprecedented power. Though a big project, it was not expected to involve as much as $100 million altogether. While this was more than the cost of almost any single job under my jurisdiction, it was much less than our total over-all spending in a normal week.[2] Magnitude aside, what little I knew of the project had not particularly impressed me, and if I had known the complete picture I would have been still less impressed.

Later that morning, I saw Styer at his office in the Pentagon. He confirmed my worst premonitions by telling me that I was to be placed in charge of the Army's part of the atomic effort. He outlined my mission, painting a very rosy picture for me: "The basic research and development are done. You just have to take the rough designs, put them into final shape, build some plants and organize an operating force and your job will be finished and the war will be over." Naturally I was skeptical, but it took me several weeks to realize just how overoptimistic an outlook he had presented.

In the course of our discussion, we agreed that, because the Pentagon was so nearly finished and because I had had so much to do with it, I would continue to control its construction, despite my new assignment. There were two reasons for this. First, my sudden disappearance from the work on the Pentagon would attract much more

---

[1] Major General W. D. Styer, Chief of Staff, Army Services of Supply.

[2] The Army Corps of Engineers, with almost one million people engaged, was then completing about $600 million worth of work each month.

notice than would my absence from my other Army construction activities. Second, because of the natural interest in the Pentagon displayed by a number of Congressmen, it would be better for me to continue to carry the responsibility for that job than to pass it on to someone else who was unfamiliar with its past problems and their many political ramifications. To avoid complications I would arrange this informally.

We also agreed upon the wording of the order, to be signed by Somervell, directing me to take complete charge of the atomic project.[3]

Before I left, Styer told me that General Marshall had directed that I be made a brigadier general, and that the list of new promotions would be out in a few days. I decided at once, and Styer agreed, that I should not take over the project officially until I could do so as a brigadier. I thought that there might be some problems in dealing with the many academic scientists involved in the project, and I felt that my position would be stronger if they thought of me from the first as a general instead of as a promoted colonel. My later experiences convinced me that this was a wise move; strangely enough, it often seemed to me that the prerogatives of rank were more important in the academic world than they are among soldiers.

At the time I was brought into the picture, research on the uses of atomic energy had been going on at a gradually accelerating pace since January, 1939, when Lise Meitner explained that the uranium atom could be split. The discovery opened up two divergent paths for its exploitation. Most physical scientists realized that nuclear energy, derived from the splitting, or fissioning, of the atom, might be used either to generate power for peaceful purposes or to create super-weapons. In general, however, it was the scientists who were personally acquainted with Hitler's New Order who first became most interested in the possible military uses of atomic energy and its effect on the existing balance of political power. Most American-born physicists were not originally so concerned with this aspect, because they

[3] See Appendix I, page 417.

had not yet become accustomed to thinking of new scientific truths in terms of their military applications. And from the beginning, of course, some of the physicists, both European and American, who considered the potential of atomic weapons were horrified at the thought of the devastation they might spread throughout the world.

With the growth of Hitlerism, many scientists gradually came to doubt the wisdom of communicating the results of their work to scientists in the enemy's camp. Again, this was particularly true of those who had come to America to escape Nazi persecution. They had great respect for the capabilities of German scientists and were familiar with their efforts in the field of atomic energy. They were also aware of the pressures that certainly would be brought to bear on the German scientists to ensure their utmost support of their country's military program. The American-born scientists, in the main, did not have so acute an appreciation of the danger of the situation as did their foreign-born colleagues.

Nevertheless, the Americans and British made a concerted attempt to achieve voluntary international censorship of information relating to atomic energy. The effectiveness of this arrangement was hindered at the start by the refusal of Joliot-Curie to participate; however, it provided a sound foundation for the strict governmental measures that followed the outbreak of war in Europe.

At the same time, the group of refugee scientists in America became a focal point of the attempts to apprise officials in the federal government of the dangers and the prospects that atomic physics held for the United States. Discussions of developments in this field took place between representatives of the Navy Department and Dr. George B. Pegram and Dr. Enrico Fermi, of Columbia, as early as March, 1939. However, Fermi expressed some skepticism at this meeting and the United States Government did not become seriously interested until October of that year when Alexander Sachs, a Wall Street economist and a personal friend and adviser of President Roosevelt, called upon him to obtain his support of the scientific research then under way. Sachs had been following the possibilities of atomic energy for some time and felt that the government should actively aid in the development. To achieve this he discussed the matter with the Columbia group and with Einstein. The latter agreed to sign a suitable letter to the President if Sachs would prepare it. This he did, had it signed and carried it to the White House. This letter emphasized the seriousness of the problem. The President, impressed by Sachs' arguments, appointed an Advisory Committee on Uranium to report to him on the situation.

The Uranium Committee, as it was called, consisted of representatives of the Bureau of Standards, the Army and the Navy. It met from time to time with various physicists and chemists to discuss the development of both atomic power and atomic weapons. On the basis of these discussions, the committee recommended that the Army and Navy make available a modest sum for the purchase of research materials. The work of the committee gained momentum when it was learned, in April of 1940, that the Kaiser Wilhelm Institute in Berlin had undertaken an extensive research program involving uranium.

When the National Defense Research Committee (NDRC) was created in June of 1940, under the chairmanship of Dr. Vannevar Bush, the Uranium Committee was made one of its subcommittees and embarked upon an ambitious research program. The work was carried out through contracts with universities and private and public institutions, using at first funds transferred from the Army and Navy, and later, NDRC-appropriated monies. November, 1941, saw sixteen projects, totaling about $300,000, under way.

During the previous spring and summer, the entire nuclear program had been carefully reviewed from both the scientific and the engineering points of view, with emphasis on the military feasibility of the project. From this review, after noting the optimism of British scientists engaged in the same field, Bush concluded that the United States' efforts to develop military uses of atomic energy should be expanded. He discussed the problem with the President and got his approval to enlarge the program, improve its organization, obtain special funds and interchange information with the British. Mr. Roosevelt created at that time a Top Policy Group, consisting of himself,

Vice President Wallace, Secretary of War Stimson, the Chief of Staff, General Marshall, Dr. Bush and Dr. James B. Conant.

In November of 1941, Bush decided that the uranium project was growing to be of such importance that it should be outside of NDRC control. Accordingly, it was placed directly under the Office of Scientific Research and Development (OSRD), of which NDRC was a part. Bush, by then the head of OSRD, set up at the same time a planning board to initiate the engineering of pilot plants and full-scale production facilities.

Virtually all laboratory research until this time had been aimed at achieving a controlled chain reaction, using U-235, a rare isotope of uranium which comprises less than one percent of the metal in its natural state. This isotope has the property of fissioning readily—a property which the far more abundant form of uranium, U-238, does not display. But it soon became apparent that unless unprecedented quantities of this material could be produced in a much purer state, a U-235 chain reaction would be impossible. The basic problem was to arrive at an industrial process that would produce kilograms of a substance that had never been isolated before in greater than sub-microscopic quantities. The processes then being considered were all designed to take advantage of the very minor physical difference between U-235 and U-238.

The way for a major breakthrough was open as a result of studies that suggested the theoretical feasibility of transmuting U-238 into a highly fissionable new element, plutonium, which might then be separated from the parent uranium by chemical means. The hope was that this would be easier to do than to isolate or concentrate the rare U-235 by physical means. The group headed by Dr. Glenn T. Seaborg at the University of California undertook to prepare extremely small amounts of plutonium, and in March of 1941 succeeded in creating the first submicroscopic amounts of Plutonium-239. Later that month the California group confirmed the theory that under neutron bombardment plutonium atoms fissioned as readily as atoms of U-235.

The desirability of investigating more carefully the military possibilities of plutonium was emphasized by Dr. Ernest O. Lawrence, of the University of California, and the following December, after further studies had been made, the Uranium Committee of the OSRD seconded his proposal. This decision was supported by calculations of the amounts of plutonium required, estimates of its military effectiveness, and roughly drawn timetables of how long it would take to produce it, assuming that the proposed production process was feasible. Later that same month, an intensive program of research on plutonium, supported and sponsored by the OSRD, was begun at the Metallurgical Laboratory[4] at the University of Chicago, under the direction of Nobel Prize winner Dr. Arthur H. Compton, who had achieved eminence by his discovery of the Compton Effect and in pioneer studies of cosmic rays. The purpose of the research was to develop the knowledge needed to design, build and operate a plant for the conversion of uranium into plutonium.[5]

In the meantime, other laboratories in colleges, universities and a few industrial plants were trying to find some method of physically separating U-235 from U-238 that would be practical from the standpoints of both economy and time. The entry of the United States into World War II caused the abandonment of all projects aimed at developing atomic energy as a source of power and gave added impetus to the efforts to build an atomic bomb. At the same time, Bush and a number of others in policy-making positions began to realize that vital as continued laboratory investigations were, even more pressing problems were developing in the fields of engineering and construction. Although they had created a planning board to cope with problems of this nature, it was fast becoming apparent that a much more powerful organization would be required. It is to their everlasting credit that Bush and his colleagues had the discernment to recognize the limitations of their own organization as well as the moral fortitude to admit them in the national interest. Very few men, confronted with a similar situation, would have done so.

[4] Metallurgical Laboratory was a code name chosen to conceal the nature of the work being done there.
[5] See Appendix II, page 418.

Consequently, when the Top Policy Group met on December 16, 1941, Bush recommended that the Army Corps of Engineers carry out the construction work, and asked that a competent Army officer become thoroughly familiar with the project. General Marshall assigned this task to Major General W. D. Styer.

The following March, in reporting favorably to the President on the plutonium project, Bush recommended again that the Corps of Engineers handle the construction that it would entail. Shortly afterward, Dr. Conant reviewed the entire nuclear program. In his report to Bush, he expressed the prevailing opinion that there were five basic production methods, each of which held out equal chances of success. U-235 could be separated by means of the centrifuge, diffusion and electromagnetic processes; while plutonium could be obtained from either the uranium-graphite pile or the uranium–heavy-water pile. All these processes appeared to be nearly ready for pilot plant construction and possibly for the preliminary design of production plants.

On June 17, 1942, Bush forwarded a detailed report on the status of the program to the President. In it he pointed out that it was possible to make a nuclear weapon that could be employed decisively in combat. He went on to discuss means of producing it and ended by saying that, under ideal conditions, it might be made ready in time to influence the outcome of the present war. President Roosevelt approved Bush's report.

The next day, Colonel James C. Marshall was directed by Styer to form a new engineer district to carry out the Army's responsibilities in the development of atomic energy. Marshall was told that he could count on the full support of the War Department and that his needs would be considered paramount to those of any other program.

The coming of autumn in 1942 brought a general reorientation of the project. No longer would it be conducted in the laboratory on a purely theoretical basis, for our scientists had now accumulated sufficient theoretical knowledge to permit the preliminary engineering of possible production processes. Of course, nobody knew whether any of the methods they were considering would work, but, assuming that at least one of them would, they could now make some guess as to the major items that would be needed.

The basic American military requirements were twofold: to provide our armed forces with a weapon that would end the war and to do it before our enemies could use it against us. To fulfill these needs we would have to move ahead with the utmost speed.

When the Corps of Engineers started its work, its job was simply to build and operate the production plants. The problems involved in the development of the bomb and its delivery were for the time being largely ignored.

Nor was the full magnitude of the project generally appreciated. No one thought of it as entailing expenditures running into the billions of dollars. Not until later would it be recognized that chances would have to be taken that in more normal times would be considered reckless in the extreme. Not until later would it become accepted practice to proceed vigorously on major phases of the work despite large gaps in basic knowledge. Not until later would every other consideration, whether the advancement of knowledge for the sake of knowledge or the maintenance of friendly diplomatic relations with other powers, be subordinated to achieving the project's single overriding aim. Not until later would all concerned grow accustomed to the idea that, while normally haste makes waste, in this case haste was essential.

For the time being, work on the project continued in an orderly, straightforward, thoroughly conventional manner.

On June 18, immediately after receiving orders to set up his special district, Colonel Marshall had informed General Eugene Reybold, the Chief of Engineers; General Robins, the Deputy Chief of Engineers and Chief of Construction; and me, the Deputy Chief of Construction, of his new assignment. He did not reveal many of the details of his work, but he did ask that we give him every possible assistance. When he first told us of his carte blanche instructions, Robins and I were rather skeptical about how long it would be before his priority would be overridden by some other super-important project. We had seen this happen before. Yet, in spite of our mental reservations, we spared no effort in supporting him fully.

Marshall selected Lieutenant Colonel K. D. Nichols as his deputy and together they began to organize the district. After visiting Dr. Bush to learn the general background of the project, Marshall went over with me his urgent need for competent assistance in solving the many engineering problems that confronted him. He said that his work would be of considerable magnitude and would involve extremely complex problems of design. After some discussion, I told him that Stone and Webster appeared to be the engineering firm best fitted to meet his requirements: they were accustomed to working with scientific people—far more than most engineering firms; they were a large firm, capable in both engineering and construction; and they were performing well on all their contracts with the Corps of Engineers. Marshall agreed and promptly began to lay the groundwork for bringing Stone and Webster into the project to take over a major portion of the engineering studies and the preparation of plans, as well as to carry out much of the construction that would follow.

On June 25, Marshall, Nichols and Styer met with the Uranium Committee[6] of the OSRD. In the course of this meeting, it became clear that neither the centrifuge nor the diffusion methods for the separation of Uranium-235 were ready for development and that it might be some time before any plants to utilize these processes could be designed. They estimated that the total power requirements of the plants to be constructed in the immediate future would be 108,000 kw and, based on this estimate, they planned for a firm power demand for 150,000 kw by the end of 1943.[7]

Many of the OSRD's contracts with the laboratories were being delayed for lack of money, so Colonel Marshall arranged for the transfer of $15 million of Corps of Engineers funds to the OSRD, in order that existing research projects could continue uninterrupted after June 30, the end of the fiscal year. His immediate money worries taken care of, Marshall's most pressing difficulties involved the procurement of scarce materials and equipment required by the labora-

---

[6] By then reconstituted and called the S-1 Committee.
[7] By 1945 our power requirements totaled almost 1,000,000 kw.

tories. Shortages were growing more critical daily and, unless relief were obtained promptly, all research would soon come to a standstill.

Then the question of a name for the project came up. Toward the end of June, Reybold called Robins, Marshall, Nichols and me to his office and told us that he had conferred with Somervell and Styer, and that they had agreed that the new establishment—that is, the various plants where the bomb's ingredients would be produced—was to be called the "Laboratory for the Development of Substitute Materials" or "DSM." I demurred on the grounds of security, feeling that the name was bound to arouse curiosity. Several other names were suggested, but Reybold decided that, since Somervell had already agreed to it, no change would be made. Before we parted, it was decided definitely to hire Stone and Webster as the over-all architect-engineer-managers, and to use the Corps of Engineers to acquire whatever land might be needed by the project.

On June 29, officials of Stone and Webster met with Robins, Marshall, Nichols and me. Robins and I questioned them in great detail to determine whether they could place a force of adequate quality and sufficient strength on the project without impairing their performance on a number of very important Corps of Engineers and other war contracts which they had already undertaken. (I wanted to do all I could to help Colonel Marshall, but at the same time I did not want his project to impede any of the construction work for which I was responsible.) Their answers were satisfactory in every respect, and they were hired on the spot.

Earlier, at Marshall's request, I had initiated within the Office of the Chief of Engineers a survey of the United States to assist in the selection of a suitable site for his production plants. The plan then was to concentrate all facilities required for the production of fissionable materials at a single site. We assumed that a very large tract would be required. Consequently, we sought out a relatively undeveloped area where land was still available at a reasonable price. It had to be located well away from both coasts, so that there would be no threat of direct enemy interference. We knew that there would be a heavy requirement for electric power. Travel between the site and the Washington, New York and Chicago areas had to be reasonably convenient and not too time-consuming. There must be an assured and quite considerable supply of water. As was the case with the acreage and the power, we could only guess at the total water requirement, but we thought that it would be large. It was essential that the site should have a climate that would permit heavy construction to be carried on throughout the year. And, finally, it was most desirable that an adequate labor force, both for construction and for operation, be available.

On the basis of these criteria, I soon concluded that the best location would be somewhere near Knoxville, Tennessee. That area seemed to meet all the requirements, and beyond that, it gave promise of being a pleasant place for the permanent operating force to live. This was most fortunate because we anticipated that many of the operating personnel would have to be imported, since the specialized skills we would need would be difficult to recruit locally. Marshall agreed with this view and, on July 1, he and Nichols, together with their assistants and representatives of Stone and Webster and the Tennessee Valley Authority, began investigating sites in the vicinity of Knoxville. Colonel T. B. Parker, the Chief Engineer of TVA, informed them at that time that the Authority could supply the power needs of the project, provided it obtained priority assistance in procuring heavy generating equipment.

A few days later, meetings were held at the Metallurgical Laboratory to give the responsible Stone and Webster people a thorough understanding of the plutonium process. Compton was concerned over a possible shortage of uranium oxide, so Stone and Webster was directed to purchase 350 tons of this material immediately. At the same time, efforts were made to select a site for a pilot reactor plant, for it had been decided that a full-scale production plant should not be built until an experimental reactor had been previously placed into operation and this had been followed by a semi-works.

The site chosen for the pilot reactor was quite close to Chicago, in the Argonne Forest, which was under the control of the Forest Preserve of Cook County. The Forest Commission, headed by Mr.

Clayton F. Smith, was most co-operative, and great pains were taken to insure that the plant would be designed in such a manner that it would not interfere with any of the Commission's plans for the future. Under the terms of the lease for the land, both the site and any buildings erected were to be returned to the Forest Preserve after the war.

On July 9, Marshall, Nichols and the Stone and Webster people met with the S-1 Committee of the OSRD. Among other things, it was agreed that Stone and Webster would subcontract the design and fabrication of the heavy-water plant to be built at Trail, British Columbia, to E. B. Badger and Sons of Boston, and the erection and operation of it to the Consolidated Mining and Smelting Company of Canada, on whose property it would be situated.[8]

Out of this same meeting a tentative construction program emerged. It called for starting the construction of the plutonium reactor piles by October 1, 1942; of the centrifuge process by January 1, 1943; of the gaseous diffusion process by March 1, 1943; and of the electromagnetic separation process by November 1, 1942. It soon became apparent that these target dates were wholly unrealistic, for basic research had not yet progressed to the point where work on even the most general design criteria could be started. I was disturbed by the vagueness of the starting dates for the several elements of the project, and urged Marshall to insist upon a detailed schedule, to be agreed to by all concerned, which he could then apply himself to meeting.

Marshall was still having financial problems. He estimated that he

[8] This program was brought into being to create a reserve of heavy water for use as a moderator in the nuclear reactors in case, for some unpredictable reason, the use of graphite proved to be impossible. The problem was that no one knew how much heavy water such a plant would be able to produce, or how much we would need—if we needed it at all. The requirements depended upon the final choice of a production method for plutonium, and this had not been settled at the time the decision about heavy water had to be made. If it were needed, however, we wanted to be ready, and so decided to provide facilities, in addition to the small installation at Trail, that would yield enough for our initial requirements. Fortunately, graphite worked and the heavy water was not needed. Our supply of it was then loaned to Canada for use in an experimental reactor.

The expansion of the heavy-water program was handled by du Pont's Ammonia Department, which at a cost of some twenty million dollars built three plants as adjuncts to existing Army Ordnance plants.

would need $85 million, yet every attempt he had made to obtain it had been turned down. We agreed that if he continued to meet refusals he should, acting through Styer, ask Bush to appeal directly to the President for help.

On July 13, the project received an AA-3 priority from the War Production Board, with the promise of AAA for special items if any difficulties were encountered. This priority was completely inadequate for the expeditious completion of a venture as complicated as this one, and serious priority troubles soon arose. I began to feel that the doubts that Robins and I had originally entertained about the support the project would receive were being confirmed.

Toward the end of the month the estimates started to come in on the most likely Tennessee site. The 83,000-acre tract which was under consideration would cost slightly over $4 million, and would require the relocation of some four hundred families. Faced with this prospect, Marshall felt that he should delay any site acquisition proceedings, pending further developments in research on the plutonium process. It must be remembered that at this time the pile in Chicago, which was to prove that a chain reaction was possible, was not yet in operation—indeed, it did not have its first test until December.

Marshall devoted considerable attention during this period to supply contracts, among the most important of which were one with Mallinckrodt for the purification of uranium oxide, and one with Metal Hydrides and one with Westinghouse for converting uranium oxide into the uranium metal needed for the plutonium process. At the same time, aided by Nichols, he had been striving to perfect his organization and to define the responsibilities and interrelationships of its several elements, to recruit competent personnel, find suitable office space, and in general to take care of myriad other necessary details.

Marshall brought in, largely on a temporary basis, a number of regular officers to get things under way while he was recruiting the key nonscientific personnel. Most of the latter were men who had been civilian employees of the Corps of Engineers for years; they were familiar with government operations and their records and abilities were well known. These men were given reserve commissions in the Army, and formed a trained and competent nucleus that proved to be of tremendous value throughout the entire project.

On August 11, Marshall handed me the draft of a general order to be published that day announcing the formation of the new district. In it he used the designation "DSM." I again objected to this term because I felt that it would arouse the curiosity of all who heard it. After some discussion, during which we considered the possibility of using "Knoxville," we decided upon "Manhattan," since Marshall's main office would at first be in New York City. Our choice was approved by Reybold, and so the Manhattan Engineer District, or MED, came into being.

During this month improvements to existing facilities in the laboratories at Berkeley, where research on the electromagnetic separation process was being carried on, and at Chicago were initiated. The design of the heavy-water plant at Trail got under way, but because this involved a great deal of difficult copper work, it immediately ran into conflicts with the synthetic rubber program, which carried a higher priority.

The situation during the summer of 1942 was not promising for any operation that did not hold the very highest priority. Stone and Webster estimated at the time that the construction of a pilot plant for the electromagnetic separation process would take eleven months under an AA-3 priority, the highest then available to the MED; yet the same job could be completed in eight months if an AA-1 rating were obtained.

Toward the latter part of August, considerable doubt began to crop up concerning the Tennessee site. For the most part, this indecision stemmed from the desires of the leaders of several laboratories to have the production plant situated conveniently close to their own facilities.

Toward the middle of September, Marshall met once more with the S-1 Committee in the Bohemian Grove, just north of San Francisco. A number of decisions were made at that time, the most important of which was to proceed with the Tennessee site, in spite of the contrary opinions. Almost equally important was the decision to design and to procure materials for a small electromagnetic separation plant, at an estimated cost of some $30 million, subject to cancellation at any time prior to January 1, 1943.

On September 16, I agreed to an AAA rating for the copper needed at Trail. The metal thus made available to the project would have to be taken from the Corps of Engineers' quota and would upset a number of other projects of the Corps, but I was anxious to emphasize to Marshall that he had our full support and to encourage his people to push their work to the utmost.

Little did I anticipate then that within twenty-four hours I would be the one who needed support.

CHAPTER 2

FIRST STEPS

Immediately upon returning to my office on September 17 after my conversation with General Styer, I delivered to Reybold the memorandum assigning me to the MED and informed Robins of its contents. Both of them concurred in my intention to await promotion before taking over openly, and agreed with the measures which I had proposed to minimize drawing attention to my change in duties.

Next I discussed the recent turns of events with Nichols, who, in Marshall's absence on the Pacific Coast, had been summoned to Washington by Styer the night before. I was anxious to learn the details of the problems which confronted the Manhattan Engineer District, especially in areas in which I had not been previously involved. Up to this time, I had asked no questions about the scientific difficulties and the probability of success in overcoming them, because it was not necessary for me to know them in order to carry out my responsibilities, which had involved chiefly assistance, where wanted, to Marshall in the site location and the procurement of land, equipment and materials. Now, I was particularly concerned with determining to what extent our work would be based on real knowledge, on plausible theory or on the unproven dreams of research scientists. Next, I wanted to know about the available supplies of raw materials. Finally, I went into the matter of obtaining high enough priorities to enable the work to go ahead at full speed.

I was not happy with the information I received; in fact, I was horrified. It seemed as if the whole endeavor was founded on possibilities rather than probabilities. Of theory there was a great deal, of proven knowledge not much. Even if the theories were correct, the

19

engineering difficulties would be unprecedented. The raw material situation was not certain, but that did not disturb me, for this was a tangible difficulty. Neither was I particularly disturbed by the poor priority situation, for I felt that I could better it through my own efforts or, if not, that I could at least force a decision either to grant us the needed priorities or else to recognize that the project was not urgent.

Accompanied by Nichols, I called on Dr. Bush that same afternoon, under the assumption that he had been told of my appointment. Through some oversight, he had not been informed and, consequently, was quite mystified about just where I fitted into the picture and what right I had to be asking the questions I was asking. I was equally puzzled by his reluctance to answer them. In short, the meeting was far from satisfactory to both of us. As soon as I left, Bush got in touch with Styer, who informed him of my new assignment.

Bush accepted the appointment, although he was quite disturbed. He told Styer that he felt I was too aggressive and might have difficulty with the scientific people. Styer told him that this was quite possible and that the two of them, personally, might have to smooth out a number of difficulties, but that the work would move. Bush, still far from happy about my designation, thereupon sent the following note to Harvey Bundy, who was then handling the details on atomic energy development for Secretary Stimson, expressing the fears he had developed during my brief call.

MR. BUNDY:

I visited General Styer since I feared that comments would be made before you returned. I told him (1) that I still felt, as I had told him and General Somervell previously, that the best move was to get the military commission first, and then the man to carry out their policies second; (2) that having seen General Groves briefly, I doubted whether he had sufficient tact for such a job.

Styer disagreed on (1) and I simply said I wanted to be sure he understood my recommendation. On (2) he agreed the man is blunt, etc., but thought his other qualities would overbalance.

Apparently Somervell saw General Marshall today regarding Groves. I fear we are in the soup.

V.B.

(Despite this inauspicious beginning, my relations with Bush, from that day on, were always most pleasant and we soon became, and remain, fast friends. We have often laughed about "the soup." Never once throughout the whole project were we in disagreement. He was a pillar of strength upon whom I could always rely.)

Not particularly happy over the meeting, but completely unaware of the cause of my obvious failure to get through to Dr. Bush, I returned to my old office in the Construction Division. Finding my secretary, Mrs. O'Leary, there, I told her I was being reassigned and that if she wanted to come along, I would be glad to have her. I added, in what proved to be a great understatement, that this would be a very quiet and easy job for her and she should be sure to bring along some knitting to keep herself occupied. This prediction proved valid for about two days.

When I returned home that evening I told my wife and daughter and wrote to my son, a cadet at West Point, that I had a new job, that it involved secret matters and for that reason was never to be mentioned. The answer to be given if they were asked what I was doing was, "I don't know, I never know what he's doing." To my son, I added, "If it is an officer who knows me well, and he is persistent, you can add, 'I think it's something secret.' "

My already long hours were gradually extended, but the extent of my travels away from Washington was about the same. My family continued to live as most Army families lived during the war and before, accepting the situation and asking no questions. Unlikely as it may seem to many people, they first learned of the nature of my assignment at the same moment, three years later, that the bombing of Hiroshima was announced to the rest of the world.

On September 19, Colonel Marshall arrived in Washington, and together we went over the new setup. I told him that the Manhattan District would function as it had before, with him continuing as the District Engineer. Soon afterward, Robins told us that Reybold and he had been relieved of further responsibility for the project and that I would attend to those matters that had formerly been handled by Styer.

With Marshall, I went to see Bush again. This time our discussion was cordial and uninhibited. Bush outlined succinctly the past history of the project, and, in doing so, cleared up a number of questions in my mind, and gave me many valuable suggestions. He mentioned that the Navy had been left out of the project at the explicit direction of President Roosevelt, and indicated that he considered this to be a mistake. He seemed pleased when I told him that I expected to visit the people at the Naval Research Laboratory in the very near future to discuss their work in the atomic area.

Before I left, Bush remarked that the handling of information within the project was becoming rather loose and that he hoped we could tighten things up. He seemed particularly interested in keeping recent scientific discoveries secret, especially those at Berkeley.

One of my first acts after learning that I was to have charge of the uranium project was to tackle what my recent construction experience led me to believe would be our greatest single obstacle. I did not see how we could possibly get the job done with nothing better than an AA-3 priority, and I did not feel inclined to fail by default. It seemed quite simple to me—if ours was really the most urgent project, it should have the top priority.

On September 19, I called upon Donald Nelson, the head of the War Production Board, and stated my views very simply but most definitely. His reaction was completely negative; however, he quickly reversed himself when I said that I would have to recommend to the President that the project should be abandoned because the War Production Board was unwilling to co-operate with his wishes.

When I left his office, I had in my pocket a letter signed by Nelson, saying:

I am in full accord with the prompt delegation of power by the Army and Navy Munitions Board, through you, to the District Engineer, Manhattan District, to assign an AAA rating, or whatever lesser rating will be sufficient, to those items the delivery of which, in his opinion, cannot otherwise be secured in time for the successful prosecution of the work under his charge.

Just why Nelson gave in so easily, I will never know. I would have been most unwilling to have had this difficulty brought up to the

President; the problem was mine. To have admitted frustration so early would have been most distasteful. And while I still had little liking for my new assignment, it was mine to carry.

In any event, as a result of my visit to Mr. Nelson, we had no major priority difficulties for nearly a year.

A couple of days later, Marshall and I went to the Naval Research Laboratory, where Rear Admiral H. G. Bowen and Dr. Ross Gunn, the technical adviser to the laboratory explained their experimental work on the separation of fissionable materials. This was my first direct contact with any laboratory work in the nuclear field, but I did not find myself at too great a disadvantage, for to an engineer their process, liquid thermal diffusion, was straightforward and not too mystifying.

Gunn had served on several of the committees that had previously controlled the uranium project, and it was quite evident that he was not particularly pleased over his omission from recent committee reorganizations. Bowen assured me, however, that his people would like to co-operate and co-ordinate their efforts with the Army. He was quite open in discussing their tests and the results they had obtained, although, so far as he then knew, I was only the Army's representative with the OSRD for atomic development.

The Navy group was extremely small. Its only scientific men were Gunn and one other, a most capable physicist, Dr. Philip Abelson. Their research in thermal diffusion was being conducted on a very small scale, and I gathered that Gunn was not able to devote his full time to it. All in all, my first impression of the Navy's uranium research was not favorable, primarily because of the evident lack of urgency with which it was being prosecuted.

On September 23, I became a brigadier general and officially took charge of the project. That afternoon I attended a meeting in an outer office of the Secretary of War. Those present were Secretary Stimson, General Marshall, Dr. Bush, Dr. Conant, General Somervell, General Styer, Admiral Purnell, Mr. Harvey Bundy and I.

The purpose was to decide on the form and make-up of the policy supervision of America's atomic effort. It was obvious that the responsibility for making atomic policy was too great to be thrust upon one man alone, yet every member of a group selected for this purpose must be able to devote as much of his personal time to the project as might be necessary. This automatically eliminated from consideration the policy group already in existence, for three of its members, Vice President Wallace, Secretary Stimson and General Marshall, could not spare the time that adequate supervision of the project would require.

At Mr. Stimson's request, I outlined the general way in which I expected the MED to operate, emphasizing that we intended to avoid empire-building by utilizing to the utmost existing facilities of other government agencies. After that, the Secretary said he felt there should be a new military policy committee made up of extremely able men from the OSRD, the Army and the Navy. There was general agreement to this. He next suggested that the committee should consist of nine or possibly seven men. I objected quite vigorously on the grounds that such a large committee would be unwieldy; it would cause delays in taking action; and some, if not the majority, of its members would tend to treat it as a secondary responsibility, to the detriment of our progress. I felt that a committee of three was ideal and that any more members would be a hindrance rather than a benefit. I pointed out that I could keep three people reasonably well informed on our major problems, and, furthermore, that I would be able to obtain advice from them much more readily than I could from a larger group. In the end, my views were accepted.

There was one obstacle, however, and that was that everyone recognized that both Bush and Conant should be members. Secretary Stimson solved this problem when he proposed that the committee consist of Bush as chairman (with Conant as his alternate), Rear Admiral W. R. E. Purnell of the Navy, and me. Everyone seemed well satisfied with this solution and especially liked having both Bush and Conant on the committee.

There followed some general conversation, in the midst of which, and with no small amount of embarrassment on my part, for I was by far the most junior person present, I asked to be excused if I were no longer needed, for I wanted to catch the train to Tennessee and inspect the proposed production plant site, so that the land acquisition could proceed. With this, the meeting broke up. I was a bit relieved when Somervell told me several days later that my request could not have been better timed, because it convinced everyone that he had not overemphasized my urgent desire to get a job moving.

Later in the afternoon, Mr. Stimson decided that Styer, instead of me, should be the Army member on the Military Policy Committee. I did not learn of the change until after my return from Tennessee, when I was told of it by Styer, who seemed just a bit embarrassed, and he assured me that he had not had anything to do with it. I replied that I knew he had not, and that in view of his position and consequent ability to help us, I was glad that he would be on our side of the fence.

The change was not the least objectionable to me, and, as time passed, its advantages to the project and to me became increasingly evident. Among these were having available in Styer the counsel of a most capable engineer officer, and in having him familiar with our problems, and ready at all times to assist us in securing the support of the various elements of the Army Services of Supply.

Colonel Marshall and I met in Knoxville early the next morning after the conference in Mr. Stimson's office, and together we went over the proposed site as thoroughly as the existing roads permitted. To my great pleasure, it was evident that it was an even better choice than I had anticipated.

The site Marshall had selected was near the town of Clinton, about seventeen miles from Knoxville. The highways from Knoxville would be satisfactory in every way for our early operations. Railroad service was adequate, and an adequate source of water existed in the small river which flowed through the area. It was then typical rural Tennessee country, containing only a few schools of moderate size. Farms were scattered throughout. There were no towns or villages and the population was relatively small and dispersed.

Originally, the entire site went under the name of "The Clinton Engineer Works," the title deriving obviously from the near-by small town of Clinton. The name "Oak Ridge" did not come into general use until the summer of 1943, when it was chosen for the new community's permanent housing area, built on a series of ridges overlooking part of the reservation. To avoid confusion, as well as to lessen outsiders' curiosity, the post office address was Oak Ridge, but not until 1947 and after the establishment of the Atomic Energy Commission was that name officially adopted, in lieu of the Clinton Engineer Works. As in the cases of Hanford and Los Alamos, our first consideration in the selection of names was to find the one least likely to draw attention to our work.

As soon as I had gone over the site, procurement of the land—about 54,000 acres—began. As always happens when the government takes over any sizable area, some owners suffered real hardships by their dislocation. This is inevitable, despite the fact that they are paid the full value of their property, as established by the government appraisers and accepted by the owners or, if not accepted, as fixed by federal judges and juries after condemnation proceedings.

It became necessary at this time to have a Presidential proclamation issued, setting up certain restrictions on this area. The proclamation came out in due course, but because of the nature of our operations, I did not consider it wise to give it any wider circulation than was absolutely necessary. For that reason, I told Colonel Marshall that it should be brought to the personal attention of the Chief Executive of the state, Governor Cooper, before it became known to anyone else.

Unfortunately, I did not discuss the way in which this should be done, since I assumed Marshall would handle the matter personally. As it turned out, I should have, for Marshall, who was most adept at dealing with high public officials, because of other pressing matters did not see the Governor himself, but sent another officer in his stead to perform this delicate task. To the Governor this was a serious breach of protocol, and it was compounded by the facts that the emissary was not a senior officer and that it was his first experience in such matters. The Governor must have been somewhat displeased in the first place by having such a large installation placed in Tennessee without his having been consulted or informed when the matter was under consideration. He was further disturbed by the relatively junior rank of the officer who finally brought him word of our movement into his state. Security limitations prevented any disclosure of our purpose and our lack of definite plans made it impossible to give him any idea of the possible size of the installation. About all that could be done was to present the Governor with a copy of the proclamation establishing the federal reservation. On the whole, we got off to a most inauspicious start with our new neighbors, and the ensuing resentment plagued us for several years.

My experience in this case highlights what everyone who issues general instructions, or what in military terms are called mission orders, must always remember. While normally instructions will be followed much more intelligently if they are general rather than spelled out in great detail, inherently they are subject to misinterpretation, and when this happens, a great deal of effort must be expended in picking up the pieces. Nevertheless, mission orders must be used whenever there are many unknowns. Our project was full of unknowns, which was the principal reason why I habitually employed mission orders. Marshall and later Nichols followed the same practice. Only in the rarest instances were direct, detailed instructions issued, and even then they usually dealt with fairly simple matters for which specific instructions were perfectly appropriate.

I decided to make my headquarters initially in Washington; I could always move later if it seemed desirable. However, as time went on, I came to realize that Washington was the ideal place for me to be, for there I was able to keep in contact with the War Department and the other government agencies, and to smooth the way for our people in the field. This was particularly true during the later phases of the work, after we became involved in international negotiations, and in military intelligence, and when we began to deal so closely with the Secretary of War, with Air Force Headquarters and with the overseas commands. Another major advantage was that distance alone prevented me from becoming involved in too many details, which is so dampening to the initiative of subordinates.

The office out of which I was to control the Manhattan Project

was set up while I was on my first visit to Tennessee. It consisted
of two rooms, 5120 and 5121, in the new War Department Building (now the State Department Building), at Twenty-first and Virginia Avenue, N.W. I moved as soon as I returned.

As I consider Washington today, it seems incredible that these
accommodations were as limited as they were. My secretary, Mrs.
O'Leary, who was soon to become my chief administrative assistant,
and I occupied one room. The only furniture added was one, and
later another, heavy safe. Alterations were limited to those essential
for security and consisted of sealing the ventilating louvers on an
outside door, which was kept locked and bolted. One other door,
leading to an adjoining conference room, was also locked permanently, so that the only access to the room which I occupied was
through my outer office. This room, at first used by visitors, eventually
accommodated three assistant secretaries and file clerks.

After several months we took over another small room. By a year
later, we had grown to a total of seven rooms, of which two were
occupied by district personnel working under the direction of the
District Engineer on procurement matters. This arrangement lasted
until shortly before the bombing operations, when we took over a
few more rooms for our public information section, which had to
be ready to start functioning with the news release on Hiroshima.

It was undoubtedly one of the smallest headquarters seen in
modern Washington. Nevertheless, I fell far short of my goal of
emulating General Sherman, who, in his march from Atlanta to the
sea, had limited his headquarters baggage to less than what could be
placed in a single escort wagon.

Our internal organization was simple and direct, and enabled me
to make fast, positive decisions. I am, and always have been, strongly
opposed to large staffs, for they are conducive to inaction and delay.
Too often they bury the leaders' capacity to make prompt and intelligent decisions under a mass of indecisive, long-winded staff studies.

Initially I did plan on having an executive officer, and I selected
first one and then a second highly competent man for that position.
Yet, before either could even begin to take over his duties, I had to

reassign him to fill a pressing need in the field, one at Hanford and the other at Los Alamos. I soon realized that as long as we were under such pressure I would always find it necessary to assign to the field anyone whom I might consider acceptable as an executive officer in my headquarters. Consequently, I abandoned all further attempts and relied instead upon my chief secretary, who became my administrative assistant. With her exceptional talents, and her capacity for and willingness to work, Mrs. O'Leary more than fulfilled my highest expectations.

After our work at Oak Ridge got under way, and shortly after Nichols succeeded Marshall as District Engineer, it became essential to move the district headquarters to that point, though we continued to carry on many activities through our four suboffices in New York. Having our offices dispersed throughout the city was most advantageous because it obviated many of the exasperating difficulties in transportation which consume so much time during the working day. It also served to compartmentalize our people and reduced the opportunity for cross-chatter between the various segments of our activities.

In mentioning Marshall's replacement by Nichols, I should explain that in August, 1943, the Chief of Engineers asked me if I could relieve Marshall for a key assignment overseas, which would mean his promotion to brigadier. Since the project was by that time well organized, I did not feel I should refuse, and appointed Nichols in his stead. This was an excellent choice and one I never regretted. He and I had been in the same battalion in Nicaragua for two years, some twelve years before, and he had shown much promise there. But the decisive factor in his selection was the extraordinary grasp of the technical and scientific details of our work that he had demonstrated in the year he had been with the MED.

My agreement to Marshall's relief was in line with my belief that the project should be managed with as few regular officers as possible. Those that Marshall had brought in at the beginning had been relieved without too much delay, as soon as they could be satisfactorily replaced and an overseas assignment was open.

I believed strongly that in time of war every possible regular officer should be in the combat area. I was undoubtedly influenced in this belief by my personal knowledge of the disappointment suffered by many regular officers who were kept in this country during World War I, with no chance of combat experience. In my own case, I was already a cadet when the war started, and remained at West Point until a few days before the Armistice. Had my own experience been different, I would quite probably have had a considerable number of regular officers assigned to the project throughout its duration.

As I look back now with a full appreciation of the tremendous import of the development of atomic energy, I think it was a mistake not to have had them. Our country would have been much better off in the immediate postwar years if we had had a group of officers who were thoroughly experienced in all the problems of this type of work—not only in problems of atomic energy but in all the manifold problems involved in technical and scientific developments that have played such an important part in our national defense since 1945.

While I am on the subject of my own mistakes, I perhaps should add that there was another consideration, similar to this, to which I did not give adequate attention. That was the necessity of having replacements available if either Nichols or I died or became disabled. Many serious problems would have arisen if anything had happened to either of us, and it was not proper for me to have placed such great reliance, fortunately not misplaced, upon the physical and mental ability of both of us to stand up under the strain, to say nothing of the possibility of accidental death or injury, particularly since we did so much flying.

This was brought very vividly to my attention in December of 1944, when Mr. Churchill suggested that I should come to London to talk over our problems, and particularly our progress, with him and other members of his government. In discussing his request with Secretary Stimson, I said that while I would like very much to go to England, I was afraid that it might take me away from my work for a considerable period of time, especially if something developed that would make it impossible for Mr. Churchill to receive me immediately on arrival.

Mr. Stimson told me that if I went, I could not go by air, because of the hazards involved. When I said, "Well, I don't see what difference that would make," he replied, "You can't be replaced." I said, "You do it, and General Marshall does it; why shouldn't I?" He repeated, "As I said before, you can't be replaced, and we can." Harvey Bundy, who was also present, said he had heard that I had previously urged flying when air safety dictated otherwise, and then asked, "Who would take your place if you were killed?" I replied, "That would be your problem, not mine, but I agree that you might have a problem."

I went on to say that if anything happened to Nichols, I felt that I could continue to operate, though it would mean a very strenuous period for me personally, but that if it were the other way around, while Nichols was thoroughly capable of taking over my position, I thought because he was not so familiar with my responsibilities as I was with his that he could not do both my job and his.

Mr. Stimson said, "I want you to get a Number Two man immediately who can take over your position, and with Nichols' co-operation, carry on in the event that something happens to you." He added, "You can have any officer in the Army, no matter who he is, or what duty he is on."

I drew up a list of about six officers who I thought would be satisfactory, keeping in mind that it would be all-important for the man selected to be completely acceptable to Nichols, since success would depend on the utmost co-operation between them. I particularly wanted someone who would not attempt to overrule Nichols in any of his actions or recommendations until he had had time really to understand what the work was all about, and I doubted whether it would be possible for anyone to accumulate the essential background for this before the project was completed.

Having made up my list, I discussed the matter with Nichols. I asked him to look over the names and to strike from the list anyone whom he would prefer not to have in such a position. He struck several names. I always suspected he struck the first one just to see if I really meant what I had said, because it was the name of a man

whom I had known for many years, and who was a very close friend.
When he struck that name, I did not bat an eye, but merely said,
"Well, he's out."

After he had crossed off the names of the men he considered unacceptable, I asked him if he had any preference among the remainder. He replied, "You name him and I'll tell you." I said that I
felt that the best one on the list was Brigadier General Thomas F. Farrell, and Nichols replied, "He would be my first choice, too."

Farrell was a former regular officer who had been Chief Engineer
of the State of New York. Early in 1941, he came back on active military duty as my executive officer. He was then stationed in India, but
was on leave in the United States. I asked him to come to Washington
to see me and told him that I intended to ask for him, but that I did
not want him if he had any objection to coming. He did not, fortunately, and soon joined us. He proved to be of inestimable value, especially throughout the trying period of the summer of 1945, when all
of our efforts attained their climax.

CHAPTER 3

THE URANIUM ORE SUPPLY

I mentioned earlier that one of my first concerns in my new job was about the supply of raw materials. The most important, of course, was uranium ore. It is sobering to realize in this connection that but for a chance meeting between a Belgian and an Englishman a few months before the outbreak of the war, the Allies might not have been first with the atomic bomb. For the most important source of uranium ore during the war years was the Shinkolobwe Mine in the Belgian Congo and the most important man concerned with its operation was M. Edgar Sengier, the managing director of Union Minière du Haut Katanga or, as it is usually called, Union Minière.

In May of 1939, Sengier happened to be in England, in the office of Lord Stonehaven, a fellow director on the Union Minière Board, when Stonehaven asked him to receive an important scientist. This turned out to be Sir Henry Tizard, the director of the Imperial College of Science and Technology. He asked Sengier to grant the British Government an option on every bit of radium-uranium ore that would be extracted from the Shinkolobwe Mine. Naturally, Sengier refused. As he was leaving, Sir Henry took him by the arm and said most impressively: "Be careful, and never forget that you have in your hands something which may mean a catastrophe to your country and mine if this material were to fall in the hands of a possible enemy." This remark, coming as it did from a renowned scientist, made a lasting impression on Sengier.

A few days later, he discussed the future possibilities of uranium fission with several French scientists, including Joliot-Curie, a Nobel Prize winner. They proposed a joint effort to attempt the fission of

uranium in a bomb to be constructed in the Sahara Desert. Sengier
accepted their proposal in principle and agreed to furnish the raw
material and to assist in the work. The outbreak of World War II
in September, 1939, brought this project to a halt even before it
began.

Tizard's warning and the obvious interest of the French scientists
emphasized to Sengier the strategic value of the Katanga ores, which
were of exceptional richness, far surpassing in that respect any others
that have ever been discovered.

Sengier left Brussels in October of 1939 for New York, where he
remained for the rest of the war. From there, he managed the operations of his company, both inside and outside the Belgian Congo, and
after the invasion of Belgium in 1940 had to do so without the benefit
of any advice from his fellow directors who were in Belgium behind
the German lines.

Before his departure from Brussels, he had ordered shipped to the
United States and to Great Britain all available radium, about 120
grams, then valued at some $1.8 million. He had also ordered that
all uranium ores in stock at the Union Minière-controlled refining
plant in Oolen, Belgium, be sent to the United States. Unfortunately,
this order was not complied with promptly; later, owing to the German
advance into Belgium, it became impossible to carry it out.

Toward the end of 1940, fearing a possible German invasion of
the Belgian Congo, Sengier directed his representatives in Africa to
ship discreetly to New York, under whatever ruse was practicable, the
very large supply of previously mined uranium ore, then in storage
at the Shinkolobwe Mine. All work at the mine had stopped with the
outbreak of the war and the equipment had been transferred to vitally
important copper and cobalt mining operations for the Allied war
effort. In accordance with Sengier's instructions, over 1,250 tons of
uranium ore were shipped by way of the nearest port, Lobito, in Portuguese Angola, during September and October of 1940, and on arrival
were stored in a warehouse on Staten Island.

At a meeting in Washington in March, 1942, sponsored by the
State Department, the Metals Reserve Corporation, the Raw Materials

Board and the Board of Economic Warfare, Sengier was invited to submit a report on the nonferrous metals resources of the Belgian Congo. That same day, in the course of a private conversation with Thomas K. Finletter, then Special Assistant to the State Department on Economic and International Affairs, and Herbert Feis, also of the State Department, who were urging him to double the Belgian Congo's production of cobalt, Sengier pointed out that Union Minière had available material that was much more important than cobalt; namely, uranium. Apparently this did not make any impression on the State Department representatives. A little later, in April, Sengier brought the matter up again, emphasizing that much valuable material was stored in a warehouse in New York. This time a little interest was aroused and there was some talk of transferring the ore to Fort Knox, where it would be stored with the national gold reserve.

On April 21, Sengier tried a third time, writing to Finletter: "As I told you previously during our conversation, these ores containing radium and uranium are very valuable." Still nothing happened.

As is now well known, the State Department was not informed of the atomic project until shortly before the Yalta Conference in February of 1945, when, for compelling reasons, I asked that Secretary Stettinius be told of it. Just why the State Department had been kept completely in the dark, I do not know, except that it was in accordance both with President Roosevelt's policy of personally conducting international relations and with his disinclination to bring any unnecessary persons into atomic affairs. And it was well known, of course, that there was friction in the higher levels of the department. Nevertheless, it is hard to understand why, with Sengier so insistent on the value of the uranium ore, and knowing that the ore contained radium, the State Department officials did not make a serious effort to determine the real value of the ores. Anyone with even a superficial knowledge of the metals field would have been extremely interested in them because of their radium content. And anyone who was reasonably well read in current publications should have been interested in uranium per se; for the newspapers and magazines by that time had printed a number of articles, such as "The Atom Gives Up" by William L.

Laurence, published in the September 7, 1940, *Saturday Evening Post*.

In my discussions with Colonel Nichols on the afternoon of September 17, the day that I was designated to take charge of the Manhattan Project, one of the first matters that we talked about was the adequacy of the ore supply. As we reviewed the situation, our prospects did not appear to be too satisfactory, except for the possible Sengier ores. The first intimation of their existence had come to the MED only ten days before in a telephone call to Nichols from Finletter, the purpose of which was to ask about the urgency of a shipment from New York of some uranium black oxide by African Metals to Canada for refining. Nichols had postponed any definite reply pending investigation, which he promptly began. He soon concluded that African Metals did have sizable stocks of rich ore in the New York area. The S-1 Committee was informed at its California meeting on September 14, and had recommended the acquisition of all available Sengier ore.

Nichols and I agreed that there should be no delay in getting in touch with Sengier, who we understood was the key figure, at least in the United States, in the Belgian Congo uranium picture. Nichols already had an appointment to see him the next morning to explore the ore situation. We knew nothing at that time of Sengier's previous futile approaches to the State Department.

The next day, when Nichols opened the conversation, Sengier was somewhat guarded in his reply, recalling how the State Department had consistently ignored his repeated proddings. After inspecting Nichols' credentials, he said: "Colonel, will you tell me first if you have come here merely to talk, or to do business?" Nichols answered diplomatically, as always, that he was there to do business, not to talk. Sengier then really pleased him by saying that it was true that over 1,250 tons of rich ore were stored in some two thousand steel drums in a warehouse on Staten Island.

A delighted Nichols left Sengier's office an hour later carrying with him a sheet of yellow scratch paper on which were written the essentials of an agreement to turn over to us at once all the ore in the

Staten Island warehouse and to ship to the United States all the richer uranium ore aboveground in the Belgian Congo.

A simple handshake proffered by Sengier had sealed the bargain. The exact wording of the written contract would be settled later.

This was typical of the way in which a great many of our most important transactions were carried out. Once the seller understood the importance of our work (and there was no need to explain it in this case), he was invariably perfectly willing to deliver his goods or his services on our oral assurance that fair terms and conditions would be settled at a later date. We always promised that he would not be out-of-pocket for any expenses incurred if for some reason final agreement was not reached. And we always kept that promise.

Union Minière's ore was of tremendous value to us, not only because of its quantity but because of its richness.[1] With it under our control, we were able to proceed with our atomic development without fear of running out of the basic material, uranium, during the critical war years ahead.

Sengier's revelation of September 18 made us realize how vastly important he was to the Allied cause, and from that time on we helped him in every way that we could. All the details of our various agreements were kept as secret as possible, including the handling of payments. Here, it was necessary for Sengier to tell the head of his American bank why he was opening a special account, to which would be credited the funds derived from the sale of materials identified under a special number. In order to safeguard this information, it was arranged that the reports of the Federal Reserve Bank would not mention any of these transactions. There was to be a minimum of correspondence on the subject, and the auditors were directed to accept Sengier's statements without explanation.

---

[1] At the start of our purchases, the hand-sorted tonnage derived from the mine contained an average of over 65 per cent uranium oxide. This seems almost incredible when it is realized that ores from the Colorado Plateau and Canada, which contain two-tenths of one per cent ore, are of marketable quality, and the South African uranium ores derived from gold-mining operations have a uranium oxide content of the order of .03 per cent.

CHAPTER 4

THE PLUTONIUM PROJECT

It is essential for the reader to keep in mind the truly pioneering nature of the plutonium development as well as the short time available for research, to appreciate the gigantic steps taken by both scientists and engineers in moving as rapidly as they did from the idea stage to an operating plant of commercial size. It was a phenomenal achievement; an even greater venture into the unknown than the first voyage of Columbus.

The laboratory investigations had to be conducted in the face of incredible handicaps. At the laboratory in Chicago, we were seeking to split atoms, and in the process to transmute one element into another —that is, to change uranium into plutonium. The transmutation of an element involves the conversion of its atoms—the smallest known submicroscopic particles capable of existing alone which are not susceptible to further subdivision by chemical means—into atoms of another element possessing different chemical and physical properties. In effect, the scientists were reviving the classical, but always unsuccessful, search of the ancient alchemists for ways to convert base metals, such as lead, into gold; and the continuing, but theretofore unsuccessful, attempts of more modern chemists to change the character of elements. The precedents of history were surely all against us.

To carry out the transmutation process, even on a laboratory scale, and at an almost infinitesimal rate of production, a reactor, or as we often referred to it, a pile, of considerable size is necessary; for full-scale production, obviously, a much bigger pile is needed. The laboratory unit, it was estimated, would require, among other items, some forty-five tons of uranium or uranium oxide. Such amounts were not available in sufficient purity until late in 1942. Even then, the laboratory unit would not be able to produce enough plutonium to permit normal laboratory research on its recovery—that is, on ways to separate it chemically from the basic uranium and the other radioactive materials that would also be produced.

In June, 1942, when the Corps of Engineers came into the picture, the necessary research on plutonium production and recovery had scarcely begun. There was no experimental proof that the hoped-for conversion would actually occur; it was predicated entirely on theoretical reasoning. Not until December 2, 1942, did we have any such proof, and this was weeks after we had decided to go ahead at full speed on the plutonium process, and many days after we had started to prepare the plans for a major plant. On October 5, 1942, I paid my first visit to the Metallurgical Laboratory at the University of Chicago, where Arthur Compton and I spent the morning inspecting the laboratory facilities and discussing with a number of scientists the work on which they were engaged.

That afternoon I had a meeting with Compton and about fifteen of his senior men. Among them were two other Nobel Prize winners, Enrico Fermi and James Franck, together with the brilliant Hungarian physicists Eugene Wigner and Leo Szilard, and Dr. Norman Hilberry, Compton's assistant. The purpose of the meeting was to give me an idea of the extent of their knowledge about the plutonium process, and the anticipated explosive power of an atomic bomb, as well as of the amount of fissionable material that a single bomb would require. Of particular importance to me was an understanding of the gaps in knowledge that remained to be filled. I wanted to be sure also that everyone recognized the intermediate goals that had to be achieved before we would attain ultimate success, and that I, too, had a clear understanding of these goals. I was vitally interested in just how much plutonium or how much U-235 would be needed for a reasonably effective bomb. This was all-important, for it would determine the size of our production facilities, not only for plutonium, but also for Uranium-235.

Compton's group discussed the problem with me thoroughly, backing up their postulations mathematically and eventually arriving at the answers I needed. In general, our discussion was quite matter-of-fact, although much of it was highly theoretical and based on completely unproven, but quite plausible, hypotheses on which all the other participants seemed to be in complete agreement.

As the meeting was drawing to a close, I asked the question that is always uppermost in the mind of an engineer: With respect to the amount of fissionable material needed for each bomb, how accurate did they think their estimate was? I expected a reply of "within twenty-five or fifty per cent," and would not have been greatly surprised at an even greater percentage, but I was horrified when they quite blandly replied that they thought it was correct within a factor of ten.

This meant, for example, that if they estimated that we would need one hundred pounds of plutonium for a bomb, the correct amount could be anywhere from ten to one thousand pounds. Most important of all, it completely destroyed any thought of reasonable planning for the production plants for fissionable materials. My position could well be compared with that of a caterer who is told he must be prepared to serve anywhere between ten and a thousand guests. But after extensive discussion of this point, I concluded that it simply was not possible then to arrive at a more precise answer.

While I had known that we were proceeding in the dark, this conversation brought it home to me with the impact of a pile driver. There was simply no ready solution to the problem that we faced, except to hope that the factor of error would prove to be not quite so fantastic. This uncertainty surrounding the amount of material needed for a bomb plagued us continuously until shortly before the explosion of the Alamogordo test bomb on July 16, 1945. Even after that we could not be sure that Uranium-235 (used in the Hiroshima bomb) would have the same characteristics as plutonium (used in the test and later against Nagasaki), although we knew of no reason why it should be greatly different.

The day's discussions did leave me with a very high opinion of the scientific attainments of the Chicago group. It was obvious that it would not need major strengthening in any scientific area, and that the existing organization was more than adequate as a basis for any build-up that Compton might find necessary.

After the meeting, Compton and I resumed a discussion we had begun earlier with Szilard on how to reduce the number of approaches which were being explored for cooling the pile. Four methods—using helium, air, water and heavy water—were under active study. It was essential that we concentrate on the most promising and more or less abandon work on the others. By the end of the afternoon we settled on helium cooling. But within three months this decision was changed. The design problems early encountered in the comparatively small air-cooled reactor at Clinton indicated that the handling of any gaseous coolant in the much larger Hanford reactors would be very difficult. And as the operation of the Fermi test pile in December had proved that in a properly designed uranium pile water could be used as a coolant, it was adopted for the plutonium reactors we built at Hanford.

I left Chicago feeling that the plutonium process seemed to offer us the greatest chances for success in producing bomb material. Every other process then under consideration depended upon the physical separation of materials having almost infinitesimal differences in their physical properties. Under such circumstances, the design and operation of any industrial processes to accomplish this separation would involve unprecedented difficulties. It was true that the transmutation of uranium by spontaneous chain reaction into usable quantities of plutonium fell entirely outside of existing technical knowledge; yet the rest of the process—the chemical separation of the plutonium from the rest of the material—while extremely difficult and completely unprecedented, did not seem to be impossible.

Up until this time, only infinitesimal quantities of plutonium had been produced, and these by means of the cyclotron, a laboratory method not suitable for production in quantity. And by quantity production of plutonium, I do not mean tons per hour, but rather a few thimblefuls per day. Even by December, 1943, only two milligrams had been produced.

Several possible methods for the chemical separation of plutonium from uranium had been studied during the previous year by Compton's group, but there was no agreement on the most feasible process. Nobody had any real concept of the kind of equipment that would be needed.

I returned to Washington convinced that our first efforts should be applied to the plutonium project and that our other problems would have to be resolved later. This was in accord with the general philosophy I had followed throughout the military construction program and to which we adhered consistently in this project; namely, that nothing would be more fatal to success than to try to arrive at a perfect plan before taking any important step.

Even before I talked to Compton's group in Chicago, I had begun to realize that we were involved in an enormously bigger undertaking than I had previously understood and that it was unreasonable to expect Stone and Webster to carry the full burden of all the engineering, even with subcontractor assistance, to say nothing of all the construction. It was also clear that the operation of the plants would be so complex that no governmental agency could handle it under the usual and necessary rules and regulations. Moreover, the plants would be so large that we ought to have a different operator for each one. We needed going concerns of sufficient size to provide the solid nucleus of management upon which we could build successful operating units.

In removing some of the load from Stone and Webster, I concluded it would be most logical to reassign the plutonium effort, on which work had not yet started. Next, I had to decide whether to place the various responsibilities for the engineering, the construction and the operation with one firm or with several. To me, a single firm carrying the threefold responsibility seemed by all means preferable. For one thing, it would lessen the problems of co-ordination that would fall into my lap, which by that time was becoming rather crowded with major problems. After I had studied all the possibilities, I concluded that only one firm was capable of handling all three phases of the job. That firm was du Pont.

Its engineering department, always of the highest quality, had become even more vigorous through the tremendous construction program it had been carrying on for the Army. It was thoroughly experienced in both design and construction and was accustomed to dealing with highly technical processes. A further attraction of du Pont, from my point of view, was that for several years I had been working most successfully with its engineers on the Army construction program.

A number of us had come to realize by that time that the operation of the plutonium plant would be extremely difficult, and that it would require highly skilled technical management, thoroughly experienced in large operations. A thorough knowledge of chemistry and chemical engineering was important, and this du Pont had. The need for a sound knowledge of atomic physics was much less vital, and for this we could lean on the Chicago laboratory.

Before arriving at any final decision to bring du Pont in, I discussed the problem with the members of the Military Policy Committee and particularly with Conant. There was no dissent.

My next step was to prepare the group at the Metallurgical Laboratory for the changes that were about to take place. This did not prove to be at all easy. When I broached the subject to Arthur Compton, he agreed at once, saying that he knew Stone and Webster were overburdened and were way out of their field of experience, and that it would be a great relief to have du Pont in the picture. However, he warned me that we would encounter opposition, some of it quite strong and quite influential, from some of the people in his laboratory.

He told me that in the previous June he had assembled his staff and proposed bringing in an industrial firm to take over responsibility for the production phase of the plutonium project. The suggestion had resulted in a near rebellion, particularly among those whose entire experience had been in academic institutions. They simply did not comprehend the immensity of the engineering, construction and operating problems that had to be overcome. Whenever attempts were made to explain them, they brushed them aside as inconsequential. After the furor had subsided, Compton announced that he expected to go ahead with his idea.

He said that while his position had been accepted then, he had no doubt that there would be many objections, voiced and unvoiced, and that the selection of du Pont—the very symbol of large industry—

would be particularly opposed. He went on to assure me that personally he was very much in favor of my proposal and, moreover, that he felt that du Pont was by far the best choice that could be made.

On the other hand, a number of his scientific people, particularly those who had been trained in Europe, where scientific and engineering education were more closely linked than in this country, had the idea that all design and engineering for the project should be accomplished under their personal direction. Some even went so far as to say that they could also supervise the construction. When I visited the laboratory on October 5 and again on October 15, I was told by several different persons that if I would provide them with from fifty to one hundred junior engineers and draftsmen, they would then themselves design and construct the plutonium plant, rapidly and without delay. They added that the plant could then be turned over to a private company for operation, or possibly be run under the Civil Service. The absurdity of such a proposal is apparent when it is remembered that this was the plant where our construction forces reached a peak of forty-five thousand and was so difficult an undertaking as to strain even the great resources of du Pont, with the full power of, and considerable aid from, the government and much of America's industry behind it.[1]

Needless to say, I did not share their views, and their completely inflexible attitude made it virtually impossible to explain to them satisfactorily just why their ideas were unacceptable. Some of the most determined members of this group continued to maintain their position even after the magnitude of the Hanford works became apparent to everyone, and a few die-hards continued to argue the point even after the project had been completed.

Some months later Conant told me he thought that it might help if I would officially appoint Dr. Richard C. Tolman and himself as my scientific advisers. Conant, though it was not generally known, was already really serving in that capacity, and Tolman, a distinguished physicist, the graduate dean of the California Institute of Technology

---

[1] Before the Hanford works were finished, du Pont employed over ten thousand subcontractors.

and a vice chairman of the NDRC, was familiar with our aims and had assisted us from time to time.

I immediately adopted Conant's suggestion, for the soundness of his reasoning was obvious. He felt that this step would alleviate, at least in part, the resentment that some of the scientists felt about having a nonscientist control their work. I also knew that it would be invaluable to me to have two such distinguished able scientists to turn to for advice and assistance, particularly in dealing with scientific personnel. As soon as I secured Tolman's acceptance, the dual appointment was announced to the various scientific groups with which we were involved. It had the helpful results predicted by Conant.

Nevertheless, there was still a small group of scientists, mostly, though not entirely, European born, who felt that they should be given complete control of the entire project. Long before the Army was brought into the picture, they had expressed dissatisfaction over the fact that Bush and Conant were in controlling positions. They seemed to feel that no one who was over forty years old, no matter how distinguished a scientist he might be, could possibly understand the intricacies of atomic energy. This was quite absurd, for it was not and is not extraordinarily difficult for anyone who will apply himself to learning them to understand the basic principles of atomic physics.

Later another angle developed that did not tend to ease matters. Several members of this group had filed patent claims on various phases of atomic energy.[2] These did not seem to me or to our patent adviser, Captain R. A. Lavender, USN, Ret., to be valid under United States patent laws, since none of them had been reduced to practice and there was no certainty that they could be. Consequently we did not show any enthusiasm toward speedy financial adjustment for these claims. This could have provided a conscious or subconscious cause for irritation. In any event, I did not envy Compton's having to contend with the problem on a day-to-day basis.

Much has been said since the war about friction within the Manhattan Project between its military and scientific members. Such friction is natural and to be expected when two groups with entirely

[2] See Appendix III, page 418.

different backgrounds and training are thrown together. At Chicago the situation was made a bit more difficult than usual because the unique array of scientific talent that had been collected there was imbued with an active dislike for any supervision imposed upon them and a genuine disbelief in the need for any outside assistance. I should emphasize that while some of these frictions were annoying, none of them interfered with the successful conclusion of the project. We were not engaged in a popularity contest, but in an extremely serious undertaking. At no time did I ever have reason to doubt the intense devotion to the accomplishment of our goal of the Chicago group—indeed, of the entire Manhattan Project—and to me that was all that ever mattered.

On October 30, I called Mr. Willis Harrington, a senior vice president of du Pont, and asked him to come to see me in Washington to discuss a highly secret matter of the utmost importance. I apologized for being unable to come to Wilmington, and asked him to say nothing about my call or his trip. He wanted to know whether he could bring with him Dr. Charles M. A. Stine, also a vice president of du Pont and a distinguished chemist; I assured him that this would be quite all right. The next day in my office in Washington, Conant and I explained to them the entire atomic situation as it then existed, giving our views on the major problems we faced and the urgency under which we were working. We made no effort to hide our uncertainty about the feasibility of the entire project. We ended by telling them that we needed and wanted the assistance of the du Pont Company in developing the plutonium processes.

We asked for their views on our chances of having a large-scale plant in operation within a reasonable time, assuming that the necessary technical data could be supplied by the scientists at the University of Chicago. We gave them only an absolute minimum of information about the other processes, the contemplated design of the bomb, and how the final product would be used in military operations; for those matters fell outside the job that we were asking them to do.

Harrington and Stine both protested vigorously that du Pont was experienced in chemistry not physics, had no knowledge or experience in this field, and that they were incompetent to render any opihion except that the entire project seemed beyond human capability. I replied that the stakes were high and while we were inclined to share their feeling, we were going ahead anyway, that we needed the best advice and help we could get, and that we thought we had come to the right place for it. In any event, until the process was conclusively proved to be unfeasible, the design, construction and operation of a plutonium plant must be initiated and carried through to the fullest extent of our ability.

I then asked them whether, in their opinion, there was any bar to du Pont's undertaking the task. They replied that they did not think so, but that before they could give a firm answer, they would have to talk with Mr. Walter S. Carpenter, Jr., president of the company, and the other members of their Executive Committee, as well as with some of their chemists and engineers. I told them that, while the entire project was of the highest order of secrecy, they were authorized to discuss it with any member of the du Pont organization in whose integrity and discretion they had confidence. I asked that they keep the number of such persons to a minimum, cautioning them about security, and to keep a list of their names, to be sent to me in case du Pont did not eventually become associated with the project.

This was the procedure we generally followed in bringing new organizations into the picture. We talked to one or two responsible officials, preferably men already known to us, whose judgment and security-mindedness we had no reason to doubt. During the preliminary conversations they were given only a minimum of information. If the opening moves did not develop as we had anticipated, the matter was dropped and they were asked to forget that anything had ever been said. Otherwise, they were requested to explore the situation within their company as prudently as possible, giving adequate warning to those with whom they talked. The urgency of the project did not allow time for us to conduct any detailed security checks in advance of negotiations; instead, we relied upon the discretion and patriotism of American industry. We considered this a good risk and we were never disappointed.

During the following week the top officials of du Pont discussed our proposal, and at their request eight of their key employees were permitted to visit the Chicago laboratory, where they went over the status and plans of the project with Compton and his associates, and received all available theoretical and experimental data.

On November 10, I went to Wilmington to see Carpenter. My purpose was to convince him that du Pont must take over the entire plutonium project. When I entered his office I knew that I had a staggering proposal to put up to him. I knew that I was asking du Pont to embark upon a hazardous, difficult and perhaps impossible undertaking, at a time when it was already straining under the terrific war burden it was carrying. I knew, too, as Harrington and Stine had emphasized, that the work lay in a general area in which du Pont was entirely without experience. Also, I recognized that I was too new to the project to feel completely at home in discussing its many ramifications with a highly experienced chemical executive, even though he would not admit to any knowledge of nuclear physics or radioactive chemistry.

The du Pont Company was not the least bit anxious to accept the grave responsibilities it would have to carry under a contract for the entire plutonium project. Its reasons were sound: the evident physical operating hazards, the company's inexperience in the field of nuclear physics, the many doubts about the feasibility of the process, the paucity of proven theory, and the complete lack of essential technical design data. To these there were added the extreme difficulties involved in designing, constructing and operating a full-scale plant without prior laboratory or semi-works plant experience.[3] Moreover, du Pont's existing military commitments were already producing shortages among the engineering and operating personnel whom we so sorely needed.

[3] Normally the development of any vast industrial process extends over a period of years. Its usual sequence is, first, laboratory research, followed by the design, construction and operation of a semi-works. Only after the semi-works plant is in successful operation is the design of the commercial plant begun.

I had made it very clear both to Harrington and Stine that the government considered the project to be of the utmost national urgency, and that this opinion was shared by President Roosevelt, Secretary Stimson and General Marshall. When I repeated this to Mr. Carpenter, he suddenly asked if I personally agreed. Unhesitatingly, I told him that I did, and without any reservation.

I said that there were three basic military considerations involved in our work. First, the Axis Powers could very easily soon be in a position to produce either plutonium or U-235, or both. There was no evidence to indicate that they were not striving to do so; therefore we had to assume that they were. To have concluded otherwise would have been foolhardy. Second, there was no known defense against the military use of nuclear weapons except the fear of their counteremployment. Third, if we were successful in time, we would shorten the war and thus save tens of thousands of American casualties. (I have always believed it was for these reasons, and particularly the last, that Carpenter and his colleagues on the du Pont Executive Committee agreed to undertake the work in spite of all the hazards it entailed for their company.)

I went on to point out that these considerations required that the plutonium project get under way at the earliest possible moment on a crash basis and without regard to normal procedures. As our discussion drew to a close, I told him that we recognized fully the chances of failure and, particularly, the unknown dangers that might be involved in the operation of the completed plant. While most courteous and interested throughout, he remained noncommittal.

Later that morning, we joined the du Pont Executive Committee. Others present included Nichols, Compton, Hilberry, and some of the group from du Pont who had visited the Chicago laboratory.

I repeated to the Executive Committee the points I had previously made in talking with Carpenter. The du Pont representatives again emphasized their company's inexperience in the field of physics, particularly in nuclear physics. They pointed out that even in one of their own fields of specialization they would not attempt to design a large-scale plant without the necessary data that could be accumulated only by a long period of laboratory research, followed by semi-works operation: for example, it had taken them many years to get nylon into mass production; yet the nylon process was simple compared to what we were asking of them. They added that, regardless of any assumption of responsibility by the government, du Pont's moral obligation to its own employees might well preclude their going into any project that involved such extraordinary and unpredictable health hazards as the operation of a plutonium plant.

I should make it clear that reactor theory at this time did not overlook the possibility that once a chain reaction was started, it could, under some conditions, get out of control and increase progressively to the point where the reactor would explode. If highly radioactive materials were blown into the atmosphere and spread by winds over a wide area, the results could be catastrophic. We knew, too, that in the separation of the plutonium we might release into the atmosphere radioactive and other highly toxic fumes which would constitute a distinct hazard for operating personnel. It was not surprising, therefore, that du Pont should entertain grave doubts about the desirability of joining us in our work.

However, they went on, because of the extreme importance and the urgency of the work—and here they recapitulated my arguments— they felt that they could not refuse to undertake it if the government asked for their company's assistance. They stressed, however, that this was only their opinion as members of the Executive Committee, and that any final commitment by the du Pont Company on a matter of such importance could be made only by its Board of Directors.

The du Pont officers then presented their estimate of the situation regarding the plutonium process. This estimate was based on the impressions of their employees who had just spent three days in our Chicago laboratories. In essence it said:

There was no positive assurance of success for the following reasons: A self-sustaining reaction had not yet been demonstrated.

Nothing conclusive was known about the thermal stability of such a reaction.

None of the pile designs under consideration at that time was believed to be workable.

The feasibility of recovering plutonium from a highly radioactive medium had not been demonstrated.

Making every possible favorable assumption concerning the various stages in the development of the process, production would be limited to a few grams of plutonium in 1943 and not much more in 1944. Given a workable plant, not until sometime in 1945 could production possibly reach the planned rate, and there were many doubts whether this rate could ever be achieved on any basis.

No valid opinion on the practicability of the process under investigation in the Chicago laboratories could be reached without comparing it with the Uranium-235 processes under investigation at Columbia and Berkeley.[4] For the same reasons that had led the Army to seek a critical appraisal of the Chicago process, there should be an examination and comparison of the alternative processes.

In all this we concurred. Two days later Carpenter informed me that du Pont would take the job. He said the decision of the Executive Committee, subject of course to the approval of the Board of Directors, was unanimous.

As the directors entered the room at their next Board meeting, they were asked not to look at the faced-down papers on the table in front of them. Carpenter explained that the Executive Committee was recommending that du Pont accept a contract from the government for a project in a previously unexplored field so large and so difficult that it would strain the capacity of the company to the utmost. He added that there were elements of hazard in it that under certain conditions could very well seriously damage if not well-nigh destroy du Pont. He said that the highest officials in the government, as well as those who knew the most about it, considered it to be of the highest military importance. Even its purpose was held in extreme secrecy, although if any Board member wished to he was free to read the faced-down papers before voting. Not a single man, and they were all heavy stockholders, turned them over before voting approval— or afterwards—a true display of real patriotism.

[4] At Columbia, scientists under Harold C. Urey were working on a gaseous diffusion method of separating U-235 from U-238; in the laboratory at the University of California, under Ernest O. Lawrence, another group was trying to do the same thing by an electromagnetic process.

The successful accomplishment which followed was due to the tremendous capacity and determination of the entire du Pont organization which had been built up through many years of intelligent management.

On November 18, Conant and I had a thorough discussion with Stine and Crawford H. Greenewalt, then one of du Pont's experienced chemical engineers, now its president. As Stine occasionally had before, they expressed a fear that we were asking them to take the most difficult of the processes under consideration—a process that would be unlikely to succeed. When I told them that we were assigning them the one that we thought was the most likely to succeed, their faces and their words reflected their disbelief. Stine commented, "If that is the case, you have even more nerve than we thought, and that is saying something."

To allay their misgivings, Conant and I told them that we would appoint a reviewing committee to investigate and compare the various processes for the production of fissionable material. Furthermore, we would be happy to have, in fact we would insist on having, some du Pont men on the committee, and we would want these men to inform the du Pont management of their findings.

The committee was a well-balanced and competent group,[5] thoroughly capable individually and collectively of understanding the very complex proposals and arriving at sound conclusions. Conant and I considered its mission to be twofold: first, to give us the benefits of their judgment; and second, to give du Pont assurance that we had not misled them in any way, either intentionally or inadvertently.

Earlier in November, we had been faced with a serious problem involving the location of the first experimental test pile. The original plan had been to place it in the Argonne Forest, some fifteen miles out of Chicago, where special facilities were being built to accommo-

[5] The committee consisted of: Dr. W. K. Lewis of MIT, Chairman; Roger Williams, T. C. Gary and C. H. Greenewalt of du Pont; and, originally, Dr. E. V. Murphree of Standard Oil Development Corporation. Unfortunately, owing to a subsequent illness, Dr. Murphree was unable to take part in the review.

date the pile and its accompanying laboratories. The already insufficient time available for this construction was cut even further by some labor difficulties which, while not particularly serious, led to delays.

In the meantime, work had begun on a small pile under the West Stands of Stagg Field at the University of Chicago. This pile was to be used to perform exponential experiments to determine the feasibility of the larger test pile. An exponential experiment, as its name indicates, is one from which, using measurements of the results obtained under varying conditions, the results to be expected under vastly different conditions can be calculated. When the supply of pure graphite necessary for the construction of a self-sustaining pile became available somewhat sooner than had been anticipated, Compton raised the question: "Why wait for Argonne?"

There was no reason to wait, except for our uncertainty about whether the planned experiment might not prove hazardous to the surrounding community. If the pile should explode, no one knew just how far the danger would extend. Stagg Field lies in the heart of a populous area, while the Argonne site was well isolated. Because of this, I had serious misgivings about the wisdom of Compton's suggestion. I went over the situation with him, and told him of my feelings, but I did not interfere with his plans, nor did I display outwardly my concern by being present during the initial test. I learned then that nothing is harder for the man carrying the ultimate responsibility, in this case myself, than to sit back and appear calm and confident while all his hopes can easily be destroyed in a moment by some unexpected event over which he has no direct control.

At this time, the exact status of responsibility for the operations of the Chicago laboratory was still a bit hazy to some, but not to Bush, Compton and me. Compton was in direct charge. The over-all responsibility was now mine. Chicago was still under contract to the OSRD, although we were rapidly taking over all the supervisory responsibilities, and there was still a feeling among the scientific personnel that their first loyalty was to the OSRD. This was perfectly understandable, and it never caused any trouble beyond a few irritations to the MED representatives in the various laboratories.

Several experiments carried out while the pile was under construction indicated that the chances of any untoward happening were very slight. With these results in hand, Compton obtained the agreement of his senior men in Chicago that it would be safe to carry on the self-sustaining reactor experiment at Stagg Field. These same preliminary results led most of them to expect that the reaction would be self-sustaining.

It did not seem possible that, with the control system to be used, there could be an accident. For that reason, I did not interpose any further objection, although through our area engineer at the laboratory, Captain A. V. Peterson, I kept in close touch with the research efforts there to be certain that I would be promptly informed if any adverse developments should arise that might make it advisable for me to stop the work.

The building of the West Stands pile went ahead full tilt. Compton was very anxious to demonstrate the chain reaction to the reviewing committee we had appointed before it completed its study and report, feeling this might greatly influence du Pont's decision about entering the project. I was not too much concerned about this, however, for we already had the agreement of Walter Carpenter and the company's Executive Committee; yet I did recognize, of course, that a successful demonstration of the pile would make the final conclusion of our negotiations easier.

Although the committee was in the Chicago laboratories on December 2, 1942, when the Fermi experimental atomic pile was first placed in oper: tion, the only committee member to witness the actual demonstration was Greenewalt. He was invited by Compton on the ground that he was the youngest and would be able to talk about it for the most years. I was en route east from the Pacific Coast at that time, so Compton could not even inform me of their success. He did telephone Conant at Harvard to pass on the now famous message: "The Italian navigator [Fermi] has just landed in the new world. The natives are friendly."

The December 2 test proved that a controlled chain reaction could be achieved, but it gave no assurance that it could be used to produce plutonium on a large scale. Neither did it give us any assurance that a bomb using plutonium or U-235 would explode. In the reactor the chain reaction was based on slow neutrons, *i.e.*, ones slowed down by graphite or other means. In the bomb, the neutrons would be fast, for because of technical limitations there could be no moderators. Nevertheless, the committee, basing its opinion on what it had seen and heard during its inspections, reported favorably on the plutonium process. Although it did not come up with anything that Conant and I did not already believe, its work did have a reassuring effect upon us as well as upon du Pont. From that time on, whatever doubts that company may have had about its prospects of success were kept within the family. Yet this same normal evaluative process had the unexpected result of causing our most steadfast supporter in the Chicago laboratories to waver momentarily.

On December 23, Compton sent me a copy of a letter he had written, on which was inscribed in his hand: "This letter is addressed to Conant, but I am equally anxious that you give it immediate and careful attention." In his letter, Compton was quite positive. He stated that the production of plutonium following the procedure then in hand was feasible; that there was a 99 per cent probability that it would be successful; that the probability of a successful bomb was 90 per cent; and that the time schedule, assuming continued full support, would see delivery of the first bomb in 1944 and a production rate of one bomb per month in 1945. This was by far the most optimistic estimate that I ever received prior to the explosion of the first bomb some thirty months later; and it was not at all justified by the existing knowledge. As it turned out, he was way off as to the time required, but then this was before all our difficulties were fully understood and before our scientific people came to realize that we were playing for blue chips and that they had to produce.

Compton pointed out that his opinion was in sharp contrast to that of the du Pont engineers. He implied that Stine saw only about a one per cent probability that the production process was feasible. From what Stine had told me on several occasions, I knew that Compton was mistaken. Compton went on to say that assuming success there, Stine's forecast of the production schedule was approximately one year behind what Compton predicted.[6] Compton said that his people had taken into account all the factors considered by du Pont and felt that their own appraisal was correct. He then went on to give a very spirited pitch about the quality of his own people.

He mentioned that he had originally favored du Pont for engineering and operating the process because of its extensive experience, excellent engineering organization and reputation for co-operation with other research organizations. In all of his subsequent dealings with the company he had found no reason to change his initial opinion. However, since du Pont seemed skeptical of the outcome, it would be handicapped from the start, and on that basis Compton preferred not to have it brought any further into the picture.

He closed by proposing that, if du Pont had doubts about the job, we call in General Electric and Westinghouse to engineer and build the power units, keeping du Pont only for the extraction process and for operating the power units after they were built and functioning smoothly. As a Westinghouse engineer for three years, and a G.E. consultant for seventeen, he felt that if he went to these organizations with confidence, they would accept his judgment of the feasibility of the process; and further, that if convinced of its high military importance, they would do everything in their power to see the project through. In his opinion, a co-operative engineering enterprise by these two companies could be arranged, which would be preferable to similar work undertaken by a skeptical and reluctant du Pont. I was not at all impressed by Compton's letter. I knew it did not truly represent the du Pont attitude or Stine's position. I knew that Compton's proposal with respect to the combination of Westinghouse, General Electric and du Pont was entirely impractical if not impossible. I knew, too, that no matter how big the odds were against success, Stine and all of his associates would go all out to achieve it.

Moreover, by this time we had the review committee's report in hand and Conant and I were confident that du Pont was securely in the

[6] Stine was right, as events proved.

fold. Conant's reassurance on this score removed Compton's misgivings as quickly as they had arisen and so we finally achieved the close and wholehearted co-operation between the Chicago scientists and the du Pont engineers upon which our eventual success was so dependent.

In our agreement with du Pont, as in all war work where time was of transcendent importance, we made use of the letter of intent. Once the contractor acknowledged its acceptance, it became a binding agreement. The letter contract with du Pont covered the design, construction and operation of a large-scale plutonium plant. It was understood that all du Pont's work would be based on technical information to be furnished by the Metallurgical Laboratory, and that the government assumed all responsibility for the results of the endeavor, as well as for any damages that might be incurred in the course of the work. This last provision was necessary because of the nature of the entirely unpredictable and unprecedented hazards involved.

Normal insurance coverage was impossible because of the need to maintain security. While we could have disclosed the normal risks involved to a single insurance representative, there would have been little point in it, for reinsurance on large risks requires that adequate knowledge be in the hands of many groups, which would seriously have endangered our security. Moreover, the unusual hazards were such that no group of insurance companies could possibly have written the coverage, even after complete disclosure. First, no one had any reasonable idea of what the hazards might be or the likelihood of their occurring. Second, no one could predict the duration of the effects of the hazard, or, in many instances, even when the effects might first appear. Third, no one could possibly predict the extent of the damage if a major catastrophe occurred.

For all these hazards the government assumed full responsibility. To facilitate the handling of claims not resulting from a major catastrophe a special fund was established. This fund was placed under the control of du Pont so that it could continue to be available for many years. All claims were to be approved by the government before payment.

Ever-present in our thinking was the sad example of the luminous watch-dial painters of World War I. Here the effects did not become apparent for many years. The delayed reaction to excessive radiation also hit many of the original researchers and users of X-rays. How could we be certain that radiation exposure in our installations might not have similar effects despite all our efforts to prevent them?

Du Pont refused to accept our first letter of intent because it contained the standard proviso that, in addition to being reimbursed for costs, it would receive a fixed fee to be computed in accordance with the usual governmental procedures. Mr. Carpenter said that du Pont did not want any fee or profit of any kind for this work, and wanted furthermore to be certain that the company would receive no patent rights. A new letter of intent incorporating provisions to this effect was prepared and was immediately accepted.

As the preparation of the detailed contract proceeded, du Pont expressed a desire to have it approved by the Comptroller General, particularly with respect to the provisions covering reimbursement and indemnifications, in order to make certain that the basic intent of the contract to provide full reimbursement of expenses without profit would not at some later date be upset by his office. I felt this was only fair and that it would also be to the government's advantage. Accordingly, I called on the Comptroller General, Lindsay C. Warren, and asked him to review the proposed contract. Although such a procedure was entirely contrary to the long-established practices of his office, he agreed to conduct the review.

One of his principal assistants, who was then called in to handle its details, opposed the idea very strongly, pointing out that it was contrary to all existing procedures, that it would open the door to similar requests in the future and thus would completely upset the orderly conduct of business in the office, which theretofore had consisted exclusively of passing upon the legality of payments for work already accomplished. Without further ado, Mr. Warren replied: "I promised General Groves to do it and I see every reason why we should and none why we should not."

At du Pont's request, Dr. Bush forwarded a letter to the President outlining the circumstances surrounding the assumption by the

United States of all responsibility for the unusual hazards involved in this work. Mr. Roosevelt initialed his approval on the letter and a photostatic copy of it was given du Pont.

Despite all these protective steps, there was never any question in any of our minds but that du Pont would suffer staggering losses if a major disaster should ever occur. The damage that it would have sustained could not have been measured in dollars lost; all such losses were to be borne by the government. But the damage to its reputation and, consequently, to its future welfare, would have been untold, and the directors and the members of the Executive Committee who had agreed to undertake the work would have been completely discredited.

We encountered one other snag in making sure that, though du Pont was doing the job without profit, it would not be subject to any direct financial losses. For purely legal reasons, provision was made for a fee of one dollar.

Although the expected duration of the contract was stated, as is usual, soon after V-J Day du Pont was paid the entire fee of one dollar. This resulted in a disallowance by government auditors, since the entire time of the contract had not run out. Consequently, du Pont was asked to return thirty-three cents to the United States. Fortunately, the officers of du Pont had retained their sense of humor throughout their many years of association with the government, and were able to derive considerable amusement from this ruling.

CHAPTER 5

LOS ALAMOS: I

Before the MED very little effort had been devoted to the design of an actual bomb, for until the problems presented by the separation of U-235 and the production of plutonium were well on the way to solution, it was felt that there was no pressing need to initiate detailed work on the mechanics of an atomic explosion. All that had been done by the middle of 1942 was to demonstrate theoretically the possible feasibility and effectiveness of an atomic bomb. Its probable design and size were unknown.

In addition to his other work, Arthur Compton had been assigned over-all responsibility for the physics of bomb development. As a first step in June, 1942, he had appointed Dr. J. Robert Oppenheimer to take charge of this particular phase of the project. Oppenheimer was then at the University of California at Berkeley. He began work on the problem with a small group of theoretical physicists.

As I grew familiar with the project and having, as I did, the advantage of a fresh point of view, I became acutely aware of the need to push this particular phase of the work. A number of people disagreed, feeling that the bomb could be designed and fabricated in a very short time by a relatively small number of competent men. One extremist even went so far as to tell me that twenty capable scientists could produce a workable bomb in three months. Discussions with Arthur Compton, Conant and Bush made it clear to me that these estimates were dangerously overoptimistic and that the work should be started at once in order that one part of our operation, at any rate, could progress at what I hoped would be a comfortable pace.

During our numerous talks about the organization of Project Y, as this work was later called, a difficult question arose: Who should be the head of it? I had not before been confronted with this special problem, for the directors of the other laboratories connected with the project had been appointed before my arrival upon the scene.

Although Oppenheimer headed the study group at Berkeley, neither Bush, Conant nor I felt that we were in any way committed to his appointment as director of Project Y. I did not know Oppenheimer more than casually at that time. Our first meeting had been on October 8 at the University of California, when we had discussed at some length the results of his study and the methods by which he had reached his conclusions. Shortly afterward, I asked him to come to Washington and together we had explored the problem of exactly what would be needed to develop the actual bomb.

In the meantime I was searching for the best man to take charge of this work. I reviewed with everyone to whom I felt free to talk the qualifications its director should have, and asked for nominations. Today, Oppenheimer would be considered a natural choice because he proved to be successful. Having been in charge of this particular field under Compton, he knew everything that was then known about it. Yet all his work had been purely theoretical and had not taken him much beyond the point of being able to make an educated guess at the force an atomic fission bomb could exert. Nothing had been done on such down-to-earth problems as how to detonate the bomb, or how to design it so that it could be detonated. Adding to my cause for doubt, no one with whom I talked showed any great enthusiasm about Oppenheimer as a possible director of the project.

My own feeling was that he was well qualified to handle the theoretical aspects of the work, but how he would do on the practical experimentation, or how he would handle the administrative responsibilities, I had no idea. I knew, of course, that he was a man of tremendous intellectual capacity, that he had a brilliant background in theoretical physics, and that he was well respected in the academic world. I thought he could do the job. In all my inquiries, I was unable to find anyone else who was available who I felt would do as well.

Of the men within our organization I had no doubt that Ernest

Lawrence could handle it. He was an outstanding experimental physicist, and this was a job for an experimental physicist. However, he could not be spared from his work on the electromagnetic process; in fact, without him we would have had to drop it, for it was far too difficult and complex for anyone else. I knew of no one then and I know of no one now, besides Ernest Lawrence, who could unquestionably have carried that development through to a successful conclusion.

Compton had a thorough background in physics, and he had had considerable administrative experience. But he could not be spared from Chicago.

Urey was a chemist and, though an outstanding one, was not qualified technically to head up this particular job. Outside the project there may have been other suitable people, but they were all fully occupied on essential work, and none of those suggested appeared to be the equal of Oppenheimer.

Oppenheimer had two major disadvantages—he had had almost no administrative experience of any kind, and he was not a Nobel Prize winner. Because of the latter lack, he did not then have the prestige among his fellow scientists that I would have liked the project leader to possess. The heads of our three major laboratories—Lawrence at Berkeley, Urey at Columbia, and Compton at Chicago—were all Nobel Prize winners, and Compton had several Nobel Prize winners working under him. There was a strong feeling among most of the scientific people with whom I discussed this matter that the head of Project Y should also be one.

I think that the general attitude toward these laureates has since changed. They no longer are looked up to quite so much as they were then, primarily because so many men have produced remarkable results recently without receiving prizes. However, because of the prevailing sentiment at that time, coupled with the feeling of a number of people that Oppenheimer would not succeed, there was considerable opposition to my naming him.

Nor was he unanimously favored when I first brought the question before the Military Policy Committee. After much discussion I asked each member to give me the name of a man who would be a better choice. In a few weeks it became apparent that we were not going to find a better man; so Oppenheimer was asked to undertake the task.

But there was still a snag. His background included much that was not to our liking by any means. The security organization, which was not yet under my complete control, was unwilling to clear him because of certain of his associations, particularly in the past. I was thoroughly familiar with everything that had been reported about Oppenheimer. As always in security matters of such importance, I had read all the available original evidence; I did not depend upon the conclusions of the security officers.

Finally, because I felt that his potential value outweighed any security risk, and to remove the matter from further discussion, I personally wrote and signed the following instructions to the District Engineer on July 20, 1943:

In accordance with my verbal directions of July 15, it is desired that clearance be issued for the employment of Julius Robert Oppenheimer without delay, irrespective of the information which you have concerning Mr. Oppenheimer. He is absolutely essential to the project.

I have never felt that it was a mistake to have selected and cleared Oppenheimer for his wartime post. He accomplished his assigned mission and he did it well. We will never know whether anyone else could have done it better or even as well. I do not think so, and this opinion is almost universal among those who were familiar with the wartime operations at Los Alamos.

Toward the end of 1944, because of his far from rugged constitution and the vital character of his job, I had to consider what I would do if anything happened to him and I had to select a successor. I talked over the problem with Tolman and Conant, but could arrive at no good solution. There was none that was evident.

As to his loyalty, I have repeatedly stated in recent years—in print, on TV, on radio, and before the Personnel Security Board, headed by Dr. Gordon Gray—that I would be greatly surprised if Oppenheimer had ever consciously committed a disloyal act against the United States.

Once his appointment was settled, and Oppenheimer started to consider the setup of his organization, we were faced with the problem of choosing a site for his laboratory. In considering suitable areas, we had to take into account a number of factons that did not apply to the location of our other sites, although outwardly the requirements for Project Y appeared quite similar to theirs. We needed good transportation, by air and rail, adequate water, a reasonable availability of labor, a temperate climate, to permit year-round construction and out-of-doors experimental work, and all the other things that make for an efficient operation. As before, we sought an isolated area so that near-by communities would not be adversely affected by any unforeseen results from our activities. Yet this installation would be different, because here we were faced with the necessity of importing a group of highly talented specialists, some of whom would be prima donnas, and of keeping them satisfied with their working and living conditions.

In view of our requirements, we concentrated our search on the southwestern part of the United States. Major J. H. Dudley, who made an extensive field search over the whole Southwest for us, confirmed our preliminary views that there were only two general areas that might be satisfactory. One would lie somewhere along the Santa Fe Railroad, in New Mexico or Arizona, while the other would be in California. The Navy was already interested in the most promising California site, which did not have suitable living conditions from our point of view, and which would have been extremely expensive to develop adequately. While shielded by surrounding mountains against the chance of an accidental explosion, the teeming millions of Los Angeles County were too near for us to maintain the security we deemed necessary. I gave more than the usual amount of weight to this consideration because our work there would be such that its purpose might easily be suspected from a considerable distance. A general barrier around the entire property, such as we had at Hanford and Oak Ridge, would not be practicable because of the tremendous population in the immediate vicinity. I was also certain that it would be extremely difficult and unpleasant to try to keep our scientific personnel from mixing socially with the faculty of the California Institute of Technology. Inevitably, we would have had security breaches, and there would have been just too many people knowing what we were trying to do.

We considered another site in California on the eastern side of the Sierra Nevadas, not too distant from Reno. Although it was on a line of the Southern Pacific Railroad, it would not have been easily accessible by commercial airlines, nor would the passenger service on the railroad have been satisfactory; and the heavy snows would have impeded our operations in winter.

After Oppenheimer and I had gone over the possibilities at some length, we agreed that there seemed to be nothing that suited our purposes as well as the general vicinity of Albuquerque. There was good rail service between that city and Chicago, Los Angeles, San Francisco and Washington, and all TWA flights to the Coast stopped there. Its climate was excellent, it was well isolated, and had the additional advantage of being far inland, which appealed to me because of the ever-present threat of Japanese interference along the Coast.

Because a New Mexico National Guard regiment had been captured in the Philippines we could count on a population and a state government intensely interested in furthering the war effort. The support we received was superb.

Oppenheimer owned a ranch in the near-by Sangre de Cristo Range, so we could draw upon a firsthand source of information on the general character of the country in judging whether our scientific people might find working there to their liking.

In October, 1942, I met Oppenheimer, Dudley and several others in Albuquerque to look at a site which had been selected by one of my officers. As we went along the road to the north, we drove by many small Indian farms, and I began to have misgivings about the troubles we would have in dispossessing the owners. When we reached the site itself, we found it well isolated and possessed of a fine supply of water, which was most unusual in that area. However, it was hemmed in by cliffs on three sides. Oppenheimer felt that this might have a depressing effect upon the laboratory workers, and I recognized that we would have a number of insurmountable problems if we should ever have to expand. After inspecting the site, I told Oppenheimer that, while I preferred the general area of Albuquerque, this particular property was not satisfactory for our purposes and cited my reasons. He did not argue the point, and his associates were openly pleased with my disapproval, since they had not liked the location.

In the hope of finding something more suitable, we drove over the mountains toward Sante Fe, to look at a possibility suggested by Oppenheimer. As we approached Los Alamos, we came upon a boys' boarding school that occupied part of the area. It was quite evident that this would be an isolated site, with plenty of room for expansion. In fact, I could see only two potential sources of difficulty in its physical layout. One was the access road to the hill on which Los Alamos was located. It was poorly laid out, going up the side of a gulch, and could not possibly carry the heavy traffic that would have to pass over it. A careful personal inspection led me to believe that it would not be too difficult to relocate the road and make it passable, but it was obvious that we could not quickly improve it to the point where it would be completely safe for normal driving. Its use would always involve risk, both to people and equipment.

The other problem was water. It seemed that there would be an ample supply, provided we exerted care in its use. As it turned out, we had considerable trouble because the population at Los Alamos grew far beyond what we had anticipated. We also found—which did not surprise me—that it was almost impossible to control the use of water by the residents. This was not so much because they were not under military discipline as because they were twentieth-century Americans and they are always prodigal in their use of water.

The fact that there were already a number of buildings at Los Alamos, though nowhere nearly as many as we would ultimately require, meant that we could move in our first people at once and expand on that base. This would save us months in getting started.

From the standpoint of security, Los Alamos was quite satisfactory. It was far removed from any large center of population, and was reasonably inaccessible from the outside. There were only a few roads and canyons by which it could be approached. Also, the geographically enforced isolation of the people working there lessened the ever-present danger of their inadvertently diffusing secret information among social or professional friends outside.

The only major problem left was whether the school's owners would object to its being taken over. It was a private school with students from all over the country and, had they chosen to do so, its owners could have made considerable trouble for us, not so much by making us take the condemnation proceedings into court as by causing too many people to talk about what we were doing. When the initial overtures were made to them, I was most relieved to find that they were anxious to get rid of the school, for they had been experiencing great difficulty in obtaining suitable instructors since America had entered the war, and were very happy indeed to sell out to us and close down for the duration—and, as it turned out, forever.

CHAPTER 6

HANFORD: I

Originally, as I have said, we had intended to set up the semi-works for the plutonium process in the Argonne Forest outside Chicago, and to locate all full-scale production facilities at the Tennessee site. This arrangement was basically sound. It placed the semi-works close to the laboratory carrying on the scientific research and development. It avoided the mistake of putting the people in charge of the full-scale plant too near to the laboratory researchers. Whenever this is done, there is a tendency to incorporate into the final plant every new idea before it has been thoroughly tested.

However, it soon developed that the possible hazards from the plutonium process, many of which we were unable to assess accurately, were far too great to warrant operating the semi-works in the Argonne woods. We decided, therefore, to build it at Oak Ridge, which was not too distant from either Chicago or Wilmington.

There was much discussion about who should operate the Clinton semi-works. Obviously it should be either the University of Chicago or the du Pont Company. Chicago University did not want the responsibility because it was so far removed from its other activities. Du Pont did not want it because of its complete lack of knowledge and experience with processes of this nature. Furthermore, the company felt that until it was demonstrated that the process was operable on a semi-works scale, responsibility for its development should remain with the scientists.

After considerable pressure had been brought to bear upon both, the University of Chicago finally agreed to take on the job and du Pont agreed to lend them a sufficient number of supervisors, technicians and clerical personnel experienced in plant operations to make the effort surer of success.

The solution to this problem characterizes the co-operativeness of all those who were concerned with the project. This was a major factor in our success. Virtually every organization that became involved in our work would have preferred to avoid it, but never once were individual preferences permitted to interfere with wholehearted co-operation. On the other hand, no one ever failed to come forward with his views when he considered them to be important to the success of the project.

When du Pont took on the plutonium project in November, Carpenter had suggested that, for safety's sake, it might prove desirable to locate the production plant elsewhere than at the Tennessee site. I accepted his proposal with almost no discussion, for if we built it at Oak Ridge, we would have to take about 75,000 additional acres of land. Moreover, by now we knew it would require a great deal of electric power and power would soon be in short supply in Tennessee. Besides that, I was more than a little uneasy myself about the possible danger to the surrounding population. The Clinton site at Oak Ridge was not far from Knoxville, and while I felt that the possibility of serious danger was small, we could not be absolutely sure; no one knew what might happen, if anything, when a chain reaction was attempted in a large reactor. If because of some unknown and unanticipated factor a reactor were to explode and throw great quantities of highly radioactive materials into the atmosphere when the wind was blowing toward Knoxville, the loss of life and the damage to health in the area might be catastrophic. Moreover, the interruption of important war production, particularly of aluminum, and the disruption of all normal living conditions would be a serious blow to our nation's military effort. It would wipe out all semblance of security in the project and would bring our work to a jarring halt, from which it would take many years, if not decades, to recover. If our electromagnetic and gaseous diffusion plants were working by then, they would probably be rendered inoperable. And it undoubtedly and quite properly would have resulted in a Congressional investigation to end all Congressional investigations.

I mention this last consideration only because, while we never gave any serious thought to it, it did give rise to a number of jokes during an otherwise deadly serious effort. I knew, as did Bush and Conant, as well as the President, Secretary Stimson, General Marshall and General Somervell, that if we were not successful, there would be an investigation that would be as explosive as the anticipated atomic bomb. Once, in 1944, Somervell told me with a perfectly straight face, at least for the moment: "I am thinking of buying a house about a block from the Capitol. The one next door is for sale and you had better buy it. It will be convenient because you and I are going to live out our lives before Congressional committees."

We could not afford to spend time on the research and study that would have been necessary to make certain that the operation of the full-scale plutonium installation at Clinton would be safe, as we know now it would have been. By this time, too, I was getting a better perspective on the size of the entire job, and I did not think that it would be wise, or even possible, to conduct such a tremendous effort in a single area.

Consequently, on December 14—twelve days after Fermi's dramatic proof that a controlled chain reaction was possible—I arranged for a meeting at the du Pont offices in Wilmington to ensure that the du Pont organization, the scientific people at Chicago and the MED would all have the same understanding of the then-accepted atomic theories, the known scientific and technical facts, the scientific probabilities, and the construction and operating problems. After reaching a common understanding, they were to arrive at the criteria for the plutonium plant site, with special attention to the limitations imposed by safety. The site requirements (based on helium-cooled reactors), which all agreed upon, definitely ruled out any further thought of putting the plant at Clinton and were the controlling guides in the ultimate selection of the Hanford site.

These requirements were:

1. An estimated 25,000 gallons per minute of water would be

needed, assuming recirculation of cooling water.

2. An estimated 100,000 kw of electricity would be required, with favorable load and power factors.

3. The hazardous manufacturing area should be a rectangle of approximately 12 miles by 16 miles.

4. The laboratory should be situated at least 8 miles away from the nearest pile or separation plant.

5. The employees' village should be no less than 10 miles upwind from the nearest pile or separation plant.

6. No town of as many as 1,000 inhabitants should be closer than 20 miles to the nearest pile or separation plant.

7. No main highway or railroad should be closer than 10 miles to the nearest pile or separation plant.

8. Theoretically, the climate should have no effect on the process, although its bearing on the engineering problems should not be minimized.

These specifications were far from fixed, and would be changed as our knowledge increased. They were based on what we knew at the time, which was very little indeed. They were intentionally quite conservative. Thus, they provided for the accidental explosion of a pile, although this was considered only a remote possibility, even in that early period. Yet, though far from final, they were all we had to go on in choosing our site and determining the layout and construction of the plant.

The discussion soon brought out the fact that one of the controlling factors, as far as the size of the manufacturing area was concerned, would be the number of piles we would need. Until we learned just what would be required we planned a layout for six production piles and three chemical separation plants, though the best guess at the time was that we probably would not need more than four piles and two separation plants. Of these four piles, three would be in operation, with one in reserve. The operating cycle would be three months under power, followed by one month to unload and to reload the uranium in the pile. The piles did not have to be separated by any great distance— between three-quarters of a mile and a mile would be sufficient—

but the separation plants should be at least four miles from each other as well as from any pile.

Although we were hoping for a thermal generating capacity of 250,000 kw for each reactor, we based our over-all design on 200,000. If our theories proved to be correct—and not all of them did—and if we could design and operate in accordance with them, this would provide a rate of plutonium production that we hoped would let us turn out a reasonable number of bombs in time to be of use. But, as I have already said, no one knew then just how much plutonium or U-235 would be required for an effective bomb; the most reliable estimates varied by a hundredfold from the minimum to the maximum. Never in history has anyone embarking on an important undertaking had so little certainty about how to proceed as we had then.

Incidentally, the same uncertainty obtained in nearly every phase of our work. However, with the agreement of everyone who held any degree of responsibility for the project, I had decided almost at the very beginning that we would have to abandon completely all normal, orderly procedures in the development of the production plants. We would go ahead with their design and construction as fast as possible, even though we would have to base our work on the most meager laboratory data. Nothing like this had ever been attempted before, but with time as the controlling factor we could not afford to wait to be sure of anything. The great risks involved in designing, constructing and operating plants such as these without extensive laboratory research and semi-works experience simply had to be accepted.

I did not attend the meeting in Wilmington myself, but sent Nichols and Lieutenant Colonel F. T. Matthias as my representatives. Matthias was not then assigned to the MED. However, I had been using him on a part-time basis to work on various problems involved in the Pentagon's construction and on a few special studies for the MED. From the latter, he had gained some concept of the scope and purpose of the project. I had instructed him to attend this meeting solely as an observer and, on his return, to give me his reactions to the discussion, the decisions and the people there. I told him that I was especially interested in the people.

When Matthias came back to Washington after the meeting, he was surprised to find me waiting for him at the railroad station. I drove him to his home in my car. During our ride of some twenty minutes, he gave me his impressions, and I then told him to start the next morning on a study of areas where there might be enough power for the plutonium plant. Two days later, I met with the site selection team, which now consisted of Matthias and A. E. S. Hall and Gilbert P. Church, of du Pont. We went over the requirements that had been worked out in Wilmington and considered the locations that would most nearly meet them. From many years of travel throughout the country, both as the son of an Army officer and as an officer myself, and particularly through the experience I had gained the past two years when I was head of military construction operations in the United States, I was quite familiar with most of the general areas in which the plutonium plant could be located. This, added to the information gathered by Matthias and that which had been put together by du Pont, enabled us to determine pretty well the most promising site areas before the team left Washington that afternoon.

At the head of the list, and the most likely prospect, was an area about thirty miles south of the site finally chosen at Hanford. A second possible choice was quite close to Grand Coulee Dam. Beyond these two, our prospects seemed dim. The two remaining most likely prospects appeared to lie in northern California, with power from Shasta Dam, and in southern California, with power from Boulder Dam. I made it clear to the selection team that I very much preferred the Pacific Northwest area, not only because of the availability of power from Bonneville and Grand Coulee, but because of the open winters and the long, dry, not excessively hot summers, which would permit uninterrupted construction throughout the year. During their inspection, they were to take full advantage of any knowledge, advice and assistance obtainable from the various Army District Engineer offices concerned. To facilitate this, I telephoned Colonel Richard Park, the District Engineer in Seattle, that afternoon and asked for his co-operation, for the areas in which we were most interested lay in his district.

Although they were unaware of it, my purpose in sending Matthias and the du Pont men out together to inspect the sites was to make certain that Matthias and Church, both of whom I considered truly superior, would work well together. Church had been tentatively selected by du Pont as the construction head for the job. I had originally intended to use Matthias as my executive officer in Washington, but had since decided, in accordance with Colonel Marshall's request, to put him in charge at the plutonium plant, for I was much more interested in having that job adequately covered than I was in keeping the best available man in my office.

During the site reconnaissance, it was Matthias' responsibility to see that my wishes were carried out, just as it was the responsibility of Hall and Church to make certain that nothing was overlooked from du Pont's standpoint. Matthias made all the necessary arrangements with the military authorities throughout the trip; Church and Hall were introduced as civilian employees from the Office of the Chief of Engineers because, for security reasons, we did not want to stir up undue curiosity, which the knowledge that du Pont was involved might have done.

For two weeks the reconnaissance party scoured the West from Washington to the Mexican border, and arrived back in my office on New Year's Eve, unanimously enthusiastic about the Hanford area. An important factor in their choice was the discovery that the high-voltage power line from Grand Coulee to Bonneville ran through the site and that there was a substation at its edge. Reinforcing the group's good opinion of Hanford was the fact that the next best site had been recently converted to an aerial gunnery range, and to have taken it over would have drawn undue attention to our work.

The members of the group took about an hour to give me their verbal report, and then went their several ways. They had already finished writing their formal report while traveling and in hotel rooms at night.

The real estate appraisal, which is always an important part of government condemnation procedures, was started on January 7, 1943. I went over the site on January 16, and approved it. I was pleased with the relatively small amount of cultivated land we would have to take over. Most of the area was sagebrush suitable only for driving sheep to and from summer pastures in the mountains and even for that purpose could not be used oftener than once in several years. The total population was small and most of the farms did not appear to be of any great value.

Hanford was then a very small town. It may have had two small general stores, but I recall only the one where we stopped to buy some crackers for lunch. The principal deterrent to the development of this area had been that all irrigation water had to be brought in from a point twelve miles away, and consequently was so expensive that the average agricultural venture could not succeed. There were some orchards of cherries and apricots and a few farms had large flocks of turkeys. There was also one farm, not in Hanford proper, but on the over-all reservation, where mint was raised. And mint, because of the war, had attained a very high price. One ranch across the Columbia River seemed quite attractive, but it was not particularly large. Most of the farms in the area, however, conveyed the distinct impression that the owners were having a pretty hard time making them pay.

The soil appeared to be mostly sand and gravel, which is almost ideal for heavy construction. The plateau on which the plant would be built was only a few miles from the Columbia River, which had a superabundance of very pure and quite cold water. The site was well isolated from near-by communities, the largest of which was the town of Pasco, and if an unforeseen disaster should occur, we would be able to evacuate the inhabitants by truck.

On February 8, the Secretary of War's formal directive was issued, authorizing the acquisition of the necessary land. Direct responsibility for the acquisition of the site was vested in the Corps of Engineers.

In setting up the land requirements we divided the area into three sections. The first of these, where the plant was to be located, was taken over by the government and was cleared of all persons who were not directly connected with the operation of the plant. The rest of the land was needed to ensure safety.

The second section was also taken over by the government, but certain portions of it could be leased by the original owners if they so desired, or by the adjacent property owners. The only stipulation here was that no one would be allowed to live on this land.

The third section was either purchased in its entirety and then leased back to the former owners, or easements were obtained, giving the government certain rights, particularly that of evicting any inhabitant at any time without warning and without having to give any reason. In all these real estate activities, there was the definite understanding that there would be no abnormal increase in the population. Of course, this did not mean that a family could not have another child, or that a farmer could not hire an additional hand. It did mean, though, that no one could open up any new operation like a tourist court or a camping ground. Our purpose here was to avoid complications in case the area had to be evacuated. This third section of the tract caused us much trouble before we were through with it.

The Hanford site was one of the largest procurements of land handled during the war, or at any other time. The total acreage taken was something less than a half-million acres.

The suitability and the exact boundaries of the site were studied as thoroughly as time permitted during January. It was essential that there be no afterthoughts, for once the actual purchasing or condemnation of a particular parcel of land was started, it would be confusing and expensive to make changes.

As we learned more about how the plant would probably operate, we became concerned over the possible effect of air stratification and other meteorological conditions on the dilution of the various waste gases. For this reason, and to avoid any future discussions, we insisted that the site must have the definite approval of the Metallurgical Laboratory.

There was one serious mistake made in our handling of the land procurement. From my first inspection, I knew that we would not have to occupy much of the area until the crops could be brought in the following summer, for we would need most of the intervening period for drawing plans and obtaining materials and mobilizing our working forces. We had no plans; we did not know what the plans would be like; we had to start from the very beginning, and it would be a long time before construction would really get under way. Because of this, I felt that we should not ask for possession of land under cultivation immediately. It would have been perfectly proper for us to do so and the condemnation could have been put through without any particular trouble. But I felt that it was only fair to give the owners plenty of time to settle their affairs and get relocated and, besides that, by withholding any immediate action we would give the farmers another crop-growing season in support of the war effort.

Actually my decision in this case cost the government a considerable amount of money, for reasons I did not foresee. Growing conditions that spring and early summer proved to be astoundingly good, so that the crops were better than they had been at almost any time since farming began in that area, and were extremely profitable to the growers. When it came time for the courts to settle the land values, the juries decided on much higher values than had been anticipated, or, in our opinion, were fair.

Compounding the problem was the fact that the juries were almost exclusively from Yakima, which was the center of a very rich agricultural section. Land values there were far greater than in Hanford, and it was only natural for any Yakima jury to think of farming at Hanford in terms of their own highly productive land. This bad situation was made virtually impossible by the attitude of the federal judge before whom the cases were tried. For reasons known only to him, he allowed the juries to feel that the government was not coming into these cases with clean hands, and that the whole proceeding was a means whereby the government could condemn private land for the contractor.

Despite every effort we could not stem the tide, and the government had to pay what I have always thought were exorbitant prices for the land.

Long before all this confusion in the courtroom began, the first du Pont employee, an engineer named Les Grogan, arrived in Hanford on February 28, 1943.

CHAPTER 7

HANFORD: II

Before the plans for the separation plants for the plutonium process could be completed, we desperately needed a supply of uranium that had been exposed to radiation—that is, uranium which had been subjected to neutron bombardment in a pile and thus contained plutonium. This was a compelling reason for the decision to build the small semi-works plant at Clinton.

The Clinton plant had to be designed and built with a minimum of preliminary study and thought—far less than would have to be devoted to the reactors in the final production plant. Our major consideration was speed. Although the works would be relatively small, we had to apply the same basic principles and incorporate in them many of the facilities necessary for the large-scale production and separation of plutonium that we would use later at Hanford.

It was different in a number of details: The Clinton pile was air-cooled while those at Hanford used water; it was designed to operate at a rate of 1,000 kw, a small fraction of the amount used by the piles we built later at Hanford. Air-cooling was used because it made for simpler, and therefore more rapid, construction. While air-cooling was entirely practical for a pile of this size, it could not be used for the larger reactors because of heat transfer problems. Later, it was found possible to increase the power of the Clinton pile through a number of modifications.

At the time the design of Hanford was begun, plutonium had been separated from uranium only in infinitesimal quantities and by laboratory methods. The most feasible process was largely a matter of opinion and the equipment needed was even more conjectural. Yet design had to be carried forward on the basis of what information we had.

Because of the overwhelming importance of time, we had to ignore the normal methods of orderly development. Design, procurement and construction had to proceed concurrently with the selection of the separation process, with its development, and with the growth of the required basic scientific knowledge. The primary purpose of the Clinton semi-works, which it achieved, was to produce a sufficient amount of irradiated uranium to give us the essential information upon which to base the Hanford separation process. This despite the fact that the Hanford installation had to be designed and much of it built before we had the benefit of any of our experience at Clinton.

Within du Pont, all work on plutonium was placed under the jurisdiction of a special section of the Explosives Department, known as TNX, headed by Roger Williams. The company's Engineering Department was delegated responsibility for engineering under TNX for whatever design and construction might be required. The Chief Engineer was Everett G. Ackart; the head of construction, and later Assistant Chief Engineer, was Granville M. Read; the head of design was Tom C. Gary; and, as I have said, Crawford H. Greenewalt represented du Pont at the Metallurgical Laboratory in Chicago. Greenewalt's was a particularly difficult and important assignment, for he served as the bridge between the hard-driving, thoroughly competent, industrial-minded, scientific engineers and executives at Wilmington and the highly intelligent and theoretically inclined scientists at Chicago. This meant he had to shuttle back and forth between Chicago and Wilmington and later Hanford, easing tensions and calming tempers and, at the same time, seeing to it that needed scientific decisions were promptly reached at Chicago. He was eminently qualified for this liaison assignment, for he was a well-trained chemical engineer and had had more than twenty years of experience in the du Pont organization. He did a superb job.

Information on research developments at Chicago was supplied to du Pont, largely through Greenewalt, in several ways. Copies of all research reports were sent to du Pont. Du Pont people made frequent visits to Chicago and to Clinton to discuss the latest results of research, to review the preliminary designs and to obtain specific detailed information. Some du Pont people were stationed at Chicago and Clinton throughout the design period to provide a link between theory and practice. Every major decision involving details of layout and process had to be concurred in by Compton's people, to ensure that the plant would be built on the basis of the latest and most reliable technical knowledge. This was constantly changing. In addition to the usual final review and approval by the district of all drawings, those that dealt with the process details were formally approved by the Chicago laboratory before they were released for construction.

Security necessitated special handling of the multitude of drawings, reports and correspondence that passed between the various offices; it also made communication by telephone and telegraph difficult, so we had to take extra care to avoid administrative delays. To speed up construction, working drawings were broken down to disclose as little information on the over-all project as possible. By this means, it was possible to treat many of them as unclassified. Such measures were particularly necessary when drawings were sent to subcontractors or used by construction workers at the site.

Even before du Pont came into the project, the laboratory at Chicago had begun work on the preliminary design of a helium-cooled pile that would operate at a much lower power level than that then being considered for the main plant. After receiving a detailed report from the laboratory, du Pont continued work on this design during December, 1942, and January, 1943; at the same time it began an intensive study of the relative advantages of helium- and water-cooled piles.

Our plans for basing the Hanford design upon helium-cooled piles were finally abandoned in February of 1943 when everyone agreed that helium-cooling, while possibly more attractive from a theoretical standpoint, would present a great many practical difficulties in handling and purifying the large volume of gas that would become radioactive and which, because of its radioactivity, would make the maintenance of equipment difficult, if not impossible. A major consideration in dropping the helium-cooled pile was the problem of designing and maintaining the necessary pressure-sustaining enclosure for the pile. Among other difficulties was that of loading and unloading such a unit under pressure. When it developed that the water-cooled pile would be easier and cheaper to design and build, all work on the helium method was stopped.

Although at the time the Hanford site was selected we had expected that the piles would probably be cooled with helium, we had also recognized that water-cooling might be used, and had sought a site which would be suitable if this proved to be desirable. Later, after we had decided on water-cooling, we discovered that not only was it necessary to have large quantities of cold water, but that its purity was of the utmost importance. We were just lucky that the Columbia River water did not contain dissolved chemicals in sufficient amounts to necessitate more than normal treatment. Even so, as the first pile was being finished, fears arose that if we were to keep the reactors working, we might require cooling water of a purity equivalent to that of distilled water. For that reason, we considered building a large deionizing plant for the second reactor so that at least one of the three piles would certainly be operable at all times.

I was discussing the advisability of this with G. M. Read of du Pont, one night at Hanford, when Dr. Hilberry came into the room. I asked him for his views, and he replied that he did not think we would need the deionizing plant, but if we did, we could not do without it. I turned to Read and said, "Go ahead and build it." Hilberry then asked what it would cost and I told him that it would be somewhere between six and ten million dollars. He replied, "I'm glad I didn't know that when I gave my opinion." Such quick decisions were not too frequent and they were always preceded by as much research, study and thought as could be devoted to them without delaying the completion of the project. Nevertheless, there were many decisions that had to be made when the unknown factors far outweighed the known. We built only the one deionizing plant and fortunately never found any need for it, for had it proved necessary, we would have had to build two more in a hurry, and would have lost considerable production while they were under construction.

We expected the cooling water leaving the pile to contain radio-active materials with relatively short half-lives. (The half-life of a radio-active isotope, or material, is the time required for it to lose half of its radioactivity.) The design provided for conducting this water underground to a retention basin for its final radioactive decay. We took this step to avoid any possibility of injury to fish in the Columbia River. We were certain the dilution would ensure the safety of the human population downstream.

Shortly after the Hanford site was selected, I had talked to Robins, who had built the fish ladders and elevators at the Bonneville Dam, and outlined the measures we were taking to protect the salmon in the Columbia River. He made a lasting impression on me at that time when he said, "Whatever you may accomplish, you will incur the everlasting enmity of the entire Northwest if you harm a single scale on a single salmon." As it turned out, we did not.

The concrete side walls of the retention basins were designed to extend high enough above the ground to prevent anyone within a critical distance from being exposed to radiation. To avoid any turbulence in the river, the discharge lines were brought into the main stream at an angle to provide for a converging flow, and, to prevent fish from swimming up the discharge pipe, a minimum velocity of over twenty miles per hour was planned. In addition, all effluent was monitored continuously by instruments to make certain that the radioactivity was at all times within entirely safe limits.

There were four outstanding factors that controlled reactor design. These were: first, the hazards to life and health if the radioactive gases should leak out or if the shielding failed mechanically or was insufficient, thereby exposing a portion of the area to radioactive emission; second, the amount of heat liberated while the piles were operating at capacity, which might result in a spontaneous temperature rise beyond controllable limits, if pulsations or interruptions in the cooling water should occur; third, the characteristics of the specific materials within the pile that required cooling water of the highest quality obtainable; and, fourth, the completely unknown effects on the strength of construction materials of the continuous high bombardment of neutrons.

To obtain a maximum yield of plutonium, processing of the first uranium through the pile had to be completed several months before we began separating the plutonium from the leftover uranium and the other fission products. Each pile unit was made of carefully machined, very pure graphite blocks with built-in aluminum tubes which were charged or loaded with uranium in the form of small cylinders or slugs. Since the piles were water-cooled, we were greatly concerned about the effects of corrosion, for it was estimated that the failure of only a small percentage of the tubes could cause the complete failure of the pile.

All design was governed by three rules: 1, safety first against both known and unknown hazards; 2, certainty of operation—every possible chance of failure was guarded against; and 3, the utmost saving of time in achieving full production. The complications were many, for many pieces of equipment weighing as much as 250,000 pounds each had to be assembled with tolerances more suitable for high-grade watchmaking. It was through the assistance and the strength of the industrial companies of America that du Pont was able to solve the hundreds of difficult design and material problems that had to be mastered.

It is hard now to realize how difficult some of our developmental problems were. The aluminum tubes illustrate how complicated even the most simple item could become. Seven months of persistent effort, principally by the Aluminum Company of America, were required to perfect a metal of the proper characteristics so that a satisfactory tube could be produced in quantity.

Although experimental development was started promptly it was only a few months before the first pile operation started that we were able to perfect the very special techniques required for the canning of the uranium slugs which went into the reactors.

The shielding for the pile was another problem. Ten months of work went into this before we could even begin to build it, with three more months before the first unit was completed. First, scientific requirements had to be adjusted to available, usable materials; then the program had to be outlined for design and procurement. In the course of this, a special high-density pressed-wood sheet was developed in collaboration with an outside supplier. Then special sharp tools and operating techniques were required to cut the various shapes from the standard manufactured widths. Next came the procurement of the thousands of tons of steel plates and the millions of square feet of pressed wood. At the same time the very detailed specifications for the assembly, prescribing the closest of tolerances, were written. Some sixty manufacturers were invited to bid and refused, presumably because of the complexity of construction and the close tolerances required, coupled with the short time available to develop sound fabricating techniques; but after methods were developed and prototypes fabricated at du Pont's shops in Wilmington eventually satisfactory suppliers were found.

We also encountered an extremely difficult problem in the welding of the steel plates surrounding the piles. This work had to be almost perfect. An average superior job would not do. It took months to perfect the techniques required. We created a special super-classification of welders with premium pay. To hold this classification, a welder was required to take special training. He then had to take practical examinations at regular intervals to make certain that the quality of his work would remain up to what we needed.

For one part of the reactor ordinary materials had to be converted into unusual shapes with extremely close dimensional tolerances for the sizes and weights involved. There were a number of most unconventional assemblies, such as the control and safety rods, special instruments, slug-charging and discharging mechanisms, a heavily shielded elevator cab, and the entire cooling system. These items all required the utmost in careful design and diligent inspection of every minute detail.

In order to insure an adequate water supply for each pile, completely independent water facilities were provided, each with duplicate lines. At the same time, all individual units were cross-connected. Arrangements were made for driving the water pumps by either electric motors or steam turbines, so that in case of a power failure from either source, a safe amount of water would still be provided. In addition, there were emergency elevated tanks with automatic cut-ins, in case the normal supply failed. These elaborate precautions were necessary to permit curtailed operation for the time needed to correct any source of trouble.

The piles themselves were surrounded by heavy shields of steel, pressed wood and concrete, to protect the operators from the extreme radioactivity that accompanies the formation of plutonium. The energy of this radiation is equivalent to that of hundreds of tons of radium.

Each pile was located within an area of one square mile. At first they were six miles apart. If additional piles became necessary, we planned to intersperse them between those already in existence, so that the distance between them would then be three miles.

Design of the equipment for the chemical separation plants had to keep abreast of, and in some cases ahead of, the development of the process itself. Fortunately, the two separation processes that seemed to offer the best prospects employed virtually the same equipment and piping layouts, so that it was possible to go ahead with a design the first stages of which would be suitable for either process. Almost every one of the major design decisions for Hanford had to be made before the Clinton pile was in operation.

Originally eight separation plants were considered necessary, then six, then four. Finally, with the benefit of the operating experience and information obtained from the Clinton semi-works, we decided to build only three, of which two would operate and one would serve as a reserve. I should like to point out that these separation plants were designed when we only had sub-microscopic quantities of plutonium. Here again, each plant was provided with its own water system and steam plant and the other service facilities needed for independent operation. Each plant was a continuous concrete structure about eight hundred feet long, in which there were individual cells containing the various parts involved in the process equipment. To provide protection from the intense radioactivity, the cells were surrounded by concrete walls seven feet thick and were covered by six feet of concrete.

In use, the equipment would become highly radioactive and its maintenance and repair would be difficult, if not impossible, except by remote control. Consequently, periscopes and other special mechanisms were incorporated into the plant design; all operations could thus be carried out in complete safety from behind the heavy concrete walls. The need for shielding and the possibility of having to replace parts by indirect means required unusually close tolerances, both in fabrication and in installation. This was true even for such items as the special railroad cars that moved the irradiated uranium between the piles and the separation plants. The tracks over which these cars moved were built with extreme care so as to minimize the chances of an accident. Under no circumstances could we plan on human beings directly repairing highly radioactive equipment.

After being discharged from the reactors, the uranium slugs were kept under water continuously, then sent on the specially designed railroad cars to an isolated storage area. There they were immersed in water until their radioactivity had decreased enough to permit the separation of their plutonium content by chemical treatment.

Following the removal of the plutonium the residues were still highly radioactive. They still had to be handled by remote control. These waste materials included the process solutions from the separation plant. They were finally placed in steel-lined, reinforced concrete tanks buried in the ground to guard against their possible injurious effects. Special provisions also had to be made to take care of the heat that was constantly being generated by this waste material. In the beginning sufficient waste storage capacity was installed to handle one year's operation, but as the piles continued to operate, additional storage had to be provided. We had always thought that it would be possible by intensive research to eliminate much of this radioactive problem in the future. We also hoped to recover the uranium remaining in the existing solutions and to reduce the bulk of the radioactive waste materials, thus making them easier to handle.

In designing the plants at Hanford and elsewhere, all possible care was taken to safeguard the health of the people who would be working in them.[1] Inevitably, this consideration enormously increased the time—which was most important—and the cost, but we strictly fol-

[1] See Appendix IV, page 420.

lowed the policy that where knowledge was lacking, every imaginable precaution must be observed. While placing great reliance on the Chicago laboratory for the adequacy of design from the standpoint of safety, du Pont did not delegate its responsibilities in this area. It sent its own medical and health personnel to Chicago for indoctrination and training, and also borrowed from the University an experienced roentgenologist and a health physicist. Thus, it was able to strike a proper balance between operating needs and safety requirements.

Experience gained at the Clinton laboratories indicated that the danger to anyone outside the immediate operating area would be much less than we had originally feared, but that the danger from the toxicity of the final product was considerably greater. Not too much was known about the human body's tolerance for neutrons. But their danger was realized and all the necessary steps were taken to ensure the safety of everyone who might be subjected to them.

Radioactivity was always a serious and extremely insidious hazard, for without special instruments it could not be detected. There were three types of radiation involved: alpha, beta and gamma. Alpha and beta rays are high-speed particles of very short range and limited penetrating power. Gamma rays are long-ranged and possess great penetrating power. They produce changes in the human body by ionizing the atoms within the body, thereby destroying or damaging body tissue. Those body cells which multiply rapidly, such as bone marrow, are most easily affected by gamma radiation, while the slower-growing cells are relatively unaffected. Beta rays affect primarily the tissues that are close to the surface of the body. Gamma rays, on the other hand, affect all body tissues and an overdose was certain to be disastrous, for no medical cure for its effects was known.

The National Advisory Committee on X-Ray and Radium Protection had established a tolerance dose for gamma rays at one-tenth of one roentgen per day. Because this was not definitely known to be safe, the tolerance dose at Hanford was set at one-hundredth of a roentgen per day. This dose could be absorbed in a short period of time, or over an entire day, so long as it was not exceeded within a period of twenty-four hours. It was calculated that one foot of lead, seven feet of concrete or fifteen feet of water would provide adequate protection from the maximum radioactivity to be expected during the operation of one of our reactors.

Each part of the plant where radiation could be a factor was studied separately and designed to make certain that the safety precautions were adequate. For example, a great majority of the process pipes in the separation plant were buried in concrete and designed to prevent the escape of radioactivity along straight paths. After operations started we found that the safety measures had been wisely taken.

In addition to all the other precautions, du Pont designed a control system for the piles that we thought would be safe no matter what happened. It consisted of three distinct elements: first, the control rods could be moved either automatically or manually into the side of the pile; second, safety rods were suspended above the pile so that, in an emergency, they could be instantaneously released; and third, as a last resort, arrangements were made to flood the pile with moderating chemicals. This last device was designed for remote operation from a shielded control room. If this safety device had to be employed, the pile would no longer be usable.

Besides the two main features of the plant—the piles and the separation works—we built a special pile to test the graphite, uranium and other materials that would be used in the main reactors. Similar to them in principle, it was operated at very low power so that it did not require extensive cooling and shielding, and its construction was therefore much less complex even than the pile at the Clinton semi-works.

In the same area that contained the test reactor there were facilities for extruding the uranium billets into rods and then machining the rods into slugs and finally for canning the slugs. There were also extensive laboratories and a semi-works separation plant for the study of the separation process, using radioactive materials.

To save time, all the uranium rods needed for the first loading were extruded off the site, but practically all machining and canning were done in this area. Because of our lack of knowledge and our feeling that the loading operation might be quite hazardous, a number of new instruments were devised and installed. Since even the machines and tools for their manufacture did not exist, we had to design and to assemble many of them at the plant and to provide for necessary maintenance, including periodic rebuilding. For this purpose we set up a large instrument shop.

We were fortunate that Hanford was served by adequate electric power. Grand Coulee was able to make 20,000 kva available immediately and could supply our entire power needs by September, 1943. Bonneville had no surplus power at the time, but new generator installations were in progress. Although the bulk of this increased capacity had been committed to the aluminum industry, we were certain that if an emergency should cut off Grand Coulee, we would receive an adequate amount of power from Bonneville.

Within the Hanford site, we had to build over fifty miles of 230,000-volt transmission lines and four step-down substations. Because of the copper shortage, the War Production Board thought that we should use aluminum cables. We encountered considerable difficulty and delay in obtaining any decision from the WPB on this matter and it was not until July, 1943, that procurement could begin. Then only the most vigorous expediting enabled du Pont to obtain the material in time to meet its construction schedules.

Added to all these problems was the very urgent one of providing adequate living accommodations and the essential community services for all the Hanford workers. The small village of Richland had been acquired as part of the site and was used as a base upon which permanent living quarters could be built. Provisions were made for the permanent housing of some fifteen thousand people. The main administration buildings, the central service facilities for the plant, and all the other structures normally required by such a project were located in the village. Among other things, we had to build laboratories, storehouses, shops, change houses, fences, electric, steam and water lines, sewers and storage tanks, as well as hundreds of miles of roads, railroads and distribution lines.

In all that has been written about the Manhattan Project, little attention has been given to what life was like for the thousands and thousands of people who worked at one or another of our wartime plants. Life in each had its own unique aspects but certain factors were common to them all—isolation, security restrictions, spartan living conditions, monotony. It was perhaps hardest, in many ways, on the women. Hanford affords a good illustration of what I mean. Here we employed several thousand women in various capacities—as file clerks, stenographers, secretaries and so forth.

To look after their welfare and to see that they were as content as possible under the circumstances, we were fortunate enough to secure the services of a remarkable woman, Mrs. Buena M. Steinmetz, then Mrs. Maris and the Dean of Women at Oregon State College. It was her responsibility, as Supervisor of Women's Activities, to deal with the problems that are bound to arise when a large population of women are housed as a group in an isolated area under rugged conditions, with few of the amenities of normal life.

Admittedly, our concern with morale was not entirely altruistic, for a stable clerical force was essential. We simply could not afford a constant turnover. The trouble was that employees found it easy to get jobs in Yakima, Seattle and other near-by cities where living conditions were far pleasanter. The turnover hazard started on arrival.

Recruited for a highly paid wartime job in the "great Northwest," and transported at considerable expense, often across the whole United States, many of these girls and women arrived at Hanford with unrealistic expectations. Disillusionment sometimes set in almost as soon as they got off the train at the railroad station in southeast Washington, in the middle of the night. Weary from long travel in day coaches crowded with wartime traffic, expecting to get into their quarters at last and have a bath and a good night's sleep, they found instead that they still had another forty-five-mile trip by bus to Hanford. There they had to bunk the rest of the night in a reception center, without their luggage and with the barest comfort facilities, and then face a day of employment routine before they could seek the relaxation of bath and bed. For many of the new employees the greatest disappointment was to find that Hanford was out in a sagebrush "desert"—when they had dreamed of Washington's famed evergreen forests.

To offset their natural disgruntlement, Mrs. Steinmetz held an induction session for each group of new arrivals to impress them with the overriding importance of the work being done at Hanford and to help them realize that in their jobs they would be contributing directly to an unprecedented effort to bring the war to an end. Since most of the women had someone they loved in the armed forces, this was an effective appeal. Inevitably the effect was not permanent in every case, but whenever one of the women decided to leave the project, Mrs. Steinmetz would endeavor to find out why she was leaving. In this way she picked up many clues that were useful in bettering the morale of the other employees.

In addition Mrs. Steinmetz served as a kind of wailing wall, maintaining day and evening office hours during which the women could come to her with their complaints and troubles and sorrows, knowing they would find comfort and sound advice. Also, in each of the barracks in which the women lived there was a responsible hostess, or "house mother," who gave them a further feeling of security and support.

The du Pont administrators at Hanford, she recently wrote me, "were unlike any team of administrators I have ever known in their complete commitment to the job to be done. Despite the pressures on them and the burden of making vital decisions daily, their doors were always open to me, and they apparently always had time to consider any problem I thought important enough for their attention."

What never ceased to amaze her, she added, was the promptness with which they acted. For example, the area between the gate in the barbed-wire fence that surrounded the women's compound of twenty or so barracks, and the barracks themselves, was covered with river gravel as a protection against mud and dust. But the gravel in short time wrecked the shoes of anyone who had to walk on it frequently—and shoes were rationed. After a while there was a ground swell of protest. On one of Read's visits to Hanford he asked

Mrs. Steinmetz, "What do you need?" "Sidewalks or asphalt," she answered, thinking that *sometime* they might get *something*. To her astonishment, trucks were spreading asphalt the next day. A small thing, perhaps, but it kept us from losing employees we could not afford to lose.

Concerned also with a wholesome program of activities for the families of men employees, who lived in trailer housing, Mrs. Steinmetz on another occasion arranged for the use of an old vacant church building in Hanford as a community center, and asked to have some necessary repairs made. Then she called a mass meeting of all the women at the project who would be interested in either Red Cross or Girl Scout work. The turnout was large, and hardly had the meeting got under way than a crew of workmen swarmed over the building; some of them mounted to the roof and proceeded to rip off the old shingles and clap on new ones, while others descended to the basement with a portable concrete mixer and poured a new floor for what was to be the community center kitchen. In spite of, or perhaps encouraged by, this unexpectedly prompt response to their needs, the women carried on amidst the din. Incidentally, the Red Cross and Girl Scout programs organized that day are still in existence at Richland.

Wartime communication was inadequate at best, and one of the greatest values of the Red Cross at Hanford—and the main reason that Mrs. Steinmetz persuaded it to organize a chapter there in the first place—was that it gave the employees, far from their homes as they were, an assurance that important messages from or concerning members of their families, in the services or out, would reach them promptly. It was a link with the outside world, and kept them from feeling totally cut off.

Another cause for dissatisfaction at first was that there were no clothing shops at Hanford, and only one inadequate beauty salon for thousands of women. To remedy this state of affairs, a good women's apparel store was persuaded to install a branch, and a special five o'clock bus was put on the route to the town of Pasco, forty-five miles away, which made it possible for the women employees to have dinner in town occasionally, do some shopping, get their hair done, or see a movie and so forth. There was a late return bus for women only, the need for which Mrs. Steinmetz determined by making the late bus trip back to Hanford herself.

There was a small library on the site, and of course regular services were held for both Protestants and Catholics—those for Catholics being conducted in a large circus tent that had been put up to serve as a theatre for moving pictures. But restlessness arising from plain boredom was not easy to cope with at best. One popular measure was the arrangement made with the commander of the Navy Air Force contingent at Pasco for any of the women who wanted to attend the regular dances there, with Navy buses supplying the transportation.

From the problems of reactor design to the health of fish in the Columbia River and the condition of women's shoes covers a considerable range of problems, and obviously they were not of equal importance. But they all mattered in the job we were trying to do.

CHAPTER 8

OAK RIDGE

Apart from the plutonium plant at Hanford, the heart of our effort to produce material for a fission bomb was Oak Ridge. Here were located all our uranium separation plants—the plants designed to separate the easily fissionable Uranium-235 from the more abundant but much less fissionable isotope, Uranium-238.

There were a number of ways we thought this could be done, but for practical reasons, to suit our immediate purposes, they were whittled down to two, the electromagnetic process and the gaseous diffusion process.[1] The construction for these was authorized late in 1942, at about the same time that we gave the go-ahead signal for the plutonium process which was put at Hanford. In 1944 we decided to build a thermal diffusion process plant, which was also placed at Oak Ridge.

A full discussion of the Oak Ridge plants and the research and theory behind them would take volumes. There is room here to give only a sketchy idea of that enormous, security-hedged complex known as the Clinton Engineer Works and to mention briefly a few of the main troubles and headaches that plagued us in our attempt to turn out material for a weapon that might end the war.

We had decided at the start that the several uranium process plants at Oak Ridge should be well separated, so that in case a disaster struck one it would not spread to or contaminate the others. For that reason, the electromagnetic and gaseous diffusion plants

[1] Scientific research into the centrifuge and liquid thermal diffusion processes had not progressed as rapidly as it had into the other processes adopted for production in the fall of 1942. To avoid further extending our already strained resources we decided to put these aside.

were located in valleys some seventeen miles apart. Later, when the thermal diffusion plant was built, we had to disregard this policy and put it quite near the steam-generating plant for the gaseous diffusion process, in order to take advantage of its supply of extra steam.

The electromagnetic plant (Y-12, to give it its code name) was built in a restricted area of about 825 acres in the central-south-eastern part of the reservation, approximately five miles from the commercial district of Oak Ridge, which was the town site of the over-all development.[2] From the standpoint of employment, this plant was the largest in the Clinton works. It was the first on which construction was started (February, 1943) and the first to go into operation (the first units were ready in November, 1943). Indeed, for almost a year it was the only plant that was operating, and until December 31, 1946, the only plant that was turning out the final product—that is, the fully enriched uranium needed for an atomic bomb.

In every way, this process was one of the major efforts of the MED. Its purpose was to separate Uranium-235 from uranium as it occurs in nature and to do so in sufficient quantity, and of the concentration necessary, for use in atomic weapons. It is a physical rather than a chemical process, although a great deal of chemistry is involved in the handling of the material. Basically, electromagnetic separation of isotopes is based on the principle that an ion describes a curved path as it passes through a magnetic field. If the magnetic field is of constant strength, the heavier ions will describe curves of longer radii. Therefore, the various isotopes of an element, since they differ in mass, can be isolated and collected by such an arrangement.[3]

To apply this principle on a large scale, it was necessary to carry out a vast program of physical and chemical research. Research in many allied fields, such as biology, metallurgy and medicine, was also required. We then had to design, build and operate an extremely

[2] See Appendix V, page 424.
[3] The basis for the process grew out of A. J. Dempster's use in 1918 of a simplified mass spectograph, and E. O. Lawrence's development of the cyclotron during the thirties.

large plant with equipment of incredible complexity, without the benefit of any pilot plant or intermediate development: to save time we had early abandoned any idea of a pilot plant for this process. Always we were driven by the need to make haste. Consequently, research, development, construction and operation all had to be started and carried on simultaneously and without appreciable prior knowledge.

We would never have attempted it if it had not been for the great confidence that we, particularly Bush, Conant and I, had in the ability and drive of Dr. Ernest O. Lawrence of the University of California. Rather early in the American effort, Lawrence had proved to his own satisfaction that electromagnetic separation was feasible, but he stood almost alone in this optimism. The method called for a large number of extremely complicated, and as yet undesigned and undeveloped, devices involving high vacuums, high voltages and intense magnetic fields. As a large-scale method of separating Uranium-235, it seemed almost impossible. Dr. George T. Felbeck, who was in charge of the gaseous diffusion process for Union Carbide, once said it was like trying to find needles in a haystack while wearing boxing gloves. It seemed likely, though, that it could fill our need for more than microscopic samples of U-235 for experimental purposes at Los Alamos. The main reason we embarked upon this project, however, was that through it we hoped to achieve large-scale production.

As it turned out, our decision to go ahead with it was fully justified, for, as we had hoped, it enabled us to get the essential early samples of U-235 for Los Alamos and, later, the necessary U-235 for the Hiroshima bomb. Without it we would also have been seriously delayed in the design of the plutonium bomb.

Long before the essential research was well started and before the equipment could be designed, we had to start designing and constructing the building to house it. Stone and Webster was in charge of this operation. The research on which all design was based was carried out in the Radiation Laboratory of the University of California, under the direction of Dr. Lawrence; and to operate the plant we selected Eastman Kodak, whose subsidiary, Tennessee Eastman, was an extremely competent organization with much experience in chemical processes.

There were only three electrical suppliers in the country whom we considered suitable for the manufacture of the type of equipment in the quantities needed for the electromagnetic project. To avoid overloading them, we divided our requirements among them: General Electric produced power supply equipment, Allis-Chalmers the magnets, and Westinghouse the process bins and allied parts.

Building and operating such a plant as this presented many new industrial problems as laboratory experiments using raw materials measured in grams were expanded to handle tons. Proper liaison between research, design, manufacture, construction and operation people was absolutely essential to success, and fortunately it was excellent. All the companies involved had such representatives as were needed at the Berkeley laboratories; more than fifty key specialists were transferred from Berkeley to Tennessee Eastman; and another large group of engineers and physicists from the laboratory was maintained at Oak Ridge, where they gave invaluable assistance with the installation of the equipment and its operation.

From the beginning, we realized that the plant would have to be enormous, and also that it would be very costly. The first estimate for construction alone was for an unrealistic sum of between $12 and $17 million; soon afterward this was increased to $35 million. These figures were for a plant much smaller than the one we finally built. In its first report to President Roosevelt early in December, 1942, the Military Policy Committee estimated the cost of the entire project as of the order of $400 million. At that time we thought that over $100 million would be needed for this process as a whole. In fact, all these estimates were only guesses, made before anyone had a clear concept of what would be needed for an operating plant and what its productive capacity should be. Exclusive of the value of silver borrowed from the Treasury for electrical conductors, the construction costs, by December 31, 1946, totaled $304 million; research cost $20 million, the engineering $6 million and operation $204 million. The cost of operating power was almost $10 million.

We knew before we started building that two separate stages might be required to complete the uranium separation to the necessary degree of enrichment. We called them alpha and beta. Alpha would start with natural uranium and turn out a product much richer in U-235, though it might be short of the requirements for a weapon. If sufficiently enriched, it would be used in the bomb; but if not, we would achieve the necessary concentration by feeding it through a beta stage. Initially, the equipment for the alpha separation process was arranged in the shape of large ovals, each containing ninety-six magnets and ninety-six bins or tanks. Almost immediately the quite appropriate term "race track" was applied to the setup, by one of the Berkeley scientists, I believe.

Because beta would use feed material from alpha; because its design would be a modification of alpha, dependent on alpha experience, and because originally we were not sure it would be needed, emphasis was placed on completing the alpha installation first. Eventually, however, the Y-12 plant comprised five alpha buildings, of nine racetracks, and three beta buildings, of eight racetracks with thirty-six bins each, as well as numerous chemistry and other auxiliary buildings. All were large—two of the alpha buildings, for example, were 543 feet by 312 feet—and each contained a fantastic labyrinth of equipment and piping. Much of the equipment was considerably bigger, closer in tolerance and more demanding as to accuracy than any similar equipment ever designed. Much of it was of revolutionary design. Much of it was required in large quantities. Because of material and labor shortages, the always overriding necessity for speed and the badly overloaded manufacturing plants of the suppliers, all of it was built under trying conditions.

Setting up a plant of this size in the short time we had demanded an extremely well-organized and co-ordinated field force. Stone and Webster interviewed some 400,000 people for construction jobs and brought together a large force of experienced construction men from all parts of the country to fill the key positions.

Tennessee Eastman, for its part, immediately began training its key people, and sent a number of them to Berkeley to gain experience in the operation of the experimental units which were being set up in the University of California Radiation Laboratory.

At the same time, a drive was started to recruit the necessary labor for the operation of the plant. Originally we had thought we would need a work force of 2,500. This was a sad underestimate, resulting from our inability to anticipate how complex and difficult the job would be and how many units would be needed. Eventually we had over 24,000 on the payroll. The great difficulty was that our personnel procurement program had to ensure the operation of a plant for which the needs were unknown, but the planning had to be completed and the organization ready as the various units became operational.

In many cases during the training of workers, the announced aims of their operations were completely distorted in order to avoid any unnecessary disclosure of classified information. However, their jobs were always described in such a way as to impress the workers with the importance of being alert and careful.

To begin with, classes were held in Knoxville, at the University of Tennessee. Subsequently, they were conducted at a special school at Oak Ridge and later in the separation plant itself. To give training and develop operating techniques for the beta process without losing valuable alpha products, the beta race tracks were run for some time on the basic feed material before using the enriched product from the alpha tracks.

One of our chief difficulties was a shortage of electrical workers. This became so acute that we had to turn for help to Under Secretary of War Patterson. Out of this appeal came an agreement known as the Patterson-Brown plan (Edward J. Brown was president of the International Brotherhood of Electrical Workers). It provided for the payment to employees of round-trip transportation and subsistence, a guarantee of no loss of seniority rights and a job on return to their former employers after completing at least ninety days' service at the project. Provision was also made for the official recognition of employers who released men in response to our appeal. This plan was a lifesaver, as was the co-operative attitude of Al Wegener, an official of the Brotherhood.

There was an almost complete absence of labor trouble, despite the fact that as many as four crafts were often involved in setting up a single piece of apparatus. The total time lost on the job from work stoppages, including jurisdictional disputes, was less than eight thousand man-hours as compared with the almost 67 million man-hours worked on the electromagnetic plant.

To me our excellent labor relations were a great satisfaction. The credit is due to the District organization, to the contractors, to Colonel C. D. Barker, in charge of labor relations in the office of the Chief of Engineers, to the patriotism and at times the forbearance of the employees, and to the co-operation of the union leaders of the trades involved.

I was particularly impressed by the attitude of one young union leader in connection with an effort to organize the powerhouse workers at Oak Ridge. We could not permit or even consider the unionization of the operating forces of any of the plants turning out U-235 because we simply could not allow anyone over whom we did not have complete control to gain the over-all, detailed knowledge that a union representative would necessarily gain. Neither could we permit the discussions between workers that would be bound to occur in union meetings. And obviously we could not have security officers present to monitor their meetings. Also, some information would inevitably filter back to the International Brotherhoods of the various unions.

When this organizing effort was begun at Oak Ridge, many were already union members but were necessarily inactive. If the powerhouse was organized there would surely be serious attempts and a great deal of agitation to unionize the entire gaseous diffusion process, which would soon enter into production. Naturally, strikes for any reason were out of the question.

To forestall any difficulty, I asked Patterson for his aid. We arranged a meeting at the White House with James F. Byrnes, who was then the Administrative Assistant to the President, and Fred Behler, the organizer of the union in question. He listened attentively to our reasons why unionization would be detrimental to the work we were doing. He then asked me about the importance of the work. I gave him my views and they were reinforced immediately by Mr. Patterson.

Behler then said, "If we don't organize now, we'll never be able to, because there will certainly be a plantwide election after the war, and as a craft union we'll be snowed under."

I replied that I expected that after the war the need for security would be less important and that there undoubtedly would be a plant-wide election, as he said.

I was not only pleased but extremely proud of him as a representative of American workers when he said, "General, in view of what you have told me about the importance of this work and your feelings that any attempt to unionize would be injurious to the country's welfare, I want to assure you that we'll make no effort to organize these men; we'll discourage any effort that is made, and we will do this with the full realization that this means that ultimately these men will not belong to our union."

Our one case of persistent labor trouble occurred at Hanford. There we were unable to get what we considered to be a proper output from our pipe fitters. Efforts to correct the situation through appeals to the local union officials were ineffective. The matter was so serious that G. M. Read of du Pont and I arranged to meet Mr. M. P. Durkin in Chicago. Durkin, later Secretary of Labor, was at that time head of the International Union. At this meeting we told him of Hanford's great importance to the war effort, emphasizing that every day's delay in its completion could well mean hundreds of Americans killed and wounded. We appealed for his assistance in increasing the production of these workers and in securing more of them. He did not seem to be at all interested in our pleas, and our meeting was not a success.

Later, when our needs grew even more pressing, we were unable to find enough pipe fitters to maintain our schedule. Investigation showed that there simply were not enough in the United States to fill the demands. The solution we adopted was to locate a considerable number of pipe fitters, all union members, who had been inducted into the Army. These men were given the opportunity to be furloughed to the inactive reserve on condition that they would accept employment at Hanford as civilians at the going rates of pay.

When they arrived they were kept together as a group so that their output would not be held down by the pressure of any union officials or of the men already working there. In a direct comparison on identical work, they produced about 20 per cent more than the other men. Pressure was brought on them to slow down, but they refused. A typical comment was: "I'm not working as hard as I did in the Army, nobody's shooting at me, I'm being paid a lot more and, what's more important, I've a lot of friends in my old outfit that I hope to see come back alive." As time went on, the other men were apparently shamed into greater effort, with the result that their output went up about 10 per cent.

Not long after he became general manager for Stone and Webster at Oak Ridge, in 1944 F. C. Creedon[4] told me he wanted to hold a special meeting of all his supervisors, down through foremen and including even some straw bosses, and wanted me to talk to them. He felt this might stimulate morale and thus increase the construction output.

Creedon was not a man one would expect to favor appeals of this kind. To him, as to me, they would be embarrassing and normally seem a waste of time. Since he thought it would help, I felt the chances were that it might, and agreed to do it.

On my next visit to Oak Ridge I talked for five or ten minutes to some two thousand of these men. I was not introduced by name but merely as the general in charge of the work for the War Department. The reason for this was to avoid drawing attention to me personally; this was our policy throughout the project until security no longer required it. (My wife once commented that I was undoubtedly the most anonymous major general in the history of the United States Army.)

As simply as possible, I told the group that, as the officer in charge, I could state positively, both officially and personally, that their work was of extreme importance to the war effort, and that my views were a true reflection of those of the Chief of Staff, General Marshall, of

---

[4] Formerly with the Rubber Administration and before that with me on the construction of Ordnance facilities.

Secretary of War Stimson and of President Roosevelt. I added that they could see for themselves how important it was from the terrific effort we were making, our obviously enormous expenditures in money and labor, and our evident ability to obtain materials that were in critically short supply. I said nothing about what we were working on or our hope that its success would quite possibly end the war. There was no flowery oratory; I would have been incapable of it, and it certainly would not have appealed to the audience.

Creedon estimated that after this meeting the efficiency of his construction operations improved by as much as 15 to 20 per cent. I never quite believed this, but the progress reports did indicate an increase of well over 10 per cent. This was far beyond anything I had anticipated; indeed, I would have been pleased with any improvement at all. In my opinion, whatever success the talk had was a result of Creedon's understanding, as an experienced construction leader, of the mood of his men and how to improve it.

For much the same purpose, to wring out the utmost in the way of support for the electromagnetic project, we invited the presidents of Westinghouse, General Electric and Allis-Chalmers to visit Oak Ridge. The speed with which the plant could be completed depended largely on how quickly their companies could deliver the key parts, and we wanted these men to see for themselves just how their equipment was to be used and to gain a firsthand realization of the complexity and magnitude of the project. The results of the visits were quite noticeable. Because major items of equipment were not such a controlling factor at Hanford, we did not follow this procedure there.

From beginning to end, the problems encountered in building and running the Y-12 plant, in their variety and often plain unexpectedness, would have taxed the ingenuity and industry of Hercules. That they were solved was due to the magnificent leadership at all levels that we had at Oak Ridge. One of the greatest difficulties was the failure of the process equipment to arrive in the proper sequence for orderly installation. After all, there were fabulous amounts of this, including the enormous oval-shaped electromagnets, the process bins and the units enclosed in them, the control cubicles, motor generator sets, vacuum systems, chemical recovery equipment and thousands of smaller parts. An idea of the quantity of material used may be gained from the fact that 128 carloads of electrical equipment alone were received in a two-week period. Special warehouses had to be built to hold the equipment while we waited for the key items that had to be installed first. Highly secret process equipment was stored in a special area under armed guard and was not unpacked until it had been moved to the place where it was to be installed.

Actually, the work on the first race track was well under way before the structure for the opposite end of the building was finished. The moment the overhead cranes were set and the concrete roof poured, we started to unload and place the heavy magnets.

For security reasons, parts of the buildings were partitioned off when the construction reached an advanced stage, and special passes were issued to the workmen who had to enter these areas. At first the resulting confusion was extreme, but before long the men became used to the restrictions, and there did not seem to be any appreciable slowdown in the work.

One of the most serious snags we ran into was when the magnets in the first race track began to act up after a comparatively short period of operation. These magnets were very large (about 20′ x 20′ x 2′) and were much more powerful than any common magnet in use. They were encased in heavy, welded steel. After much theorizing about what could be causing the trouble, we broke one open in the hope that a visual examination would supply the answer, though this meant that the magnet would have to be shipped back to the manufacturer for rebuilding.

There were two possible sources of trouble. The first lay in the design, which placed the heavy current-carrying silver bands too close together. The other lay in the excessive amount of rust and other dirt particles in the circulating oil. These bridged the too narrow gap between the silver bands and resulted in shorting, often intermittent, which had made it more difficult to determine what, if anything, was really wrong.

To me, all this seemed entirely inexcusable, for we should have made certain that nothing like this would ever happen. The design should have provided for a much greater factor of safety. The manufacturing specifications should have made more adequate provision for rigid cleanliness. The design and erection specifications for the circulating system for the oil used to cool the magnets should have prevented the entrance of rust and dirt.

What made it even worse was the fact that so many of us failed to foresee the hazards that this design entailed. I know I should have. It was more than annoying, too, to learn too late that Lawrence had once had a similar difficulty with one of his cyclotrons. Fortunately, this error did not in the end set us back in our cumulative production of U-235, but only because it was discovered so early that the extraordinary precautions taken thereafter enabled us to eliminate many minor stoppages that otherwise would have occurred.

When the trouble was discovered, we had completed one race track, were well along with the second, and were starting on the third. Immediate and drastic measures were called for. The magnets were removed from the track and sent back to Milwaukee to be cleaned and rebuilt. The silver bands were unwound and then rewound to allow greater spacing. A special pickling plant was built and all installed piping was taken out and passed through it to eliminate every bit of rust. The equipment was then reassembled under conditions as free from contamination as we could devise. All new piping was similarly handled. The magnets were immediately redesigned. After that, we had no more trouble from that source.

Although we were certain sabotage was not involved, in our detailed review of the situation we found that it would be possible for a saboteur, who would have to be an employee on one particular assignment, to throw iron filings into a feed opening in the oil circulation system and thus put an entire section of track out of action. Steps were taken at once to station counterintelligence agents on and around these spots.

One difficulty, which was unforeseen, because we lacked experience with magnets of such enormous power, was that the magnetic forces moved the intervening tanks, which weighed some fourteen tons each, out of position by as much as three inches. This put a great strain on all the piping connected to them. The problem was solved by securely welding the tanks into place, using heavy steel tie straps. Once that was done, the tanks stayed where they belonged.

Another headache was ordering necessary spare parts in proper quantities. This was virtually impossible since, lacking operating experience, we had no intelligent basis on which to base our estimates. The estimates were guesses; some proved to be shrewd and some rather wild. What seemed to be a very generous order for some items turned out to be insufficient, and in a few cases temporary shutdowns of some equipment resulted. In other cases, the quantities ordered proved to be greatly in excess of what was needed. Extremely rare chemicals were procured in small quantities for laboratory uses. These included samarium, rhenium, yttrium and other rare earths. Other substances that had previously had very limited application were needed in staggering quantities. For example, each alpha track used four thousand gallons of liquid nitrogen every week.

One incident that delayed production on a bin in an alpha track for several days involved a mouse. In some unknown way, he got into the vacuum system, where his presence prevented the bin from reaching the necessary high vacuum. After several days of trouble-shooting failed to reveal the source of the trouble, the run was terminated and the bin opened. The remains of the mouse, a bit of fur and a tail, disclosed what had caused the trouble, but no one ever learned how he got into the system in the first place.

More serious in effect was the suicidal action of a bird which perched on an outside wire in such a way as to short the electrical system. We had to shut down an entire building, and, because of the nature of the process, it was several days before operations again became normal.

There were numerous other causes for shutdowns—electrical storms, accidental tripping of switches and accidents in serving the motor generators supplying currents in the magnets. Once, the stone used to grind the commutators of the generators broke and nicked one of them, delaying operations in that track for some time.

Because of the fabulous value of some of the materials, strict accountability was essential to avoid waste. Waste such as piping, scrap cloth, filter cloths, papers, rubber gloves, clothing and the like had to be carefully saved in order to recover the small concentrations of uranium, particularly of Uranium-235. Inventories of the alpha cycle were made every four weeks and of the beta cycle every two. Constant studies were made to find out where losses occurred.

Early in 1946 an additional safeguard was adopted—a lie detector. It was used chiefly on people who had access to the final product chemistry building, to make certain that no one had taken, or knew of anyone who had taken, material from the plant. The first tests were carried out under the supervision of the inventor of the instrument, and one of his assistants was retained at Y-12 to conduct tests whenever necessary.

The value of one of the materials we used in quantity necessitated what was virtually a separate operation in itself. Preliminary design calculations on the Y-12 electromagnetic plant in the summer of 1942 had indicated that enormous quantities of conductor material would be required. Because the demands for copper to be used in defense projects far exceeded the national supply, the Administration had decided that the need for copper should be reduced by substituting for it silver borrowed from the Treasury Department.

Colonel Marshall thereupon called on the Under Secretary of the Treasury, Daniel Bell. Mr. Bell said that he might be able to make available some 47,000 tons[5] of free silver, together with 39,000 tons more which could be released from the backup of silver certificates, if Congress authorized its use through appropriate legislation. At one point early in the negotiations, Nichols, acting for Marshall, said that they would need between five and ten thousand tons of silver. This led to the icy reply: "Colonel, in the Treasury we do not speak of tons of silver; our unit is the Troy ounce."

Under the terms of the final agreement, the silver required by the project was to be withdrawn from the West Point Depository. Six months after the end of the war an equal amount of silver would be returned to the Treasury. It was further agreed that no information

[5] Converted from Troy ounces.

would be given to the press on the removal of the silver, and that the Treasury would continue to carry it on their daily balance sheets. Our relations with the Treasury were most cordial, and Mr. Bell and the various officials of the Mint and the Assay Office were always very pleasant and helpful.

Because of the natural reluctance of any private company to accept the responsibilities for safeguarding and accounting for the large amounts of silver that were involved, the MED had to carry out this responsibility with its own forces. This meant organizing separate guard and accountability units, establishing special inspection procedures employing special consultants and arranging to convert the silver into the conductors that we so urgently needed.

We accepted the Treasury's certification of the bar weights of the silver as we took it over at West Point. Then we delivered it to a processor, who cast the bullion bars into billets which could be extruded into forms more suitable for manufacture into bus bars, magnet coils and similar items. The casting was done by the Defense Plant Corporation and by the U.S. Metal Refinery Company. For the large magnets which used the bulk of the silver, Phelps Dodge Copper Products Company then extruded the billets into strips, which were rolled into coils about the size of a large automobile tire. These coils were shipped to Allis-Chalmers, where they were wound, suitably insulated, around the steel bobbin plate of the magnet casing.

Special MED guards watched the silver at all times while it was being processed, and accompanied every shipment except that of the final magnets from Allis-Chalmers to the Clinton works. We decided that at this point we could achieve adequate security by sending unguarded railway cars over different routes on varying time schedules. The silver coils were encased in large, heavy, steel shells which were completely welded together. Although silver is a valuable commodity, to have made away with any great amount of it during shipment would have been a major task, as our experience in opening one of these shells at Oak Ridge later confirmed. Moreover, the railroads always followed our shipments carefully, and we would have known immediately if any car had been waylaid.

Despite the great total value of the silver involved and the thousands of small, easily stolen pieces of silver that were being fabricated into equipment, we sought to strike a reasonable balance between security and economy. Useless restrictions were avoided. Many of the precautions that we took were aimed primarily at concealing our interest in the silver, and included the use of coded commercial bills of lading, the direction of all shipments to nonmilitary personnel and the requirement that our officers wear civilian clothing in many of the plants they inspected. Naturally all communications, oral, telephonic and written, were carried on in accordance with our established procedures for handling highly secret matters.

Every person involved in this work was thoroughly investigated before being employed and anyone who was not properly cleared was denied access to the areas where our work was being done, which were completely shut off from the rest of the plants. Within the process areas the silver was guarded twenty-four hours a day and seven days a week.

Routine security inspections were performed by District security and intelligence personnel. General inspections of the entire silver program were also made by other District personnel not directly connected with the program, and normally we used a different inspector on each occasion. This introduction of fresh viewpoints proved of great value in disclosing possible loopholes.

Our accounting system was very detailed, for it had to reflect the disposition, including all intermediate steps, of more than fourteen thousand tons of silver, to the nearest ounce. To assist us in this work, and as an added precaution, we employed the services of a well-known New York accounting and auditing firm to maintain a complete running audit of the silver account.

No recovery operation was undertaken unless the recoverable amounts were expected to be of more value than the cost of recovery. Nevertheless, throughout the entire operation we lost only .035 of one per cent of the more than $300 million worth of silver we had withdrawn from the Treasury.

The electromagnetic process entailed a number of special hazards: uranium is toxic as well as radioactive. Some of the raw materials were also extremely difficult to handle. High temperatures and pressures were involved and many irritants such as phosgene had to be used. Liquid nitrogen was used in large quantities at a temperature of $-196°$ Centigrade. Huge amounts of electricity were used throughout the process. Each control cubicle, for example, of which there were ninety-six for each alpha track and thirty-six for each beta, consumed about as much electricity as a large radio station.

In order to ensure complete compliance with the established Corps of Engineers' safety regulations and to impress everyone with my belief in their value, I set up specific requirements soon after the work was begun. These included an immediate telegraphic report to me personally whenever a fatal or near fatal accident occurred, to be followed with a complete written report the next day, giving the details of the accident, the cause and the steps that were being taken to prevent similar accidents. We had eight fatal accidents in all of our plant operations through December, 1946. Five people were electrocuted, one was gassed, one was burned and one was killed by a fall.

The task of whipping the bugs out of equipment, overcoming failures, low efficiencies and losses, with untrained personnel, while surrounded by the dirt and din of construction work was a prodigious one, but Tennessee Eastman proved more than capable. Despite all the difficulties that had to be overcome, the first shipment of enriched uranium was sent to Los Alamos in March, 1944, just a few days more than a year after the construction of the plant was begun. The concentration was not great, but it was sufficient for urgently needed experimental work at Los Alamos.

The most crucial point came just before the Hiroshima bomb. In a letter to Oppenheimer dated July 3, 1945, Nichols gave the predicted amount of material that would be shipped to Los Alamos by July 24, 1945. Activity at Oak Ridge became even more feverish than usual. By the time of the deadline, the production exceeded by a considerable amount the schedule Nichols had promised.

The chemical side of the electromagnetic process, in fact of the entire project, has often been treated as a simple auxiliary to its more eye-catching atomic physics aspects. Actually, chemistry was the beginning and the end of each of the separation processes. Production efficiency could be won or lost in chemistry, as well as in physics. Each was indispensable. The chemists and the chemical engineers had to keep constant pace with the possibilities unfolded in the new developments in physics. They, too, had to design processes without adequate information. And the chemical facilities had to be ready in time to meet the rest of the schedule. They always were.

The gaseous diffusion process, later termed the K-25 project, was a large scale multistage process for the separation of U-235 from U-238 by means of the principle of molecular effusion. The method was completely novel. It was based on the theory that if uranium gas was pumped against a porous barrier, the lighter molecules of the gas, containing U-235, would pass through more rapidly than the heavier U-238 molecules. The heart of the process was, therefore, the barrier, a porous thin metal sheet or membrane with millions of submicroscopic openings per square inch. These sheets were formed into tubes which were enclosed in an airtight vessel, the diffuser. As the gas, uranium hexafluoride, was pumped through a long series, or cascade, of these tubes it tended to separate, the enriched gas moving up the cascade while the depleted moved down. However, there is so little difference in mass between the hexafluorides of U-238 and U-235 that it was impossible to gain much separation in a single diffusion step. This was why there had to be several thousand successive stages.

The basic scientific research on the gaseous diffusion process was done by Columbia University's SAM[6] Laboratory in New York City under the leadership of Dr. Harold C. Urey, with Dr. John R. Dunning as chief physicist. In November, 1942, the S-1 Committee of the OSRD gave it a third-place priority, behind the electromagnetic and plutonium plants. The following month a letter contract was signed with the M. W. Kellogg Company for the extensive research and development, design, procurement and related services necessary to build a plant to produce U-235 of the purity and the quantity found needed for atomic bomb production. For operational reasons, as well as for security, Kellogg set up a wholly owned subsidiary, Kellex, to handle this project. Close co-operation was promptly established

[6] Code name originally based on Substitute (or Special) Alloy Materials.

between Columbia and Kellex. Kellex took fundamental data, developed by Columbia, developed them further and applied them to the design of the production plant.

Kellex was headed by Mr. P. C. Keith, a brilliant scientific engineer. At first some thought was given to having the Kellex organization handle the construction, too, but I decided against this on the grounds that they would be fully occupied in the development and preparation of the engineering plans. Instead, we gave the building contract to the J. A. Jones Company of Charlotte, North Carolina, with whose excellent work on several Army construction jobs I was already familiar. Union Carbide and Carbon Corporation was chosen as the operator. Here the work was under the general supervision of James A. Rafferty, a vice president of Union Carbide, with Lyman A. Bliss as his principal assistant. Felbeck was in direct charge of the operation.

Our power needs for K-25 we knew would be high. For many months before, as well as during, the construction of the plant, we labored under the belief that if the plant was shut down through power failure or for any other reason—for as much of a fraction of a second—it would take many days, some said seventy, to get it back into full operation. This proved to be quite erroneous, but nevertheless it was the basis on which the plant was designed and built. And it was the main reason we decided to build a special steam-generating plant, instead of depending wholly on TVA, whose power had to come over many miles of wire and was therefore subject to interruption from both natural causes and sabotage.

There were also a number of minor advantages to be derived from a steam-generating plant near the K-25 plant, including a better means of supplying the wide range of variable-frequency power for which the plant was designed. The use of power at a number of varying frequencies was rather typical of many features of MED. We could not afford to make our task easier by using standard equipment if we knew that special equipment would work but were not sure the standard would.

To protect our power supply from sabotage, we built an underground conduit from the steam-generating plant to the gas diffusion plant. In spite of our precautions, we did have one case of attempted sabotage. This was sabotage in its crudest form—the driving of a nail through a rubber-coated electric power cable. Fortunately, it was discovered long before we had to depend on this cable for the supply of electricity. We could never discover the culprit; it could have been done either by an enemy agent, an enemy sympathizer or by a disgruntled workman, most probably the last.

Without question the most serious problem that confronted us throughout was our inability to produce until late 1944 the barrier material which was the heart of the process. This prevented the orderly installation of the production equipment. It meant that before the first unit could be put in operation, some $200 million had been spent on construction and on the purchase of special equipment, and most of this had been done before we knew even that a satisfactory barrier could be made in the quantities we would require. Yet in spite of this major unknown factor, we had to press ahead with the construction of one of the largest industrial plants ever built, comprising over forty acres of floor space.

The first separation of uranium isotopes by the gas diffusion method had been accomplished at Columbia in January of 1942, but it was not until October of that year that the first laboratory unit was operated. The barrier in this unit was about the size of a silver dollar. While other pilot units were built to test various components, no complete pilot plant or semi-works was ever constructed. The Oak Ridge plant was a first in every sense, and its design, involving many acres of barrier, was based on this small piece less than two square inches in area. Even this practical foundation soon disappeared when it became known that the material used in the first filter could never be employed in the main plant. One of the great contributions made by Union Carbide during the design and construction period was in connection with the development of a satisfactory barrier, and with the co-operation of a large number of firms mass production was achieved in ten months after manufacture was started.

The design and manufacture of the thousands of pumps needed to force the gas through the diffusion units was solved by the Allis-Chalmers Company. Keith had canvassed the entire industry, and had come to the conclusion that our one hope for obtaining the pump we needed within the time limit lay in that firm. When he first approached them they were unwilling, primarily because they were already seriously overloaded with war work. He then asked several Carbide men and me to visit Milwaukee with him, with the aim of persuading Allis-Chalmers to undertake the job. They were already playing an important role in the project in the production of the enormous magnets used in the electromagnetic process and their president was most reluctant to consider our request. Finally, after warning us that they were so overloaded with war work that he did not see how they could possibly undertake it, he consented to our talking with his chief engineer. We were amazed when, after we had described in some detail the exacting performance specifications, he replied, "Yes, we can do that. We have already manufactured pumps of the same type, but of course of much smaller capacity." The contract was accepted and perfectly performed.

On our way back from Milwaukee, we stayed overnight in Chicago. In our hotel rooms we talked at some length about another design problem: how to handle a breakdown within a particular unit. In the course of the discussion, I expressed surprise that it was thought to be a problem, since all that was necessary was to cut out the particular unit that had broken down. The difference between the makeup of the gas varied from diffuser to diffuser so slightly as to be unnoticeable and almost unmeasurable, and I asked how the diffusers could ever tell the difference. That casual question immediately suggested the answer. As so often occurs, it was a case of a simple solution occurring immediately to someone who had not been struggling for months with the problem.

To minimize the effects of gas corrosion, it was first proposed that we use solid nickel for the some hundred miles of process piping. K. T. Keller, the head of the Chrysler Corporation, which was to produce the diffusing units, pointed out that our demands in that case would exceed the entire nickel production of the world, and insisted that heavy nickel plating on the inside of the larger pipe, four inches and above, was feasible. To attempt to heavily nickel-plate the interior of the quantity of pipe we needed was an unprecedented undertaking, but it was solved by a small manufacturer in Belleville, New Jersey, the Bart Laboratories. They developed a novel method in which they used the pipe itself as an electroplating tank. The pipe was rotated during the operation in order to obtain a uniform thickness of deposit. Their success eliminated what otherwise would have been a most difficult situation.

Instruments capable of distinguishing between the isotopes of a particular element had to be developed. This requirement was also unprecedented. Then there was the problem of regulating the gas flow within the system, which was extremely long, and was ultrasensitive to pressure waves and variations. It was feared these would cause surges.

I was never so concerned about the problem of surges as were many of our people. This was because I approached it with the background of an engineer rather than of a theoretical scientist. Many of the British scientists were extremely pessimistic. They expressed the view that it would be practically impossible to regulate the flow so as to avoid fluctuations and internal disturbances, and that the resulting surges would seriously impair operating efficiency, and might even damage the barriers. Some went so far as to say that the plant would be inoperable.

We had to be absolutely sure that in the hundreds of miles of piping the total leakage of air into the system, particularly through the welds, would not exceed that which would enter through a single pinhole. This problem was solved by industrial engineers. By using helium gas with an improved mass spectrometer, we were able to detect all leaks before the individual piping assembly was installed, and because we could not permit any leakage, no matter how slight, we could not tolerate normal commercial shop welding of the pipe connections, so special welding techniques had to be developed. Once we realized the importance of complete tightness in the gas diffusion plant, we did everything we could think of to achieve it. One step was to operate a special school for over two hundred employees of the various firms involved in the manufacture of equipment or the operation of the K-25 plant.

A very serious question arose in connection with the gas diffusion program when our special scientific safety committee told us that there was a distinct chance of an accidental atomic explosion resulting from an accumulation of highly fissionable material at some point within the piping system. Because of the distinguished membership of this committee, the utmost consideration had to be given to their fears, even though I disagreed with them. In this I found considerable comfort and support in the advice of Nichols, Conant and Tolman. Each expressed his feeling that the committee was overly apprehensive, and that there was no serious danger, or in fact any appreciable possibility, that an explosion would occur. I discussed this very thoroughly with both Keith and Felbeck. They agreed that the chances were extremely remote, and the opinion of the safety committee was ignored.

The cleaning and conditioning of equipment prior to installation was vital and the closest practical approach was made to surgical conditions. This involved the complete removal of dirt, grease, oxide, scale, fluxes and other extraneous matter. Any such material, even in small amounts, could very well have caused a complete failure.

The cleaning methods were based on procedures developed by the Chrysler Corporation. The individual steps were not too unusual in industrial practice, but the combination of all of them, their rigorousness and their application to the thousands of pieces of equipment were unheard of. Depending upon the particular item, as many as ten steps were required. For some parts the cleaning was done at the factory under the supervision of Kellex inspectors. Many, however, had to be cleaned just before installation, which required the construction of a special conditioning area at K-25.

Aside from the cleaning of equipment, the general cleanliness control measures at the K-25 plant were so rigid as at first to interfere considerably with normal methods of construction. All workers changed into clean outer clothing from head to foot upon entering a restricted building. Initially everyone had to comply with this requirement. Special lockers were provided for all concerned, including Nichols and myself. We had to change like everyone else even if we were going to walk into the area and out again in a few minutes.

Everything possible was done to eliminate dirt and dust. Vacuum cleaners were used instead of brooms, and dust mops were used in order to avoid raising dust by dry sweeping. As experience was gained, it was found possible to lessen the severity of the rules to some degree.

In scheduling the construction we established five objectives:

First, the completion of one cell of the main cascade; second, the completion of one entire process building; and third, the completion of a sufficient portion of the plant to enable the production of a lightly enriched material. As early as January 1, 1944, we set the date for this as January 1, 1945. Fourth, the completion of additional portions so as to enable the production of a maximum amount of U-235 by the critical date, as yet unset, for the first bomb. And fifth, the completion of the entire plant so as to permit maximum production.

The goal of producing slightly enriched material by January 1, 1945, was not quite met, but the later goals were achieved, although with very little time to spare.[7]

The gaseous diffusion process as it was built was essentially an all-American effort. However, the British were made aware of the general details of its design. It should be realized that the first work in this area was begun in England, even though the efforts were confined to theory and to the laboratory.

Several preliminary meetings were held with several of the British representatives (W. A. Akers, F. Simon and R. Peierls) during the spring of 1942 in the United States, at which the principles of diffusion separation and possible types of plant design were discussed. The views of the British group on plant design, however, were quite different from the American, and their methods and equipment were

[7] At the peak of construction employment the working force at K-25 totaled some 25,000. The cost of the K-25 design, engineering and procurement to the end of the fiscal year 1946 was $253 million, and the estimated total costs for the completed contracts were $275 million.

dissimilar to ours. For that reason the latter discussions did not prove to be of much practical value to the United States development.

In the fall of 1943 the British sent over a strong delegation of scientists and engineers to learn of our plans on the gaseous diffusion process. They spent the period from September, 1943, to January, 1944, going over our designs in detail.

They did not agree with us on the course we were following. To many of our people the principal reason for this appeared to be because it was not based on the design theories they had originally developed. The barrier material was discussed with them in considerable detail, but their views did not influence the selection of either our first barrier, which was later abandoned, or the second.

In December of 1943, they proposed that an investigation be made of a method by which the gas would be recycled back and forth. Theoretically this had considerable merit, since it would have permitted a considerable reduction in the number of stages required. But it would have necessitated a large increase in the total barrier requirements, and the production of the barrier was still an unsolved problem, after several years of investigation. It also would have greatly increased the complexity of design of the converter and of the stage operation.

At this late date, I was strongly opposed to any major change such as the British suggested, for any one of them would have seriously delayed the completion of the plant. To me this was completely unacceptable.

The British group also suggested designing the plant as a cascade of cascades. In such a design the plant is not a simple long cascade, but rather is compartmentalized and consists of a number of smaller groups of stages, each in itself set up as a complete cascade and each connected with its neighbors. The system is quite complicated, and thus has many disadvantages. The American plan appeared to be satisfactory and as its design was progressing on schedule, we could see no advantage in changing it. We did examine their proposal carefully to make certain that our decision was sound. The discussions on this point were quite helpful in suggesting means for controlling operating disturbances.

There were many other suggestions made by the British, none of which was adopted. Their principal value lay in the stimulation the Kellex engineers derived from discussing them. This is always of value in any scientific development.

From February to May, 1944, a group from the British delegation gave us some valuable aid in solving some of our theoretical problems. This group included Peierls, C. F. Kearton, Skyrne and Fuchs. They were stationed in New York, but they were free to travel and they did.

I have mentioned earlier that the first uranium separation process I looked into on being told of my appointment to the MED was that involving liquid thermal diffusion. The basic apparatus is a column. It consists of a long, vertical, externally cooled tube with a hot concentric cylinder inside. What makes this an effective separation method is the fact that one isotope tends to concentrate near the hotter of two surfaces, and then moves upward.

From a practical standpoint, thermal diffusion was not suitable as an independent process because of the incredibly large amount of steam required. The production costs would have been staggering. A minimum rough estimate was two billion dollars, and I would not have considered this a safe figure, but would have raised it to at least three billion if I had thought the work would have to be undertaken. Moreover, the research, though it had been carried on by Philip Abelson in a most competent manner, had been extremely limited.

He had started his investigations at the Carnegie Institute in Washington, and in the summer of 1941 had succeeded in actually separating a certain amount of U-235. The interest of the Navy was aroused by its hope for a better power source, and it was not long before it began to support the work, first at Carnegie and then moving it with Abelson to the Naval Research Laboratory in Washington. Here the research went forward for several years under Navy auspices and using Navy funds. Abelson's progress was followed closely by the OSRD and later by the MED. The process and the results obtained were reviewed on several occasions by competent scientific groups. No one was particularly impressed by its possibilities.

However, in June of 1944, Oppenheimer suggested to me that it might be well to consider using the thermal diffusion process as a first step aimed at only a slight enrichment, and employing its product as a feed material for our other plants. As far as I ever knew, he was the first to realize the advantages of such a move, and I at once decided that the idea was well worth investigating.

Just why no one had thought of it at least a year earlier I cannot explain, but not one of us had. Probably it was because at the time the thermal diffusion process was studied by the MED we were thinking of a single process that would produce the final product. No one was considering combining processes. This step came much later when we decided to limit the initial enrichment goal of the gas diffusion plant and to use its product as feed material for the beta stage of the electromagnetic plant.

If I had appreciated the possibilities of thermal diffusion, we would have gone ahead with it much sooner, taken a bit more time on the design of the plant and made it much bigger and better. Its effect on our production of U-235 in June and July, 1945, would have been appreciable. Whether it would have ended the war sooner, I do not know. It would not have affected the date of the Alamogordo test, since for that we used plutonium in the implosion-type bomb. Even with an increased production rate, we could not have afforded to use Uranium-235 in a test.

A few days after Oppenheimer made his suggestion, we took steps to carry it out. By this time the Navy was building a pilot plant at the Philadelphia Navy Yard. It was nearly complete and most of the operating techniques had been planned.

In order to get a large-scale plant into operation just as quickly as possible, we decided to use a single contractor for the entire operation. The H. K. Ferguson Company,[8] an outstanding engineering

---

[8] As a nationwide engineering construction company, the Ferguson Company operated on a closed shop basis. To avoid the union problem after the plant began to operate, a subsidiary operating company, the Fercleve Corporation, was formed. The MED then entered into two separate contracts; one for construction with Ferguson and one for operation with Fercleve. This arrangement worked out to everyone's satisfaction.

firm, was given the assignment on June 26, 1944, and on the same day Admiral King, at my request, directed that copies of the Navy plans be turned over to them. Ferguson engineers also visited the Naval Research Laboratory and the Philadelphia Navy Yard and made many engineering sketches.

To expedite the design and construction, I ordered that, insofar as possible, all process features of our plant, particularly the basic column assemblies, should be Chinese copies of those at the Philadelphia pilot plant. A great deal of time was also saved by frequently using field engineering sketches instead of the customary more formal drawings.

Various sites were hastily considered, including one in Detroit, but we finally settled on an area at Oak Ridge near the K-25 powerhouse, where steam was immediately available. There was also an advantage in having the new plant next to the K-25 plant, which was to use its product. Above all, we wanted to avoid building and operating classified plants in a new area. This process was given the security designation S-50.

When we decided on its location, Union Carbide expressed concern over the possibility of an explosion caused by the high steam pressures we would be using and its effect on their gaseous diffusion operation. They were also worried lest the powerhouse water supply might be contaminated by S-50 process material, which would cause trouble in the big steam plant. I did not think there was too great a danger of a damaging explosion, and we guarded against the possibility of contamination by installing a number of special recording devices.

The basic piece of equipment was the isotope separation column, 102 of which were arranged to form an operating unit which we termed a "rack." The column was a vertical pipe, forty-eight feet long, of nickel pipe surrounded by a copper pipe. The copper pipe was encased in a water jacket contained in a four-inch galvanized-iron pipe. The copper pipe was cooled with water at a moderate temperature. The columns were arranged in three groups, each of seven racks, making a total of 2,142 columns.

The columns at the Naval Research Laboratory were few in number and had not been standardized, yet Ferguson had to arrange for their manufacture on a mass basis. Over twenty manufacturers were consulted but none of those considered qualified wanted to undertake the job. Finally Mehring and Hanson and the Grinnell Corporation agreed to attempt it. After considerable difficulty, both firms worked out production methods that permitted them to make as many as fifty columns a day apiece. On July 5, nine days after getting the assignment, Ferguson placed the first order for the manufacture of process columns, and on the ninth, the clearing of the site began.

Training of operating personnel at the Philadelphia Navy Yard got under way in August. In September our progress received a severe blow when there was a bad explosion in the pilot plant, which injured several persons and disrupted the pilot plant installations. While we thought that it was probably due to external causes (as it proved to be) rather than to the process or to faulty designs, we could not be certain without thorough investigation, which, of course, held up the work for a while.

The importance of early production had dictated severe construction schedules. While six months were considered optimistic for the construction of the main process building, I set a schedule of 120 days to begin operation. The highest priorities were used and everything was done to speed the work. Riggers, pipe fitters, welders, sheet metal workers, electricians, carpenters and operation personnel worked feverishly on immediately adjacent parts of the unit. As soon as the riggers put a column in place and the pipe fitters completed its installation, it was pressure-tested and conditioned by the operators. Sixty-nine days after the start of construction, one-third of the plant was complete, and preliminary operation began. On October 30, the first product was drawn off, and peak production was reached the following June.

At first the operation was not too satisfactory. We were badly handicapped by the complete lack of pilot plant experience, the lack of trained people and, as it turned out, by an insufficient supply of steam. We were also plagued by leaks of the process material and by high-pressure steam leaks. It was not until January, 1945, that these difficulties were entirely ironed out.

The application of high steam pressures (a thousand pounds per square inch in this case) is a risky undertaking, particularly where the equipment has been designed and installed so hurriedly. The resulting leaks could have been avoided if we had had more time for study in the design phases of the project.

On one of the racks, owing to the hurried construction, there were large quantities of high-pressure steam escaping. The resulting clouds of vapor and the noise made it difficult for the operators to function. Yet we continued to operate the rack in spite of conditions that under any normal procedures would have called for an immediate shutdown. When the plant reached full operation, this rack was taken out of production so that it could be put into satisfactory condition.

By October it was evident that too much steam was leaking out at the screwed unions at the bottom and top of the columns. We decided, therefore, to replace all of them with new welded connections. By January, all the twenty-one racks were again ready except for a few small piping changes found to be necessary on the basis of our operating experience.

As the various plants at Oak Ridge were gradually put into operation, the main problem we faced was how best to use them, in combination and separately, to get the greatest possible output of fissionable uranium in the shortest possible time.[9]

The over-all problem was extraordinarily complex, not only because of the constantly changing relationship between the quantity of U-235 produced and the degree of its enrichment, but also because the product of one process was the feed material for another, and this also was subject to change. Moreover, as additional equipment was put into use—and in the spring of 1945 this was expected to, and did, occur almost daily—the production would increase also, and it would increase still further as the enrichment of the material fed into the individual processes was augmented. We were certain, too, to have increased efficiencies in the operation of the various plants as we gained experience. But the effects of this could only be guessed at.

A sound solution was of such importance to our success that I

[9] See Appendix VI, page 427.

asked Nichols to give the problem even closer personal attention than he might normally have devoted to it. Consequently, in December, 1944, he set up a special group of scientists and engineers, for the most part officers and enlisted men, under the charge of Major A. V. Peterson. Their job was to work out the best practical plan of operation, despite the changing characteristics, performances and capacities of the different processes.

When they started their study, the exact amount of U-235 that would be required for a bomb was still not known. Peterson discussed with Oppenheimer the relative importance of quantities and enrichments, and the enrichment to be aimed at was fixed. Under his close supervision, Peterson's group was able to work out an optimum program of operations, based on the most efficient production interplay between the electromagnetic alpha, beta, gaseous diffusion and thermal diffusion plants. This schedule was constantly revised as new efficiencies were achieved and as changes occurred in the dates for placing equipment into service.

By May, 1945, we reached the conclusion that our pre-Yalta estimates of being ready early in August were reasonable and that we should have accumulated enough material for one bomb by late July. After Nichols, Peterson, Oppenheimer and I had carefully reviewed the situation, July 24 was finally set as the deadline date.

And by the end of that day, enough uranium—and a little bit more —had been shipped to Los Alamos for the manufacture of the first bomb to be dropped on Japan.

CHAPTER 9

NEGOTIATIONS WITH THE BRITISH

Vannevar Bush carried innumerable important responsibilities during the war years, but perhaps his most valuable service to the nation was his handling of the Manhattan Project's relations with the White House. He had established a close working relationship with Mr. Roosevelt long before the Army entered the atomic picture, and after the MED was formed he handled all our business with the President, directly with him or with Harry Hopkins. He was the man to whom the President looked for advice and information on the project, and no one connected with its management ever had the least desire to change this arrangement. It was a tremendous aid, not only to me but to Secretary Stimson as well, relieving us of a considerable burden.

He also, and with equal effectiveness, acted in our behalf in many other ways; and was always punctilious in keeping the other members of the Military Policy Committee and me informed about the course he was pursuing and the results he obtained. For all of this I shall be everlastingly grateful to him.

Nor shall I ever forget the masterful way in which he conducted the ticklish discussions with the British regarding the interchange of atomic information.

Before the summer of 1942, the relative amounts of work being done on the development of atomic energy in the United States and Britain were not greatly unbalanced, and it was on this basis that the interchange of information between the two countries had been instituted. Soon after I came into the project, Bush and Conant and I reviewed the situation and reached the conclusion that in the future the British effort would probably be limited to the work of a very small number of scientists without any significant support from either the British Government or industry; and that inevitably a large amount of information would pass from the United States to the United Kingdom, while practically none beyond preliminary laboratory data would pass from the British to us. In order that everyone would have a better understanding of the background of our relations with the British, I brought up the subject at the General Policy Group's meeting of September 23, 1942.

Discussions between representatives of the United States and the United Kingdom on the development of atomic energy had begun informally as far back as early 1940, while the American effort was still under the control of the original Advisory Committee on Uranium. The first attempt to put the exchanges on a formal basis grew out of an *aide-mémoire,* dated July 8, 1940, in which the British Ambassador, the Marquess of Lothian, proposed to the President that there be an immediate and general interchange of secret technical information. This paper made no mention of nuclear energy; it dealt primarily with radio and radar developments. However, the door was left open when the Ambassador said that, although the British did not want to bargain for secrets and would consequently make any or all of theirs available to the United States, "they would hope you would reciprocate by discussing certain secret information of a technical nature which they are anxious to have."

This *aide-mémoire* was discussed in a Cabinet meeting a few days later, and was approved by the President, with the Secretaries of War and Navy concurring. The State Department thereupon informed the British that the United States was prepared to enter into conversations with them on the interchange of technical information, and implied that we were in agreement with the proposal, so long as it did not interfere with our own war procurement programs. In the light of later developments, it is significant that neither the original British proposal nor our reply placed any restriction upon the type of information to be exchanged which would limit it to that having military applications only.

That fall, a British technical mission under Sir Henry Tizard, which had been empowered to disclose secret information concerning weapons, arrived in the United States to discuss the research activities of both countries. With the knowledge and consent of both the Army and Navy, the British mission had a number of conferences with the members of the NDRC, in the course of which Sir Henry proposed a full exchange of information concerning research and plans for weapons development. The desirability of such an arrangement was confirmed by an exchange of letters between Bush, for the NDRC, and the Secretaries of War and Navy, acting jointly.

In February, 1941, having received an invitation from the British, NDRC sent over a mission headed by Conant to exchange technical information. Agreements were reached there that set the pattern for liaison throughout the war. They provided that the NDRC would exchange war research information directly with the appropriate British ministries; that the British would concern themselves with research having as its immediate objectives the defense of Britain, while long-range development would, in general, be undertaken by the United States.

However, there was no official interchange of information on atomic energy until after the establishment of OSRD on June 28, 1941. The Executive Order that created it designated OSRD as the agency to "initiate and support such scientific and medical research as may be requested by the government of any country whose defense the President deems vital to the defense of the United States . . . and serve as liaison office for the conduct of such scientific and medical research for such countries." This directive, together with the Lothian *aide-mémoire,* formed the basis for a complete interchange of information between the United States and the United Kingdom during the early phases of atomic development.

During the summer of 1941, both countries exchanged reports on the subject, and several months later two American scientists, Harold Urey and George Pegram, visited England to become familiar at firsthand with British work in this field. British efforts up to that time had been concerned primarily with certain theoretical aspects of the separation of U-235 by gaseous diffusion. While they had done a little theoretical work on the heavy-water pile, they had done almost nothing on the graphite pile or the electromagnetic process—two of the three major production methods which we later employed. Recently, I was told, the British had requested and been refused representation on the OSRD S-1 Executive Committee.

This then was the general pattern of relationship between the British and American scientific groups when I brought up at the General Policy Group meeting in September, 1942, the matter of exchange of information on the scope of the work under way in the laboratories of each country and on plans for the future. Not until later, so far as I know, was there any formal agreement dealing specifically with the exchange of information on the production of atomic energy. After some discussion, it was decided to let things stand as they were until Secretary Stimson could discuss the subject with the President.

At the close of the Cabinet meeting on October 29, Mr. Stimson put the question to President Roosevelt. He told him of our accelerating progress and of his feeling that we were already doing nine-tenths of the work, and asked what foreign commitments the President had made in this field. In his diary, Mr. Stimson says of that meeting:

He said he talked with no one but Churchill and that his talk was of a very general nature. I asked him if it wasn't better for us to go along for the present without sharing anything more than we could help. He said yes, but he suggested that sometime in the near future he and Churchill and I should have a conference over this matter. He and I discussed some of the enormous possibilities and the ways of meeting the ticklish situation after the war with a view to prevent it being used to conquer the world. He, himself, was alarmed by the thought that possibly a bomb might fail to go off over enemy territory and give the thing away.

About this time, the British Government sent Mr. W. A. Akers to this country. Before the war, Akers had been a senior official of the Imperial Chemicals, Ltd., and, at the time of his visit, was acting as the head of all British work on atomic energy. While in the United States, he had several meetings with Bush and with Conant and me. He presented a number of arguments in favor of a freer flow of information about manufacturing processes, but I explained to him that, under the existing agreements, I had no authority to extend the established rules of collaboration. Bush, Conant and I met frequently during this time, as did the Military Policy Committee, to discuss Akers' proposals, and we were generally agreed that: (1) Akers might well be influenced by an undue regard for possible post-war commercial advantages for the British when speaking of interchange of information; and (2) the United States should make available only such information as would assist in winning the war. All of us felt strongly that we should not consider any change in United States policy until we had official instructions from the President.

It became apparent from our discussions with the British during this period that they would not be in a position to embark on any large-scale atomic development or production programs during the war. Disturbed because they were getting less and less information from us—for I insisted that anything we gave them must be essential to our common war effort—they put increasingly strong pressure on both Bush and me. They had always understood, they insisted, that there would be complete collaboration even beyond what was necessary for war purposes.

But they themselves had provided us with numerous precedents for rejecting their argument, for, as Conant pointed out, they had not told us of a secret bomb disposal method, or of several other of their developments in which we were interested, because, they said, this knowledge would not assist our military effort. Consequently, we felt no pangs of conscience when we decided that information on the Manhattan Project would not assist British military participation in the current war. Akers and his colleagues were greatly discomfited by this interpretation of policy.

With matters in this regard pretty much at a standstill, the Military Policy Committee, in December, 1942, submitted a report to the President, reviewing progress on the atomic energy project as a whole. The report emphasized the need for clear instructions on

future relations with the British and the Canadians. Three alternatives
were offered:

(A) Cessation of all interchange.
(B) Complete interchange not only in the research field, but in development and production, including free interchange of personnel.
(C) Restricted interchange of information only to the extent that it
can be used now by the recipient.

Alternative (C) was strongly recommended, and the President's
approval of it established the pattern for our collaboration with the
British from then on.

At about the same time, Akers wrote to Conant setting forth
once more the views he had previously given to Bush and me. He
expressed grave concern over the restrictions that had been imposed
on the interchange of information, saying: "It has always been the
intention of the British group that the closest possible liaison should
be maintained, not only in research work, but also in production."

If the British, particularly Akers, had not displayed such an
interest in, and had not insisted on obtaining, material of value solely
for postwar industrial possibilities, the existing interchange of information might not have been affected. Negotiations broke down not
because of American policy, but because the British refused to accept
our view that collaboration should be for the purpose of winning the
war—not for postwar purposes. I decided that, in the absence of any
official agreement, the Manhattan Project would not be justified in
continuing to supply information to the British beyond that permitted
by the President's approval of the Military Policy Committee's recommendations. Needless to say, the British were far from satisfied with
this state of affairs and undertook to have it changed.

It was not long before Mr. Churchill raised the question of interchange directly with President Roosevelt. On May 25, 1943, in an
attempt to reach a meeting of the minds, a conference was held at
the White House, attended by Harry Hopkins, Dr. Bush, and Lord
Cherwell, who, throughout most of the war, was Mr. Churchill's
scientific adviser. Lord Cherwell made it very clear during this dis-

cussion that the principal reason the British wanted information on our project at this time was for postwar military purposes. He denied that they had postwar commercial ambitions in this field. He went on to point out that unless military information for postwar use were furnished to the British, they might find it necessary to divert some of their war effort to developing the necessary information for themselves.

Mr. Churchill's efforts to remove completely all restrictions were almost crowned with success when, on July 20, the President wrote to Bush saying: "While I am mindful of the vital necessity for security in regard to this, I feel that our understanding with the British encompasses the complete exchange of all information." This new position of the President constituted a complete reversal of his approval of the recommendations of the Military Policy Committee scarcely six months before.

Dr. Bush did not receive Mr. Roosevelt's note at once, for he had gone to London on other business. While there, he was asked by Mr. Churchill to confer on the subject of exchanging information. Bush replied that both Secretary Stimson and Mr. Bundy were also in England and that he felt they should take part in any such discussion. The meeting was held at Number 10 Downing Street on July 22.

At the time of this meeting, Bush had no knowledge of the President's willingness to enter into a complete and free interchange with the British. Mr. Roosevelt's note of July 20 had been received by Dr. Conant, acting then as the head of the OSRD, who transmitted its contents to Bush, but the cable was so garbled in transmission that the American representatives in London did not have a correct understanding of the President's instructions. Before the meeting Conant went on record, saying: "A complete interchange with the British on the S-1 project is a mistake. The proposition put up officially by the American Government, I firmly believe, was in the best interests of the war effort, the United States and the eventual peace of the world. I can only express the hope that the President did not revise his decision on a matter which may have such important bearings on the future of the United States without proper understanding of the potential possibilities of the weapon we are now engaged in developing, nor the difficulties of our enterprise. . . . In my opinion, the reopening of the exchange with the British without reservation (as contrasted to our restricted offer of some months ago) cannot in any way assist the war effort and will greatly diminish our security provisions here in the United States. Whatever time and energy those concerned with the S-1 project devote to British interchange (outside of the areas we have already offered to open) will be a pure waste of time as far as the job of winning this war is concerned." I completely shared his opinion.

The British were represented at the July 22 meeting by Prime Minister Churchill, Sir John Anderson (the British Cabinet officer handling atomic energy matters) and Lord Cherwell. The United States representatives were Secretary Stimson, Bush and Harvey Bundy. Mr. Churchill opened with a plea for a complete re-examination of the interchange problem. He emphasized the fact that Britain was not interested in the commercial aspects of atomic energy, but instead was vitally concerned with being able to maintain her future independence in the face of the international blackmail that the Russians might eventually be able to employ. After considerable discussion Churchill modified his position to agree to an interchange of only that information which would support the war effort, finally removing all British objections to our position.

He suggested an agreement between himself and the President that would contain the following provisions:

1. A free interchange would be established to the end that the matter would be completely a joint enterprise.

2. Each government would agree not to use this invention against the other.

3. Each government would agree not to give information to any other parties without the consent of both.

4. Each government would agree not to use atomic energy against other parties without the consent of both.

5. Information passing to Great Britain concerning commercial

or industrial uses would be limited in such manner as the President might consider fair and equitable in view of the large share of the expenses being borne by the United States.

I have never been able to find any justification for the notion advanced from time to time that the fourth provision held some hidden meanings. The fifth provision was the result of an effort by the British Government to remove any shadow that might have been cast upon their good faith by the vigorous representations of Akers.

After further discussion in the conference, Secretary Stimson agreed to present the Prime Minister's proposals to the President.

Shortly afterward, Mr. Churchill wrote to Mr. Stimson to tell him that he had received a message from the President, and suggested that the British send a representative to Washington "to discuss arrangements for the resumption of collaboration." For this purpose, the Prime Minister had designated Sir John Anderson, who would bring with him a draft of the heads of agreement which Mr. Churchill had proposed at the July 22 meeting. I do not know when Mr. Stimson told the President of his conversations in London, but probably not before his return to the United States on July 31, so once again (through his close personal relationship with the Prime Minister) Mr. Roosevelt had been precipitated into a matter on which he was not fully informed.

When Sir John arrived, he sent to Bush for review an expanded version of the draft heads of agreement governing future collaboration. The principal area in which Mr. Churchill's proposal had been expanded lay in the fifth point, where Sir John had attempted to prescribe in some detail the machinery for achieving American-British collaboration in research and development. Sir John's fifth point provided that:

a. There should be established in Washington a Combined Policy Committee. Its duties were described.

b. There should be complete interchange of information and ideas on all sections of the project between members of the Policy Committee and their immediate technical advisers.

c. There should be full and effective interchange of information and ideas between those in the two countries engaged in the same sections of the field of "research and development."

d. "In the field of design, construction and operation of large-scale plants, interchange of information shall be regulated by such *ad hoc* arrangements as may, in each section of the field, appear to be necessary or desirable if the project is to be brought to fruition at the earliest moment. Such *ad hoc* arrangements shall be subject to the approval of the Policy Committee."

Bush replied that the first four points in the draft were concerned with matters of international understanding which lay far beyond the problems of interchange and should thus be for the consideration of the President and the Prime Minister. He therefore limited his comments to the fifth point, indicating that the General Policy Group was in agreement that the proposed procedure was practicable. He reiterated our understanding of interchange of information as meaning that it would be pursued wherever the receipt of such information would definitely advance the project as a war measure.

Sir John replied immediately, confirming Bush's interpretation that, while provision "b" was intended merely to authorize the British to discuss with their immediate scientific advisors such information as they might have, provision "d" would really govern the interchange. He went on to state that "it will not be for the Combined Policy Committee to interfere with the control of the American programme by the Corps of Engineers of the United States Army. My thought is that the members of the Combined Policy Committee should have such information as may be necessary to enable all of us to be satisfied that we are making the greatest possible contribution towards bringing the project to fruition at the earliest possible moment."

On August 7, Bush notified the President that a basis for interchange with the British had been reached. Although he referred to Mr. Roosevelt's letter of July 20, which directed full interchange, he forwarded to the President the draft of the British proposals which placed interchange on a much more equitable basis.

Throughout his negotiations with Sir John, Bush had been most meticulous in consulting with all members of the General Policy

Group and the Military Policy Committee. However, after August 7, when the matter was referred to the White House, until the nineteenth, when the Quebec Agreement was signed, we were all pretty much in the dark. The Military Policy Committee was not consulted by the President during this time, nor, so far as I know, were any of its members. Most of us were extremely concerned lest the President enter into an undesirable arrangement. At General Marshall's request, Bundy asked Secretary Stimson to tell the President that Bush and Conant felt strongly that the "agreement should stand on a reasonable basis of *quid pro quo,* and exchange should be limited to the exchanges of information which will help expedite the S-1 development" in order to "avoid at all costs the President's being accused of dealing with hundreds of millions of taxpayers' money improvidently or acting for purposes beyond the winning of the war."

I was not informed of Churchill's presence in this country just prior to the Quebec Conference in August, 1943, nor did I learn of the conference itself until it was about to get under way. When I finally did hear of it, I knew that atomic energy matters were bound to be discussed between President Roosevelt and the Prime Minister. While their discussion would probably be centered on the interchange of atomic information, there was no way to be certain that it would be confined to this area.

Bush had been keeping the President informed of our efforts verbally. Our only written report to him had been delivered early in the previous December. The need for a new report was obvious. I started to prepare one personally, and with Nichols' assistance, on the project. On the thirteenth of August, the Military Police Committee agreed that this report should be sent to the President at once.

In order that further information would be immediately available, I had Nichols take the report to Quebec on the War Department's courier plane. When he delivered it to General Marshall, he was told that it was probably too late, as an agreement had already been signed, but that if anything else came up, it would be useful.

I do not know whether President Roosevelt consulted any Americans at Quebec on atomic energy. I doubt very much that he did.

All we were able to learn of the proceedings there was that, on August 17, the President signed the Quebec Agreement, which matched almost word for word the version that had been presented by Sir John Anderson to Dr. Bush in Washington a few weeks earlier. In that, I feel, we were very fortunate—fortunate that the British were not aware of the President's instructions of July 20, fortunate that the President's directions to Bush were garbled in transmission, and fortunate that Bush had possessed sufficient initiative and courage and skill in negotiating, after he did learn of them, to modify them to the extent necessary to protect our national interest.

The Quebec Agreement established the official basis for the relationship of the United States to the United Kingdom throughout most of our wartime nuclear energy programs.

Under the Agreement, a Combined Policy Committee was set up which was to meet in Washington and to supervise the joint efforts of the United States, the United Kingdom and Canada.[1] Its operations went smoothly at all times and there were never any serious differences of opinion among its members over the conduct of its business during the war. I attribute this largely to the outstanding membership and to the awareness on the part of the British delegates of the magnitude of the American contribution in comparison to theirs. The decisions of the Combined Policy Committee did not at any time interfere with the United States program. On the contrary, it supported our efforts to the fullest extent that could be desired.

Carrying out the policies governing the interchange of information, we arranged for selected British scientists to work in certain of our laboratories; we also supplied certain information, personnel, equipment and materials to the heavy-water pile at Chalk River, in Canada. In December, working closely with Colonel John Llewellin and Dr. James Chadwick, the recently appointed Chief of the British Scientific Group assigned to the Manhattan Project, I drew up the

[1] The committee, as it was originally designated in the Agreement, consisted of: Secretary Stimson (U.S.), Dr. Bush (U.S.), Dr. Conant (U.S.), Field Marshal Dill (U.K.), Colonel Llewellin (U.K.), and Mr. Howe (Canada). Colonel Llewellin was in the British Ministry of Supply and was stationed in Washington. Mr. Howe was the Canadian Minister of Munitions.

rules regulating the operations of British scientists attached to our project, and I assigned some twenty-eight British scientists to the work that was under my control. The Combined Policy Committee approved this action and appointed a subcommittee consisting of Dr. James Chadwick (U.K.), Mr. C. J. Mackenzie (Canada) and me to establish rules for the interchange of information between the group of scientists working on Canadian projects and their colleagues in the United States. The regulations which we established at this time provided the basis for all interchange until the passage of the Atomic Energy Act in 1946.

CHAPTER 10

SECURITY ARRANGEMENTS
AND PRESS CENSORSHIP

For about the first year, internal security in the MED was supervised by War Department Counter Intelligence and was thus a responsibility of the Army G-2, Major General G. V. Strong. As far back as February, 1942, he and J. Edgar Hoover of the FBI had agreed upon the various phases of security each organization would cover. The War Department's area of responsibility was to include all its civilian employees, as well as all civilians on military reservations or under military control.

On March 18, 1943, General Strong requested that the FBI discontinue its investigation of one of the scientists working at the Berkeley, California, laboratory. This was to lessen the chances that the suspect would find out he was being watched. Through its surveillance of the Communist party leaders in the San Francisco area the FBI had got wind of the project, for Russian espionage of activities in this laboratory had been vigorous for a long time. But it was not until a meeting on April 5 that the Bureau was first officially notified of its existence. At that time Strong told E. A. Tamm, an assistant to Mr. Hoover, of the Army's plans for protecting the development, and it was agreed that the Army's responsibility would include the Manhattan Project.

The MED's own security organization was then limited to a few officers and men who performed minor security tasks and carried out certain liaison responsibilities with G-2. This was in accord with our general practice of not undertaking any task ourselves that we could have competently done for us. Through frequent meetings with the responsible officers in G-2, I maintained proper contacts with this phase of the project.

It was not until much later in 1943 that a reorientation in the philosophy of the counterintelligence operations of the War Department made it impossible to rely any longer on the formerly very satisfactory centralized organization. Our only recourse was to set up our own complete security staff.[1]

For the staff control of this activity I selected Major John Lansdale, Jr., a graduate of VMI and a successful young lawyer in civil life. For some months he had been devoting almost his entire time in G-2 to our problems. He had, with the full support and approval of General Strong, set up a security organization operating outside of regular military channels. Selected officers and agents in every service command reported directly to Major Lansdale, who in turn reported directly to Strong and also to me. Thus we had been able to utilize all the resources of the Army counterintelligence organization without having to disclose through regular channels the nature of our work.

When it became necessary to move this activity into the MED a special counterintelligence group was formed, into which the existing security force was merged with little change. Later a few additions were made to this, but in the main it was kept as it was originally formed.

Throughout the entire history of the MED, both before this change-over and after, there was the fullest co-operation between the FBI and our security organization. This was vital, for the FBI had a great deal of background information that was of much value to us, and we were acquiring information that was of interest to them. There were also persons with Russian connections entirely outside our control—that is, not in the MED—who were attempting to procure information about the project. Since they were also trying to secure other military information, they were under the surveillance of the FBI. Obviously the closest co-operation was essential.

---

[1] By the end of the war the MED's force of "creeps," as they became known, numbered 485.

Our counterintelligence operations were under the District Engineer, who had a District Security Officer, first Captain H. K. Calvert, and later Lieutenant Colonel William B. Parsons. At every laboratory, plant and other facility there was a security officer with such assistants as he needed. He was responsible to the District Security Officer, our local MED representative and to the head of the installation.

Because of the nature of the work and our close co-ordination with the FBI and the Office of Censorship, more of the details than I liked had to be handled in my office in Washington by Lansdale and his assistants. I found it difficult at times to control to the required degree of smoothness the direct communications between my office and the many field offices. This on more than one occasion caused unnecessary, but not too serious, friction with the district organization.

The basic security problem was to establish controls over the various members of the project that would minimize the likelihood of vital secrets falling into enemy hands. Dr. Bush had already expressed concern over the risks incurred through the free exchange of information among the various people in the project. This flow had to be stopped, if we were to beat our opponents in the race for the first atomic bomb. Compartmentalization of knowledge, to me, was the very heart of security. My rule was simple and not capable of misinterpretation—each man should know everything he needed to know to do his job and nothing else. Adherence to this rule not only provided an adequate measure of security, but it greatly improved over-all efficiency by making our people stick to their knitting. And it made quite clear to all concerned that the project existed to produce a specific end product—not to enable individuals to satisfy their curiosity and to increase their scientific knowledge.

Nevertheless, security was not the primary object of the Manhattan Project. Our mission was to develop an atomic bomb of such power that it would bring the war to an end at the earliest possible date. Security was an essential element, but not all-controlling.

Never once was any definite country named to me as the one against which our major security effort should be aimed. At first it seemed logical to direct it toward the Axis Powers, with particular emphasis on Germany. She was our only enemy with the capacity to take advantage of any information she might gain from us.

Japan did not in our opinion have the industrial capacity, the scientific manpower or the essential raw material. Italy was in the same position, with the further disadvantage that any large plants would be exposed to Allied bombing attacks. We did not feel that information secured by Japan would reach Germany accurately or promptly, and we suspected that the Italian-German intelligence channels were not too smooth either.

I had learned within a week or two after my assignment that the only known espionage was that conducted by the Russians against the Berkeley laboratory, using American Communist sympathizers. Our security aims were soon well established. They were threefold: first, to keep the Germans from learning anything about our efforts or our technical and scientific advancements; next, to do all we could to ensure a complete surprise when the bomb was first used in combat; and, finally, insofar as we were able, to keep the Russians from learning of our discoveries and the details of our designs and processes.

All procedures and decisions on security, including the clearance of personnel, had to be based on what was believed to be the over-riding consideration—completion of the bomb. Speed of accomplishment was paramount.

Naturally, we made every effort to find out before employing any-one whether there was anything in his background that would make him a possible source of danger, paying particular attention to his vulnerability to blackmail, arising from some prior indiscretion. A new person was usually kept on nonsecret phases of the work until a hurried investigation could be completed. Since there was no assurance that secret information might not be disclosed during the original discussion of employment, we tried to be certain, even before ap-proaching him, that there was no likelihood that he would prove unqualified.

A number of foreign-born persons were employed throughout the project, although it was usually impossible to obtain any but the most meager information regarding their past activities. Several of these people were refugees from countries with which we were now at war, or from other countries whose ideologies they had been unable to stomach. Although it seemed highly probable that they could be entrusted to share secret information affecting the security of the United States, it was always possible that someone with disloyal intentions might slip through our screening procedures. Despite the fact that there was so much at stake, a number of critics within the project enjoyed talking about our "Gestapo" methods. However, we thought it was absolutely necessary to maintain reasonable security checks on those whose records of prior affiliations were not available to us or to the British Government.

When I was first placed in charge of the MED I found that a number of people in the project had not as yet received proper security clearances, though some of them had been engaged in the work for months.

Any question of the trustworthiness of any one of these people was troublesome, for he would already be in possession of valuable information. To remove him would create only a greater hazard, particularly if he thought our suspicion of him unjustified. (I remembered that Benedict Arnold's treason had been sparked by his feeling that he had been unfairly treated.) Moreover, if we were to dismiss a person without publicizing the proof, which we would not want to do, the understandable resentment of his friends and associates in the project might seriously interfere with their work.

Almost all our original scientific workers came from academic surroundings. Most of them had been in universities as students or young teachers during the depression years, when there was more than the usual amount of sympathy for Communist and similar doctrines. Almost all of them at one time or another had been exposed to Communist propaganda and had had friends who were secret or even semi-open Communists.

I realized what the temper of the times had been, even though I never had any sympathy for the philosophy or for the educated Americans who adopted it. Discussions with others experienced in this area led me to the belief that among those whose employment would be to the advantage of the United States a reasonable distinction could be made between individuals whose use might be dangerous and individuals whose use would probably not be.

We gave a great deal of weight to how closely the person had followed the party line and for how long. We were particularly interested in how closely he had followed the twists and turns of Soviet relations with Germany. In most doubtful instances this was a deciding factor.

Our problem was made much more difficult by the very limited number of qualified atomic scientists available in this country. We could not afford not to use everyone possible.

The most disastrous break in security was that resulting from the treasonable actions of the English scientist, Klaus Fuchs. Fuchs was born in Germany and had fled to England, where he completed his education. The British authorities had been informed by the Germans prior to the war that he was a Communist. For some reason they ignored this and did not even record the information where they would find it. After the outbreak of the war he was interned as an enemy alien, first in the British Isles and then in a prisoner of war camp in Canada. After some time there he was released and returned to work in England on atomic research. After his return he was made a British citizen.

Our acceptance of Fuchs into the project was a mistake. But I am at a loss when I try to determine just how we could have avoided that mistake without insulting our principal war ally, Great Britain, by insisting on controlling their security measures.

When I received the names of the first group of British scientists coming over to work in the Manhattan Project, under the terms of the Quebec Agreement, I observed that there was no mention of their reliability. I told the British official with whom I was dealing that I would have to have a statement that they had been properly cleared. The statement furnished in reply was inconclusive, in my opinion, and I asked for a more definite one. This was given me; it said that each member had been investigated as thoroughly as an employee of ours engaged on the same type of work.

Since the disclosure of Fuchs' record, I have never believed that the British made any investigation at all. Certainly, if they had, and had given me the slightest inkling of his background, which they did not, Fuchs would not have been permitted any access to the project. Furthermore, I am sure the responsible British authorities would have withdrawn his name of their own volition, before giving me his history.

If Dr. Chadwick had been in charge of the British mission at that time, as he was later, I am sure that no such deception would have been attempted. Chadwick was always most punctilious in informing me of the slightest question of background, including that of German blood. Unfortunately for the free world, Chadwick did not take over until a few weeks later.

Since Fuchs was uncovered, it has often been suggested that I should have investigated each British subject before he was admitted to the project. This would have been most presumptuous and, in fact, impossible without complete infringement of British rights and without the co-operation of the British Government, which we would not have obtained.

It was a British responsibility. As partners in the atomic field each nation had to be responsible for its own personnel. The United Kingdom not only failed us, but herself as well.

I have always felt that the basic reason for this was the attitude then prevalent in all British officialdom that for an Englishman treason was impossible, and that when a foreigner was granted citizenship he automatically became fully endowed with the qualities of a native-born Englishman. With the uncloaking in recent years of Fuchs, May, Maclean and Burgess, as well as others, I doubt if this feeling still prevails.

No one but Fuchs and the Russians know what he told them. We do know, however, what he knew and consequently what he could have told them. He knew the general progress of atomic development up until the fall of 1942, when we stopped the almost complete interchange of information. He knew the general design of the gaseous diffusion plant at Oak Ridge and many of its details. He knew the details of design for both types of atomic weapons, the gun type and the implosion type. He knew our thinking about how these could be improved, and he knew of our studies and belief in the possibility of the H-bomb, not as it was later developed, but as the much more expensive and complex weapon we then envisaged.

It was, of course, utterly impossible to make a thorough investigation of the past history, loyalty, habits, and citizenship of the thousands and thousands of construction and operating employees at Oak Ridge, Hanford and elsewhere. In general, all were investigated to some degree, but for practical reasons, the degree varied widely. The investigations of those who would not have access to any classified information, such as truck drivers, cafeteria workers, and the like, were limited to a brief check of police and fingerprint files. The background of others who were in positions where they could obtain secret information was much more carefully examined. For a few, the investigation was most thorough and went back to their infancy.

In the restricted areas, each employee had to fill out a special personnel security questionnaire before he could enter the area or handle classified documents. Where there seemed to be any question about him he was thoroughly investigated.

All fingerprints were sent to the FBI. If a record was on file, it was returned to be compared with the arrest record given by the employee at the time he was hired. Most of these arrest records involved traffic offenses or drunkenness. Any employee who had failed to give the correct arrest record when he was hired, or had failed to give a complete record on serious charges, was interviewed. Depending on his attitude when questioned, the seriousness of his arrest record, the quality of his work, his absentee record, and the need for men of his particular ability, he was either retained or discharged. Many were discharged for falsifying their records of arrests on serious charges, and many were retained. Of those retained, most were used in the general areas and not in the restricted areas. No one was hired or kept on who had been convicted of rape, arson or narcotics charges. Such persons were felt to be unreliable because of their demonstrated weakness in moral fiber and their liability to blackmail.

Press security was the other side of the coin. Here we had the invaluable co-operation and assistance of the Office of Censorship under Byron Price and N. R. Howard, editor of the Cleveland *News*. Howard was succeeded by Jack Lockhart of the Scripps-Howard papers, and it was he who was our usual contact. With his wise advice this aspect of our security problem was well handled indeed.

The general principles governing our control of information were simple. First, nothing should be published that would in any way disclose vital information. Second, nothing should be published that might attract attention to any phase of the project. Third, it was particularly important to keep such matters out of any magazine or newspaper that was likely to be read by an enemy agent or by anyone whose knowledge of scientific progress would enable him to guess what was going on.

An item carried by an Amarillo, Texas, newspaper, for example, had nothing like the potential danger of an identical item in a New York newspaper or in a national weekly magazine. For the same reason we extended the ban even to the reprinting of relevant articles that had appeared abroad. The press was always a bit restive under this restraint, but we were concerned lest the republication of such articles in this country might lead to others and to speculation by astute reporters about the purpose of the MED. We were only too aware that the piecing together of bits of published information is a prime source of knowledge to every intelligence organization.

It was in order to prevent speculative articles as well as the publicizing of any of our efforts that the press and radio had been asked to avoid the use of certain words, such as "atomic energy." Certain decoy words, such as "yttrium," were included in the list to camouflage its real purpose. This was a step we did not want to take, for it automatically pointed out to the press that the government was interested. However, Howard insisted that we simply had to do it if press security was to be maintained. Most reluctantly we agreed. As it turned out, it was a very wise move and an absolutely essential one.

We wished, too, to avoid any widespread mention of such places as Hanford or Oak Ridge and all mention of Los Alamos, as well as any reference to the MED. We also did not want any mention of my name that might arouse the interest of a foreign agent in my activities. Yet to have banned all reference in the near-by papers to Oak Ridge or Hanford would have been neither practical nor desirable, for it would only have tended to attract attention locally. We did try to keep Los Alamos entirely out of the news, but the Knoxville papers were permitted to carry items—mostly in the nature of social notes—about employees and events at Oak Ridge, though nothing, of course, that would help the average reader determine the purpose of the project or its importance. The same leeway was given to the papers close to Hanford.

We did have several unfortunate security breaks, but none of them, so far as we could ever find out, attracted any particular interest. The one with the worst potential for damage was a radio program that discussed the possibilities of an atomic explosion. The script for this had been prepared for the regular news reporter on a network program; he himself had had nothing to do with writing it. Unfortunately, in order to meet his travel schedule, he delivered it from a small affiliated station, where apparently it had not been reviewed to make certain that it did not violate press censorship rules.

From all that we could ever discover, there had been no deliberate breach of security. The information on which the talk was based came from a scientist who was not connected with the project in any way but who evidently had an inkling of what was going on, gleaned, we thought, from some of the project's scientists at the large laboratory in his city. The actual text was written for the reporter by a friend of the scientist. There was never any question in my mind but that the reporter delivered it in good faith. The failure of the radio station to stop it was attributable to plain carelessness.

Another incident that concerned us greatly was the appearance in a national magazine of an article hinting at the theory of implosion. While it did not violate any rules, it was most disturbing. A thorough investigation indicated that it resulted from the work of an alert and inquisitive reporter in another country.

There was one unfortunate happening not too long before the bombing, when a Congressman, in discussing an appropriations bill, commented on the importance of the Hanford Project. This item was picked out of the *Congressional Record* and was republished in a newspaper without any comment. I could never disabuse myself of the feeling that this newspaper did it with the deliberate intent of letting me know that our security prohibitions were not so effective as we thought.

After the war the excellent co-operation of the American press continued. Articles written by anyone with access to classified information were invariably cleared with my office. In one instance a newspaper learned of a rather large construction job in progress and wrote a series of articles about its most unusual character. Although the publisher of the paper completely disagreed with me, he canceled the articles before publication when I said that I thought it would be injurious to the best interests of the United States to publish them.

CHAPTER 11

LOS ALAMOS: II

The task for which Project Y, as the bomb development was called, was brought into being was without precedent. It would require the utmost in collaboration among civilian engineers, metallurgists, chemists and physicists, as well as military officers, some of whom would have to use the final weapons in combat. Many of the most difficult problems encountered at Los Alamos could not even be anticipated until work was well under way throughout the entire project. For this reason, scheduling at the Y site was both definite and indefinite. The bomb had to be ready as soon as sufficient quantities of atomic explosives were available, yet no one could say when this would be.

In looking for a contractor to carry out the work, I sought an organization experienced in research, which still retained an uncommitted capability to take on work. It soon became apparent that the University of California was our best prospect. The University authorities were not too eager to join us, but finally consented when I convinced Dr. Robert G. Sproul, the president, that this seemed to be the best solution to a crucial problem.

The nucleus of the Los Alamos organization came from the various groups that had been working under Oppenheimer at Berkeley. It was reinforced by others recruited with the assistance of Conant who, as chairman of the NDRC and president of Harvard, exerted a tremendous influence. The big problem in getting good people arose from the fact that the scientific resources of the country, particularly in this general area, were already fully engaged on important war work. Because they were civilians, the scientists had complete freedom in their choice of jobs. They had to be persuaded to come with us despite the disadvantages of isolation and security restrictions and their natural disinclination to move, particularly when their families were comfortably located. Many of the men we wanted were used to living in cities or near large metropolitan areas and were a bit dubious about the prospects of life in a remote, sparsely populated area. We had somewhat similar trouble with the engineering people, although they were not so concerned at being isolated.

One other major handicap was that we could not hold out any financial inducement to the people we wanted. It had been decided, after consultation with the Military Policy Committee, that, in keeping with the general policies of the OSRD, we should not offer any increase in pay to people recruited for the Manhattan Project. However, academic personnel who had formerly been paid on the basis of a nine-months work year were given increases whereby they were paid for the full twelve months of work, at the original monthly rate. Occasionally this led to inequities. For instance, Oppenheimer, who came to us from a state university, originally received much less than some of his subordinates, who had worked at large Eastern schools. This was unsound, and the difference was so marked that I eventually decided to make an exception in his case and brought his salary up to a point where it was equal to that of the others. This was done without any request or suggestion on his part, and his approval was not even asked.

Conant suggested that it would help greatly if we furnished Oppenheimer with a letter that clearly defined the responsibilities of the military and scientific organizations at Los Alamos. Although I knew that Oppenheimer himself had no doubts about the setup, I could see much merit in the proposal and accordingly prepared the letter. In order to give it added weight among the scientists, Conant signed it with me. It was written in terms that would enable Oppenheimer, without violating security, to explain to the scientists he was recruiting the background of the project and to assure them that by joining us they would not be cut off from the rest of the scientific world. The writing of this letter reflected Conant's great experience in, and understanding of, the academic world, particularly its scientific element.

The letter also stated that by a fixed date all the civilians on the staff would be given military rank. This provision was based on the practice that had been followed successfully in World War I in the development of certain chemical warfare products. Conant had participated in one of these projects, and had found the system satisfactory. As Los Alamos expanded many times beyond our expectations, we found not only that it would be unnecessary to carry out this idea but that it would be unwise and well nigh impossible even to consider it.

Stone and Webster drew up the original plans for the laboratory in accordance with specifications developed by Oppenheimer with the assistance of E. M. McMillan and J. H. Manley and with Conant's and my approval. They provided for a scientific staff of about one hundred, supported by a somewhat larger group of technical, shop and administrative employees. This was a great underestimate, for by July, 1945, the personnel had expanded many fold.

As the work got under way some amazing rumors began to circulate through Santa Fe, some thirty miles away. Typical of these was that old stand-by that we were building a home for pregnant WAC's. The near-by local population followed the construction of the enclosure's fence with great interest to see whether it was designed to keep people in or to keep them out.

During the following months and years, the speculation and stories grew even wilder, although they seldom spread beyond the local community. One woman who lived near the highway leading from Santa Fe to Los Alamos wrote regularly to the local paper complaining about the mysterious and certainly nefarious goings-on there, and urged that citizens take action toward having the matter investigated. Every day, she saw large numbers of loaded trucks headed toward Los Alamos, but the trucks that came away never carried anything. It was quite obviously some boondoggling New Deal scheme for wasting the taxpayers' money.

After a number of Navy officers had been assigned to the project, and were seen on the streets of Santa Fe, rumors burgeoned about the new type of submarine that was being perfected on the Hill, as Los Alamos came to be known locally. Although the nearest navigable body of water was many hundreds of miles away, this rumor sounded entirely plausible to a number of people.

Regardless of how incredible the stories were, the people who worked at Los Alamos never confirmed or denied anything that they heard.

Colonel G. R. Tyler, the military commander at Los Alamos, once boarded a train at the railway stop nearest Santa Fe, and in the club car sat next to a man in civilian clothes who had gotten on at the same station. The stranger at once began a one-sided, rapid-fire conversation. It was obvious that he had failed to note the fact that Tyler had boarded the train at the same time that he had for, finally, he lowered his voice and said, "If we can find a secluded spot, I can tell you something which, I think, will interest you."

Both men walked to the vestibule of the car, and stood while the man related his story. "You'd never believe the strange things that are happening on a certain mountain about fifty miles from Santa Fe. They're doing some work that is very secret and the place is surrounded by belts of tall wire fencing. In order to keep intruders out, between these belts of fences they keep ferocious packs of wild African dogs. Besides, there are thousands of heavily armed soldier guards, and I can tell you that a number of people have been killed by the guards, or torn to pieces by the animals. It's a frightful thing! However, I suppose that in wartime these things have to be." He then told of other strange happenings on the Hill, none of which were true, and concluded with, "Of course, I happen to be one of the very few residents of Santa Fe who know what they are doing up there, but I do hope that you won't ask me any questions. You see, I've given my word of honor that I will not divulge their secrets."

By this time the train was approaching Tyler's station, and as the stranger followed him to the platform he said, "Colonel, I forgot to ask you, but where are you stationed, and what sort of an assignment have you?" The officer replied, "I am stationed at Los Alamos, and

I command the military personnel there." The horrified and now extremely red-faced stranger said, "I hope that you'll forget everything that I've told you. I don't *really* know what's going on at the Hill. I merely repeated some of the things that I've heard."

Oppenheimer and a few of his staff had arrived on March 15, 1943, long before the necessary construction was completed. There was neither adequate laboratory space nor housing. For security reasons, we did not want to put this group in Santa Fe, so we bent every effort to accommodate them in guest houses near Los Alamos, transporting them to and from the site. Living conditions for these early arrivals were far from pleasant. As one veteran of those days reported:

There is no doubt that the Laboratory staff and their families faced the prospect of life at Los Alamos with enthusiasm and idealism. The importance of their work and the excitement associated with it contributed to this feeling, as did the possibility of building, under conditions of isolation and restriction, a vigorous and congenial community.

The actualities of the first months were hard for many to view in this light. Living conditions in the ranches around Santa Fe were difficult. Several families, many with young children, were often crowded together with inadequate cooking and other facilities. Transportation between the ranches and Los Alamos was haphazard despite great efforts to regularize it. The road was poor; there were too few cars and none of them was in good condition. Technical workers were frequently stranded on the road with mechanical breakdown or too many flat tires. Eating facilities at the site were not yet in operation and box lunches had to be sent from Santa Fe. It was winter, and sandwiches were not viewed with enthusiasm. The car that carried the lunches was inclined to break down. The working day was thus irregular and short, and night work impossible.

Until mid-April, telephone conversations between the site and Santa Fe were possible only over a Forest Service line. It was sometimes possible to shout brief instructions, but discussions of any length, even over minor matters, required an eighty-mile round trip.

And yet it was impossible to wait until everything was ready, for it never would have been. Days counted, and they could not be wasted.

From the beginning, there were two heads at Los Alamos—the

Commanding Officer and the Director. The Commanding Officer[1] reported directly to me and was primarily responsible for the maintenance of adequate living conditions, safeguarding government property and the conduct of the military personnel. The Director, Oppenheimer, also reported to me and was responsible for the technical, scientific and security portions of the program.

Originally, there were only two administrative assistants to the Director: Dr. E. U. Condon, from the Westinghouse Research Laboratories, and Dr. W. R. Dennis, from the University of California.[2] Not long afterward, Dr. Hans A. Bethe, a brilliant physicist from Cornell University, joined the group as director of the theoretical physics division, and stayed on until after the end of the war. Bethe, who had left Nazi Germany in 1935, was already famous for his work on hydrogen fusion as the source of energy in the sun.

Condon was not a happy choice; yet the responsibility for his appointment was primarily mine. I felt that Oppenheimer should have help in handling the many administrative details of his job, particularly to ensure good relations with the Commanding Officer, so I had urged him to select as an associate director a physicist with industrial background. With my approval he chose Condon, who at that time was an associate director in the Westinghouse experimental laboratory at Pittsburgh.

At Los Alamos, Condon did little to smooth the frictions between the scientists and the military officers who handled the administrative housekeeping details. We employed Army officers for these simply because we did not wish to use scientists who were in short supply and whom we wanted to devote their entire efforts to the research work for which they alone were fitted. To keep relations between the two groups on an even keel was Condon's major responsibility as Associ-

[1] Lieut. Col. J. M. Harmon, succeeded by Lieut. Col. W. Ashbridge, and then by Col. G. R. Tyler.
[2] Condon left in May, and Dennis in July of 1943. David Hawkins of the University of California arrived in May of 1943 to serve as a special assistant to Oppenheimer to handle liaison matters with the Station Commander, and A. L. Hughes took charge of personnel matters in June. In January, 1944, David Dow was designated Assistant in Charge of Non-Technical Activities.

ate Director, and I had expected that his industrial experience would enable him to handle the problem well. Oppenheimer, as head of the entire laboratory, naturally had to give his first attention to the scientific and technical problems.

We sent the staff out to Los Alamos as we assembled it, without delay, both to avoid losing members to other jobs and to give them more time for planning. Unfortunately, trouble arose as soon as the first scientists arrived on the site. They naturally wanted everything ready immediately and had little appreciation of the difficulties confronting the Albuquerque District Engineer, whom we had arranged to be in charge of construction at the site, as he tried to accomplish a most complex job in a remote area. Nor were they temperamentally able to let him resolve his own difficulties, the greatest of which was getting enough skilled labor. What little labor there was in the vicinity was under the control of building trades locals which did not understand, particularly at first, our sense of urgency in getting the buildings ready for use. Consequently the District Engineer found himself in that most unhappy of all positions for any builder, which occurs whenever the user is given free access to a facility while it is still under construction. I did not envy him as he contended with constant suggested changes in design.

As the scientists grew impatient, they attempted to remedy the situation themselves. The principal result of their efforts was confusion and soon we began to receive complaints from the War Manpower Commission, the U.S. Employment Service and the American Federation of Labor. Some relief from this unpleasant situation came when the small permanent key maintenance staff, hired by the University of California, arrived on the scene, furnishing the scientists with a construction force of their own for small jobs. Nevertheless, serious construction problems continued to plague us throughout the duration of Project Y, both because many construction requirements could not be determined until experimentation had been completed, at which time the new facilities were usually needed immediately, and because of the continuing expansion.

From the beginning, the procurement of equipment and materials was difficult, not only because our needs were unusual but because some all-important equipment could be obtained only by negotiation or borrowing from universities. We obtained a cyclotron from Harvard, two Van de Graaf electrostatic generators from the University of Wisconsin, and a Crockcroft-Walton accelerator from the University of Illinois. Smaller items were purchased wherever they could be found.

As time passed and the administrative and technical groups grew accustomed to each other's ways, minor mutual annoyances gradually faded away. Looking at this period in retrospect, it is difficult to appreciate how vital some of these matters then seemed to the people involved. Yet it is understandable that to men eager to get on with their work they were of utmost importance, and it is not surprising that on a number of occasions tempers erupted and feelings were hurt. The fact that almost all the people who arrived during this difficult time stayed on through the entire project is indicative of the spirit and sense of duty that motivated them, both individually and collectively.

Yet Condon remained only about six weeks and then resigned. When Oppenheimer told me that Condon was leaving, I warned him that for his own protection in the future he should insist on Condon's putting into writing his reasons for resigning. The considerations he cited in his letter of resignation[3] did not seem to justify his departure.

Condon was not a stranger to this area; in fact, he had been born at Alamogordo, New Mexico. He had lived in the West for many years, and had attended the University of California. Yet, for some reason, he did not want to stay at Los Alamos. I never felt that his letter disclosed the real reasons for his departure. He did mention his children's needs for schooling and said that he did not think that the local schools were good enough. It was my impression, however, that he was being motivated primarily by a feeling that the work in which we were engaged would not be successful, that the Manhattan Project was going to fail, and that he did not want to be connected with it.

[3] See Appendix VII, page 429.

During the latter part of April a series of conferences was held at the site to acquaint the new arrivals with the existing state of knowledge on atomic physics and to arrive at a firm program of research. One of the principal theoretical questions that had to be solved involved the time available for a nuclear reaction if an explosion were to take place. In this case, two opposing considerations came into play. The violence of the explosion was dependent upon the number of neutrons released by the chain reaction. This number increased geometrically with each generation of the chain. Yet to allow the reaction to progress through a number of generations took a certain amount of time during which the energy already released by previous generations could blow the bomb apart and terminate the chain reaction before any major detonation was achieved.

During this period guesses concerning the optimum time length of an explosive reaction—and they were guesses—varied greatly. All in all, however, we were convinced that the efficiency of any bomb we might build would be low as compared to the power that would remain untapped. This is usual in all explosions.

The most straightforward proposal for the bomb's design utilized the gun-assembly method to bring a critical mass of fissionable material together. In this method one subcritical mass of fissionable material was fired as a projectile into a second subcritical mass of fissionable material, the target, producing momentarily a supercritical mass which would explode. This principle was employed in the design of the Thin Man bomb that was dropped on Hiroshima.

Another method was proposed that utilized the effects of implosion, by directing the blast of conventional high explosives inward toward a quantity of fissionable material. The force of this blast literally squeezed the material together until it reached a critical mass and detonated. This principle was used in the Fat Man bomb which was delivered against Nagasaki. It was particularly difficult to develop because, unlike the gun-assembly method, there was no previous experience upon which to base its design. The man most responsible for the development of implosion theory in these early days was Dr. S. H. Neddermeyer, who at the beginning was almost alone in his belief in that method. He held to his conviction even though it was not at first too highly regarded by his colleagues at Los Alamos. When we discovered late in 1943 that certain previously unknown properties of plutonium made it extremely difficult to employ it safely in a gun-type bomb, under the then existing knowledge, we were very thankful for Neddermeyer's persistent belief in the feasibility of an implosion bomb and the advance work in which this had resulted.

At this time, too, there was some discussion of a fusion bomb, such as has later been developed in the hydrogen bomb, but since it was realized that any such super-bomb would require the high heat generated by a fission bomb to set it off, it necessarily had to be assigned a lower priority. On the other hand, its potentialities were so great that research and study for its development could not be completely neglected.[4] There was much discussion about the possibility of its igniting the atmosphere, but such an occurrence seemed most unlikely, not only according to the scientific theories of the time, but by common sense. There was some discussion about the explosive force such a bomb could develop, but there was no real agreement on this. Everyone recognized, however, that the fusion bomb would produce a tremendous explosion far beyond that expected from the atomic bomb.

Oppenheimer was told that he should be ready to assemble a fission bomb as soon as the necessary material was available. He was given an estimate that this would probably not be until sometime in 1945, but that in any event he would probably have at least six months' advance notice of the probable delivery date of the final ingredients for the weapon. While the bomb itself could not be tested until enough fissionable material was available to produce a critical mass, many auxiliary tests could be made with a smaller amount, of less purity, which would aid in designing the bombs and in estimating their probable power and the likelihood of success.

In the beginning, Los Alamos was purely a research organization. The work there was so scientific and so highly theoretical that, in its early stages, almost all the responsibility was carried by the

[4] Most of this was carried out by Dr. Edward Teller, the physicist from Hungary, who is now known as the father of the H-bomb.

scientists rather than by the engineers. As time went on, however, the balance shifted somewhat, and engineering, still highly scientific, became of great importance as we sought to make certain that the bomb would go off when dropped.

One of our major needs was to determine the number of neutrons per fission. We had some idea of what this ought to be in the case of U-235, although we were not certain because our estimates were based upon measurements of samples containing large amounts of U-238. For plutonium, however, the number was completely unknown. We could only assume and hope that it would not differ too much from that of U-235.

At the start of the Los Alamos Project, we did not know, because of possible production difficulties, whether we would use U-235 or plutonium, or both, for the bomb. Neither did we know whether the material would be a pure metal or a compound. The use of U-233 was also under consideration for a while. Consequently, the mechanical requirements for the bomb material could not be specified.

We faced the certainty that we would have only a minimum amount of time for research after fissionable material became available in gram or kilogram amounts. Almost all chemical investigations, important as they were, would have to be performed on microscopic samples.

We also had to get started on our ordnance program at a very early date. This program involved a new field of engineering, and would have to be a joint effort not only of scientists and engineers, but particularly of explosives experts and men experienced in ordnance.

At the Military Policy Committee's meeting in May, 1943, I asked for advice in selecting a suitable head for this work. It was a position that Oppenheimer and I had not yet been able to fill. The man we needed, I said, should have a sound understanding of both practical and theoretical ordnance—high explosives, guns and fusing—a wide acquaintance and an excellent reputation among military ordnance people and an ability to gain their support; a reasonably broad background in scientific development; and an ability to attract and hold the respect of scientists. I added that, because in the later stages of the project he would be concerned with ballistic testing and the planning for and possibly the actual use of the bomb in battle, it would be helpful if he were a regular military officer. I told the committee that I knew of no one available in the Army who could fill the bill.

After some discussion, Bush asked whether I would have any objections to a naval officer. I quickly replied, "Of course not." He then suggested Commander William S. Parsons, and Purnell added his hearty endorsement.

Parsons, a 1922 Annapolis graduate, had spent much of his commissioned service on ordnance and gunnery duties and had just completed several years of work on the development and fleet tests of the proximity fuse. Subject to my meeting and approving of him, and the Navy's concurrence in his assignment to the MED, which was promptly arranged by Purnell, his selection was agreed upon.

Late in the next afternoon, Parsons reported to me; within a few minutes I was sure he was the man for the job. During our conversation he reminded me that I had had dealings with him in the early 1930's, when I was working on the development of infrared for the Army at the time he was involved in the development of radar for the Navy. I remembered that I had been impressed then with his understanding of the interplay between the military forces and advanced scientific theory.

I discussed Parsons' proposed assignment with Oppenheimer, who met him in Washington a few days later and expressed his hearty acquiescence in his selection. The necessary orders were issued promptly and Parsons soon found himself at Los Alamos.

His reception on arrival was not notable. Because of security restrictions it was extremely difficult to enter Los Alamos unless one had business there and had been granted security clearance. Those who were qualified entered by a main gate which was about two miles from the center of activities. Security guards had special instructions to be on the alert for uniformed personnel who sought entry, and whose uniforms varied in any detail from that prescribed by regulations. The guards were all enlisted Army personnel and were but vaguely familiar with Navy uniforms.

Parsons was the first Navy officer to be assigned to the station, and appeared at the gate wearing a Navy summer uniform. His arrival was announced by a frantic guard, who telephoned his sergeant: "Sergeant, we've really caught a spy! A guy is down here trying to get in, and his uniform is as phony as a three dollar bill. He's wearing the eagles of a colonel, and claims that he's a captain."

Another case of "uniform confusion" illustrates the danger of using slang. I had telephoned to Colonel Tyler from Washington to say that a senior staff officer of the Army, with whom we had dealings, would be arriving later that day at the Santa Fe Airport. He was traveling on a tight schedule and wanted to see Los Alamos. I told Tyler to meet him at the airport, drive him to the Hill, arrange a brief conference with Oppenheimer, and if possible a minor nonnuclear explosion in the test area. Because the General had so little time, Tyler arranged for his car not to be delayed by sentries, and the customary identification procedures were to be waived.

These unusual instructions were relayed to the commanding officer of the Guard Detachment. Tyler drove to the airport where he was to meet the plane at 2 P.M.; however, the airplane arrived two hours late. The General emerged, announced that he had a later appointment the same day at El Paso, Texas, and urged that his visit to Los Alamos be expedited in every possible way.

All went well until the car was driven at fast speed past a sentry post in a wooded section in the direction of the test area. As it sped past, the sentry yelled, "Halt! ———— ———— it, STOP!" This command was accompanied by the sound of the action of a rifle bolt, which could only mean that the sentry was preparing to fire. The driver applied brakes, and slowly backed the car to where the sentry stood. In a casual and deliberate manner the sentry began examining the identification papers of all occupants. At this point, the General showed evidence of impatience. Tyler then asked the sentry, "Weren't you told to allow this vehicle to pass your post this afternoon without stopping?" "No, sir," the sentry replied. "All they told me was to allow some visiting firemen to pass without stopping, and I've been expecting people wearing firemen's helmets; you ain't seen them anywhere, have you, sir?" The grim look which had been worn by the General up to this time now relaxed in unrestrained laughter.

Parsons more than fulfilled my expectations at all times, not only in his performance of his assigned duties, but in helping to smooth out some of the frictions that are bound to arise on a project of this type. As we had anticipated, the fact that he was a regular officer, with a background in proximity fuse development, proved of inestimable value to us in the later stages of the project when we moved into the final preparations for the bomb's delivery.

Conant advised me that I could improve my working relationship with the Los Alamos scientists if I appointed a committee to review their work, regardless of whether or not any direct benefits would accrue to us from its reports. He pointed out that these people were accustomed to making their views known to similar committees appointed by their university administrations, and that our adoption of this system would meet with their approbation. A further advantage which we both recognized was that a review committee, with its fresh outlook, might be able to make a suggestion that would be eagerly seized upon, whereas if the same suggestion came from me, it might be regarded as interference.

Personally, I never found the idea of a committee particularly obnoxious so long as I recalled the opinion of a very wise and successful Chief of Engineers, General Jadwin. When some of his subordinates intimated to him that there was no need to appoint a board of consultants on the Mississippi River, since its members would have neither the knowledge nor the background in this field possessed by many officers of the Corps of Engineers, Jadwin replied: "I have no objection to committees as long as I appoint them."

With this guidance, I had no qualms about setting up a review committee. Its primary purpose was to reassure Conant and me, as well as the members of the Military Policy Committee, that the program and the organization at Los Alamos were sound.[5] As with

---

[5] The committee consisted of W. K. Lewis of MIT, Chairman; E. L. Rose, of Jones & Lamson; J. H. Van Vleck and E. B. Wilson, both of Harvard; and Richard C. Tolman, Vice Chairman of NDRC.

every committee that I appointed, its members were very carefully selected, and at least one of them, in this case the secretary, was a man who was thoroughly familiar with the project and with my views on the subject under study.

Although such an approach to committees may appear cynical, in my experience it produced excellent results. Certainly that was the case in this instance. Out of the Review Committee's work came one important technical contribution when Rose pointed out, in connection with the Thin Man, that the durability of the gun was quite immaterial to success, since it would be destroyed in the explosion anyway. Self-evident as this seemed once it was mentioned, it had not previously occurred to us. Now we could make drastic reductions in our estimates of the Thin Man's size and weight. Because the gun-type bomb thus became militarily practical at an early date, work on it could go ahead on an orderly and not too hurried basis.

One recommendation made by the Review Committee, which was subsequently adopted, was that because Los Alamos would be responsible for the successful performance of the bomb, the development of the special methods necessary to purify plutonium after it had been separated should be carried out there. Although at first this involved nothing more than study and research, it gradually grew into a production operation. I considered this perfectly appropriate for the Los Alamos site because the quantities involved were extremely small and we had to build a special installation in any event at either Hanford or Los Alamos. The committee also strongly concurred in Oppenheimer's views, which were also mine, that the ordnance development should be pushed. These changes in emphasis and missions doubled the number of people at Los Alamos, with a consequent increase in its facilities, particularly in the test areas for ballistics and explosives work.

The Explosives Division was headed by Dr. George Kistiakowski of Harvard, a distinguished chemist experienced in high explosives work. The success of the Fat Man depended upon the design and quality of manufacture of non-nuclear explosives, for both of which Kistiakowski was responsible. The work was carried out under experimental conditions and it was not possible, because of lack of time, to proceed with normal caution. Chances were taken but no one was injured.

To facilitate his control of the work at Los Alamos, Oppenheimer established a governing board consisting of himself, his division leaders, the general administrative officers and selected individuals occupying important technical positions.[6] Serving under the division leaders were the group leaders. The board had cognizance of everything pertaining to the project and its operation. There was also an advisory community council. It was a thorn in the side of the station commander, who was unable to satisfy all of its demands, yet its continual prodding often got results that added to the amenities of life. On the whole, it was a valuable adjunct, for it not only improved the morale of the community, but kept the post administration on its toes.

Problems at Los Alamos included those that can always be expected to arise in any isolated community. They were aggravated by the fact that the two dominant sectors of the group were composed of people of almost directly opposite backgrounds: scientists with little experience outside the academic field; and uniformed members of the armed services, nearly all nonprofessionals, who had little experience in, or liking for, the academic life and who were interested simply in bringing the war to a quick and successful end.

There was always some undercurrent of feeling between small segments of these two groups, though Oppenheimer, Parsons, Tyler and Ashbridge made every effort to bring them together. On social occasions, for instance, they included both civilian and military personnel. On one evening at least, it was a notable success.

This was a dinner given by Tyler and his wife, soon after their arrival at Los Alamos. Shortly before, an item had appeared in a daily column syndicated in several Eastern newspapers advancing the

[6] The membership of the original board included Robert F. Bacher, a top-flight nuclear physicist who later became one of the first members of the Atomic Energy Commission; Hans Bethe; J. W. Kennedy; A. L. Hughes; D. P. Mitchell, who had previously been in charge of laboratory procurement for the physics department at Columbia; Parsons; and Oppenheimer. Later additions were E. M. McMillan of the University of California, and George Kistiakowsky and K. T. Bainbridge of Harvard.

theory that if one wished to expedite the freezing of ice cubes in a refrigerator he might do so by filling the ice trays with boiling hot water.

In a casual way, the hostess mentioned the item, and wondered whether any of the guests knew whether the freezing of water could, indeed, be hastened in this way.

Any qualms she might have felt about a topic of conversation that would absorb the interest of the leading physicists of the United States were now dispelled. One highly eminent scientist stated that the proposal was a ridiculous one. Another said that the theory was quite possibly true. Small slide rules emerged from several coat pockets; pencils and pads of paper were requested; there were heated arguments in which some of the military guests with engineering background joined, as did some of the scientists' wives, while others looked quietly resigned, as if they had many times endured similar scenes. There is no record that any agreement was finally reached; but later it was rumored that several participants in the discussion hurried home and conducted experiments in their own refrigerators.

Despite such occasional successes it was obvious that until most of the innumerable petty annoyances could be removed, the frictions would continue. Matters were made much worse by the constant and unanticipated increase in the population. Although the water supply system had been adequate for about three times the originally estimated population, it was soon overloaded, and available housing always lagged behind the numbers of potential occupants.

Because of the difficulties the builder was having, both the quality and the quantity of the living quarters varied from time to time and at best were on the austere side. To house all the scientific and administrative staff satisfactorily, we tried to provide apartments, duplexes or separate cottages for married couples, and dormitories for unmarried people.

Aggravating the housing problem was the scarcity of household help. There were no servants other than Indian girls from near-by communities, who were brought in by bus and assigned according to need, rather than according to desire for a servant or ability to pay.

This system was designed to encourage the wives of our people to work on the project, for those who worked obtained priority on household assistance. Some of the wives were scientists in their own right, and they, of course, were in great demand, but with labor at a premium we could put to good use everyone we could get, whether as secretaries or as technical assistants or as teachers in the public school that we started for the children.

To enable the mothers of young children to work, a nursery school was organized on a partially self-supporting basis; its financial losses were carried by the government. The elementary and high schools were operated as free public schools, with all expenses borne by the project.

A hospital, run for the benefit of all residents, co-operated in the health and safety program of the laboratory. I felt it was particularly important to have this hospital staffed by people who would be able to meet every possible demand, whether reasonable or unreasonable, for we did not want anyone to feel the slightest desire to use outside facilities. Medical service at Los Alamos was entirely free, except for nominal board charges for those actually in the hospital. Apparently, we provided adequate service, for one of the doctors told me later that the number and spacing of babies born to the scientific personnel surpassed all existing medical records.

Much of the friction at Los Alamos would never have existed if the laboratory could have been placed in the heart of a major city. Most of the people were used to living in urban surroundings with the facilities and leisure activities normally found in such areas. They found life strange in New Mexico, far from every form of amusement except for the more simple pleasures that they could devise for themselves. There were no symphony orchestras or concerts, no theaters, or any lectures on matters of a cultural nature. They could, of course, go down to Santa Fe, the capital of New Mexico, but still a quite small city, which was about twenty miles away, but extended trips away from the reservation were discouraged.

In addition to the advisory council that dealt with community affairs, Oppenheimer set up a co-ordinating council whose members were drawn from the group leaders and others in higher positions. This was not normally a policy-making body; its meetings were generally informative rather than deliberate. The various divisions and groups held their own meetings and seminars at which scientific information was exchanged.

Another means of stimulating interest and progress was the weekly colloquium, which every staff member was privileged to attend, provided he had sufficient scientific education or experience to enable him to give or receive benefits in any general discussion of the technical program. In the final analysis, though, the colloquium existed not so much to provide information as to maintain morale and a feeling of common purpose and responsibility. From the standpoint of security, it presented a major hazard, and it was one of the reasons why the treachery of Fuchs was so disastrous to the free world.

As liaison between Los Alamos and the other phases of the Manhattan Project was effected, it became difficult to keep the exchange of information under control, although, generally speaking, we were fairly successful in this respect. There was, however, much dissatisfaction among the Los Alamos people concerning their lack of information about production schedules. This lack did not result so much from poor liaison as from the fact that during this period all schedules were vague, incomplete and contradictory. It was not only difficult but impossible to arrive at sensible schedules for bomb research and development, when we simply could not predict when the necessary U-235 or plutonium would be ready.

The procedure for liaison with the Metallurgical Laboratory at Chicago was fairly typical of all such arrangements. Specified representatives of the two laboratories were permitted to exchange information either by correspondence or by visits of the Los Alamos agent to Chicago. The information to be exchanged was limited to that dealing with the chemical, metallurgical and nuclear properties of fissionable and other materials. The representatives were permitted to discuss schedules of need for, and availability of, *experimental* amounts of U-235 and plutonium, but not of *production* amounts. Nor were they allowed to exchange information on the design or operation of production piles or on the design of weapons. However, three members of the Los Alamos laboratory were kept informed of the time estimates for the production of large amounts of these materials so that research could be intelligently scheduled, and I was always willing to make such additional exceptions to policy as might in Oppenheimer's opinion be required.

Our security precautions were not made any easier by the reluctance of a few of the scientists to recognize the need for putting some limitations on their personal freedom. For the first year and a half, travel away from the immediate vicinity of the site was forbidden, except on laboratory business or in case of emergency. Personal contact with acquaintances outside the project was discouraged. In the main these restrictions were accepted as concessions to the general policy of isolation. Some thought they were not strict enough, and no one was satisfied with the working definitions of "personal emergency." The removal of these limitations in the fall of 1944 was a cause of general rejoicing, and resulted from my feeling that the improvement in morale would outweigh the increased security risks.

One aspect of our security policy at Los Alamos that particularly annoyed everyone was the censorship of mail. Originally there was none. Nevertheless, shortly after the first staff members arrived at Los Alamos, rumors began to circulate that some letters had been opened. As the rumors continued to spread, Oppenheimer became most concerned and, since he had no assurance that the mail was not being opened, asked me whether I had ordered any censorship. I had not, and careful investigation of every instance where someone claimed that his mail had been opened convinced me that these claims were without foundation. However, by that time a number of the more thoughtful members of the laboratory had themselves begun to urge that we institute an official censorship on outgoing mail. This was set up in December of 1943. Its primary purpose always was to guard against the inadvertent rather than the intentional disclosure of information; deliberate traitorous espionage can never be prevented by normal military censorship. The most vital information at Los Alamos was of such a nature that it could be conveyed in a few words which could be transmitted in any number of ways. To guard against this, we had to rely on the integrity of the individual. The possibility of betrayal thus became directly proportionate to the number of people employed, and to the amount of knowledge possessed by each.

CHAPTER 12

THE COMBINED DEVELOPMENT TRUST

Although we believed that because of the splendid co-operation of Edgar Sengier of Union Minière we controlled enough uranium ore for our war needs, the Military Policy Committee realized that we must increase our supplies if we wanted to be certain that at the end of the war we would not find ourselves in the embarrassing position of having the plants, the knowledge and the skills, but no raw materials to work with. Information available at the time indicated that the Belgian Congo was, by far, the best source for us. However, the Belgian Government-in-Exile was in London, and it was possible that the British might gain a monopoly over the Belgian Congo raw materials. The United States would then find itself in a most disadvantageous position.

The best solution to this rather delicate situation, it seemed to us, would be to obtain a long-term commitment from the Belgians through the medium of a governmental agreement between Belgium on the one hand and the United States and the United Kingdom on the other. Accordingly the matter was brought before the Combined Policy Committee, and it was agreed that a tripartite agreement should be negotiated in London with the Belgian Government-in-Exile.

Shortly thereafter we began work on establishing a joint Anglo-American-Canadian agency which could enter into and administer the commercial contracts that we hoped to make with the Belgians following the tripartite agreement. In these contract negotiations, carried on in London, the United States was represented by Ambassador Winant, and Great Britain by Sir John Anderson, the Chancellor of the Exchequer.

In order that Mr. Winant would have the necessary background information and also so that I would be kept fully informed, I sent over Major H. S. Traynor from Oak Ridge to assist him. As in all other diplomatic matters dealing with atomic energy, the instructions to Winant were issued in the name of Secretary of War Stimson rather than that of the Secretary of State, Cordell Hull. When I asked him once whether he found this arrangement embarrassing, he replied: "Not at all—I am the representative in London of the President, not of the State Department."

The conduct of negotiations in London can perhaps best be described by excerpts from a recent account written by Traynor at my request.

On a day in early March 1944 a telephone call to Oak Ridge informed me that General Groves requested that I report to him the next day in Washington.

General Groves told me that it had become necessary to make immediate and personal contact with British and Belgian government officials and mine owners in London to insure availability of critically needed uranium ores in the Belgian Congo and that I was to proceed to England as quickly as possible, enlist the assistance of U.S. Ambassador John G. Winant and proceed to take whatever steps appeared to be appropriate. He said that a letter of introduction from the President would be provided as would known information on the amount and availability of ore, and legal opinions on possible courses of action with respect to intergovernmental and commercial arrangements. Ambassador Winant, General Groves added, was to be given information on the Manhattan Project —or the S-1 Project, as it was then sometimes called—to the extent necessary for him to appreciate and understand the implications of the assistance he was being asked to give.

Two objectives were clear: first, the establishment of some type of organization to effect intergovernment long-term exclusive rights to Congo ores and second, the development of a commercial arrangement that would insure prompt and uninterrupted ore shipments to the United States. All arrangements had to be equitable and legally supportable in all countries concerned and should take advantage of negotiations that were already under way between Belgian mining interests and U.S. military authorities.

On March 18, I arrived in London and I was able to see the Ambassador within the hour. I gave him a brief account of the S-1 Project, laying particular emphasis on the necessity for the prompt and complete availability of uranium ores which had already been mined in the Belgian Congo, the necessity for extracting additional quantities of ore from this source at the earliest practicable moment, and the strict security which surrounded the entire project and which would have to be observed in all negotiations.

Ambassador Winant quickly grasped the importance and urgency of the work. He mentioned that he had learned that the Germans had been engaged in a similar sort of project and had wondered what steps the U.S. had been taking in this regard. I was requested to see him again the following day—a Sunday.

The Sunday meeting with the Ambassador was arranged for one hour shortly after lunch. However, as the S-1 Project was described to him in more detail, as its importance to the U.S. war effort was explained, and as the need for swift action in the Congo was emphasized, Mr. Winant's interest became so complete that the meeting continued into the early evening hours. He mentioned that in his younger days he had had some mining experience which he believed helped him in appreciating the complexities and difficulties of the matters at hand.

One could not talk with Mr. Winant without being deeply impressed by his quick perception, his intensity of interest, and his quick grasp of implications. Courteous, considerate, mild in manner, soft and plain-spoken, he treated all people alike regardless of rank or station. Yet he seemed to exert a remarkable influence on those with whom he came in contact—both foreigner and American alike. I was told that many persons—both British and American—sought his advice on personal problems as well as matters of state. Several times he expressed great concern as to whether humanity would be able to progress socially and morally as fast as it was progressing technically and thus be able to control, for the common good, the new power from the atom. . . .

On the following Wednesday, Ambassador Winant met with Sir John Anderson. Mr. Winant stated that he had received instructions from the President to collaborate with the British Government with a view to obtaining long-term exclusive rights to uranium deposits in the Congo jointly by the U.S. and the U.K. It was agreed that it would be necessary to inform Belgian Government representatives in strict confidence that certain experimental work with uranium being carried out by the U.S. and the U.K. had reached the stage at which, while no results had been achieved which made it possible to say on the basis of tests what the ultimate developments would be, it was clear that there were possibilities of the greatest significance and that if these possibilities materialized, it would be of profound importance for the future not only of the United States, but of the whole world, that uranium should not fall into the wrong hands, that the U.S. and U.K. governments felt it their duty to do their best to obtain an option on as much as possible of the world's supply of uranium and to assume the task of seeing that it was not misused, at least until other arrangements were made after conclusion of the war. The exact nature of an agreement was left open pending further legal consideration, but a request for an option on all Belgian Congo uranium ore for a term of years was considered. It was agreed that Sir John Anderson would make the initial approach to the Belgian Government, which he did the evening of the same day.

The Chancellor reported that he talked with an official of the Belgian Government along the agreed lines. He said that he also gave assurance that nothing was contemplated that would in any way derogate from national Belgian sovereignty. He mentioned the possibility of an option on the entire output of uranium on terms to be discussed, pointing out, however, that the importance of the matter transcended commercial considerations. The Belgian official gave assurance that his government would co-operate.

A need quickly developed for legal guidance and consideration. Consequently, at Ambassador Winant's request, Brigadier General E. C. Betts, Judge Advocate General of the European Theater of Operations, was assigned by General Eisenhower to Mr. Winant for "advice on matters of interest to the War Department outside the European Theater of Operations." On the British side Sir Thomas Barnes, the Treasury Solicitor, was assigned a similar role. These were fortunate appointments as both men were highly regarded for their integrity and professional competence. Both had successfully worked together on legal problems arising from the presence of U.S. armed forces in the United Kingdom.

On March 27, a joint meeting was held between Ambassador Winant, the U.K. Chancellor of the Exchequer and senior officials of the Belgian Government stationed in London. Assurance of Belgian co-operation was again given and the Belgians stated that it would be necessary for representatives of Congo commercial mining interests to be brought into the negotiations.

Two days later General Betts and Sir Thomas Barnes proceeded to the problem of an intergovernmental body and developed a draft "Declaration of Trust."[1] This was completed within several days, and was dis-

[1] The basis for this was a brief which had been previously prepared in my office by Lieutenant Colonel John Lansdale, Jr., and Major W. A. Consodine.

patched to Washington on April 5 for the President's formal approval. At the same time, arrangements were made to put the draft before the British Prime Minister. . . .

After some exchange of views between London and Washington, and modifications including provisions for thorium procurement, the Declaration of Trust was signed by the President and the Prime Minister in mid-June, 1944.

One feature of the Declaration which, I think, made it unique among all such secret executive agreements, was its last clause, which was included at my instance, because of my strong feeling that all international agreements should receive the approval of the Senate, as set forth in the Constitution. It said:

The signatories of the Agreement and Declaration of Trust will, as soon as practicable after the conclusion of hostilities, recommend to their respective Governments the extension and revision of this war-time emergency agreement to cover post war conditions and its formalization by treaty or other proper method. This Agreement and Declaration of Trust shall continue in full force and effect until such extension or revision.

This provision was never carried out as far as I know.

The Trust Agreement established in Washington an agency known as the Combined Development Trust,[2] which, under the direction and guidance of the Combined Policy Committee, was to supervise the acquisition of raw materials outside of American and British territory. Their allocation was the responsibility of the Combined Policy Committee.

The Trust functioned to good advantage. Besides providing the impetus which led to several international agreements dealing with the control of raw materials, it increased the scope of the exploratory surveys that had been initiated by the Manhattan District, and it encouraged valuable research activities in the field of enrichment of low-grade ores.

[2] Originally appointed as trustees were Mr. C. K. Leith, a distinguished mining engineer; Mr. George L. Harrison, a special assistant to Secretary Stimson, who was becoming more and more involved in atomic affairs; and Major General Leslie R. Groves for the United States; Sir Charles Hambro and Mr. Frank G. Lee for the United Kingdom; and Mr. George C. Bateman for Canada.

At the same time that the Trust was being set up, negotiations were continuing with the Belgian Government officials. In mid-April the Belgian Minister of Colonies had sent an urgent summons to Edgar Sengier to come to London at once, as a matter of "national duty." The Belgian Government was completely unaware of Sengier's previous top secret contracts with the Manhattan Project and wanted his advice before signing an agreement with the British and Americans. We for our part were delighted, since his presence during the negotiations would be an insurance against delay and unfortunate questions.

The British Supply Council in Washington made the arrangements for his trip, though the situation was complicated by a recent regulation closing all British borders to diplomatic travel.

In spite of the care we had taken, Sengier's trip was not without its awkward moments. All boat departures were kept very secret, and so that he would not have to stay on board any longer than necessary, he was picked up just before sailing time. He had already been provided with the necessary travel papers. As he told me afterward, "At six o'clock in the evening two officials in civilian clothes appeared at my hotel apartment. They prevented me from even saying good-by to my wife, who had the impression that I was being taken off to Sing Sing prison." They delivered him on board without his having to observe any of the usual formalities.

Sengier was the lone civilian traveler among nine thousand military passengers, but nevertheless on the first day he became involved in the usual disciplinary and physical exercises. Among other things, he claims he was kept standing for two hours in the rain, with discipline as the objective. Fortunately, this difficulty was cleared up promptly and his voyage thereafter was as comfortable as conditions permitted.

On his arrival at Liverpool there was trouble again: first, a serious objection to his disembarking at all, and then a suggestion that he be put into quarantine. This embarrassment ended quickly when a British colonel in civilian clothes, who had been waiting for him, intervened and took him and his baggage to London in a jeep.

On May 8 Sengier, the Belgian Cabinet members, Winant and Anderson held a meeting at which, to continue with Traynor's report:

The Belgians reiterated their readiness to co-operate in seeing that the Congo uranium did not fall into the wrong hands and that as much as possible of it was made available for experimental work dealing with military possibilities. In addition, they stated that if experimental work produced results capable of commercial and industrial exploitation, the Belgian Government would expect to share in the benefits.

Negotiations continued for several days, and on May 12 Ambassador Winant reported to Washington that their discussions had resulted in an understanding that an agreement would be negotiated and signed under which the Belgian Government would undertake to grant to the U.S. and U.K. governments, on terms to be agreed upon, first refusal to purchase all Belgian Congo uranium.

Implementation of the undertaking was provided for in three parts:

1. A contract between an agency of the U.S. and U.K. governments and commercial mining interests to reopen the Shinkolobwe mine and supply a specified quantity of ore, with the Belgian Government guaranteeing fulfillment of the contract.

2. The Belgian Government would undertake to insure that such further amounts of Congo uranium ore as the U.S. and U.K. would require for military purposes would be made available at reasonable terms.

3. When and if military requirements were met, the question of additional requirements for industrial and commercial exploitation would take into account the desire of the Belgian Government to be assured of an equitable share in the benefits of such exploitation.

The Belgian agreement became effective in late September, 1944, with an exchange of letters among Belgian Foreign Minister Paul Henri Spaak, Sir John Anderson and Ambassador Winant.

With the signing of the purchase agreement with the Union Minière the American trustees were confronted with a serious problem. The contract was for so long a term that appropriated funds could not be counted on from War Department sources. The Constitution would prevent it.

Any such long-term commitment, if made from other funds, would require prior authorization from the Congress and, even so, would be subject to a change of heart on the part of Congress. Under the terms of the contract the trustees would be personally responsible for the payments and none of us had the necessary number of millions to take care of the obligation.

Our dilemma was solved by arranging for me to be paid by the United States $37,500,000, a sum sufficient to cover the expected obligations. I then deposited this money in a personal account at the U.S. Treasury. From this I made withdrawals as necessary and deposited the money with the Bankers Trust Company of New York.

Payments for ore were made from this account. To avoid arousing undue curiosity, knowledge of the account was limited to Mr. Sloan Colt, the president of the bank, and two other officials, who handled all transactions, including deposits and the delivery of statements.

Some time after the account was opened, I was startled by an observation by Major Consodine. He pointed out that in the event of my death, and probably of either of the other two American trustees empowered to sign checks (for two signatures were required), there could be some serious complications. The state of New York might claim inheritance tax on the balance in the bank and might resist the settlement of the individual's estate besides. When I talked to George Harrison about it, he became quite concerned. We immediately prepared a letter to the bank stating that the funds were not ours and that if any of us died or were disabled, the Secretary of War would designate the person to whom the money would be paid.

The MED's associations with Sengier continued to be as pleasant as they had been from the day of Nichols' first visit to his office. Despite the large sums of money involved, agreements were reached expeditiously and without any quibbling over legal language, usually in an hour or so. Each of us would have a scratch pad on which we wrote down the various points as agreement was reached. These heads of agreement seldom covered more than a legal-sized sheet. Then one of us would read his notes aloud for the other to check, so that the two papers would be alike. After this the notes were turned over to our assistants, who would draw up à contract in accord with the listed points. After the Trust came into being, Sir Charles Hambro normally participated with me in the negotiations.

It was difficult to arrive at a proper price. By this time it was certain that the material was of immense value to the United States, provided the bomb worked. To the seller it was of great potential value if atomic energy should prove to have either military or peacetime value. Otherwise, it was worth only the value of its radium content. And if our reactor theories were sound, the radium would lose most of its value since radioactive cobalt could largely replace it.

It did have one definite value and that was what it cost to produce. Yet even this was difficult to establish fairly, for the unit production cost was much less at Shinkolobwe than in Canada or on the Colorado Plateau. Its value had never been determined in the open market and now there was only one purchaser and one seller.

As a Belgian, Sengier appreciated fully the absolute necessity of an Allied victory. It was his broad, statesman-like attitude that made it possible for us to reach an agreement satisfactory to all.

It was a distinct pleasure for me after the war to recommend the award of the Medal of Merit, the highest civilian award made by our government, to Edgar Sengier for his great services to the United States, to Belgium and the free world in making available to us adequate supplies of Belgian Congo uranium. It was also my pleasure to present this award at a ceremony in my office in Washington. Security restrictions had not yet been lifted on this phase of the MED operations and the ceremony was private and unpublicized. It has always been a source of regret to me that Sengier's services, and particularly his foresight, could not receive full public recognition at the time.

Union Minière was not our only supplier of ore. Prior to the formation of the MED all ores had been obtained from the Eldorado Mining Company, which had a uranium mine at Great Bear Lake, not far from the Arctic Circle. Eldorado also operated a refinery at Port Hope on Lake Ontario, where uranium oxide, as well as radium, was extracted from uranium ore, and through which we eventually funneled all the Belgian Congo ore.

Later, when we found it imperative to decrease the breadth of the responsibilities originally assigned to Stone and Webster, we placed a soundly trained geologist, Captain Phillip L. Merritt, in charge of all raw ore procurement and processing through the refining stage. Under his immediate direction, and under the general guidance of Nichols, the ore situation rapidly improved.

A systematic search of the Colorado Plateau disclosed uranium-bearing wastes in the dumps at the vanadium mills of Union Carbon and Carbide and of the Vanadium Corporation of America. Contracts were let for the uranium content of these dumps, which was of considerable quantity, and for its extraction.

In the fall of 1943, Merritt went to the Belgian Congo, to make certain that there were no other known easily exploitable ores in that area. As expected, there were none. He also checked to determine whether there were not some tailing dumps that contained a substantial amount of ore; that there might be seemed most probable in the light of the richness of the ores we had previously received. Merritt's inquiry was successful and as a result we had immediately available another large amount of ore. It was not so rich as that which we had previously obtained from the Congo, but the Congo's poorest was much better than the best from Canada or the Colorado Plateau. These dumps had been built up during the years as a result of hand-sorting the richer ores. Their uranium content varied widely from 3 per cent to 20 per cent.

A typical example of the difficulties that we so often encountered in operating during the war came up when the shipping arrangements were made. There were no suitable bags to be found in the Congo, or, apparently, anywhere else. We finally found some new ones in India, and some used ones in Texas City. The latter were tin ore bags and were marked "Product of Bolivia." When they were reused, no one thought to re-mark them. Consequently their delivery was held up, for the United States Customs officials knew that the "sand" which they contained came from West Africa, not Bolivia. This matter was cleared up in short order. As a matter of fact, we never had much trouble with government regulations and so-called "red tape," probably because whenever we encountered potential difficulties, we did not resort to letter-writing through channels. Instead, a competent officer was always sent immediately to the trouble spot with orders and authority to resolve the problems.

Early in 1943, we decided that we ought to learn as much as possible about the various deposits of uranium and thorium throughout the world. It was obvious to me that we had a responsibility to see to it that in the future the United States would not lack the essential raw materials known to be, or likely to be found, suitable for the production of atomic energy. Uranium, we were certain, would be all-important and we thought that thorium might prove to be of almost equal value.

To collect the necessary knowledge we decided to use the services of some existing organization rather than attempt to organize an agency of our own for the work. We also decided that we should use a private organization rather than one within the government. The principal reason for this was the need for security, since extensive field investigations by a government agency would be apt to attract too much attention.

Union Carbide and Carbon agreed to undertake the assignment. It was a large company with adequate manpower resources; moreover, it was already engaged in the Manhattan Project as the operator-to-be of the gas diffusion plant at Clinton, as the supplier of the extremely pure graphite needed for the plutonium process reactors and as a refiner and supplier of uranium-bearing ores.

Among other things, it was to make a study of all the existing literature on the world's geology. This would require competent geologists who were also competent linguists—a requirement that proved extremely difficult to meet, particularly in the Russian language area. Somehow Union Carbide succeeded in finding good people for the job.

We had no suitable officer available in the MED to serve as the Army representative on this study. Mining experts had estimated that the work might take a number of years to complete; some estimates running as high as fifteen or twenty years. To avoid any unnecessary stretch-out of our study, I wanted a man who was experienced in the oil industry, feeling that he would be used to making quick, conclusive decisions, based, if necessary, on very limited information. I did not want anyone who would always insist on 100 per cent proof before making a move.

The records of all officers on file in the Adjutant General's office were reviewed as we searched for individuals possessing the desired qualifications. Out of the million or more cards examined came only a dozen names and most of these men were already overseas. Fortunately, the man who seemed best qualified, Major Paul L. Guarin, Corps of Engineers, was then stationed in Dallas, Texas, in the office of the Division Engineer. He had had many years of experience with the Shell Oil Company and was thoroughly accustomed to crash programs.

Before he took charge, I talked with him for twenty or thirty minutes. I went over our objectives and emphasized my views somewhat along the following lines: "When the war is over, there will be diplomatic exchanges between the victorious nations, possibly even another Versailles Conference. It seems clear that President Wilson and his staff, as well as the other participants at that conference, were not supplied with all the pertinent data they should have had. I am determined that any American negotiators after the present war will have available to them all possible information concerning the sources of fissionable materials. I estimate that we must be ready with the bulk of this information within two or three years. We are after a good sound report—not a perfect one."

We would help him all we could, I told him, but like everyone else in the MED, he had to carry his own responsibilities and make his own decisions when he could not, for any reason, obtain guidance promptly from either the District Engineer or me. I assured him he would not be subjected to second guessing so long as his decisions were reasonable.

In carrying out his responsibilities, Guarin was aided by a small group of assistants, chief among whom were Dr. George W. Bain, a Professor of Geology at Amherst College, and Dr. George Selfridge, a geologist from the University of Utah.

As the reports from Union Carbide's task force began to come through, Bain developed certain generalizations concerning the pattern of conditions under which uranium ore of sufficient richness was likely to be found and, equally important to our goal, where it was unlikely to be found.

Out of all this came several extremely important technological breakthroughs. Until this time it had never been thought that extractable amounts of uranium would be found in any hydrocarbon-bearing material, such as petroleum or coal. From his pattern studies, Bain concluded that they should be; and was proved to be right.

He had a remarkably thorough knowledge of geological formations throughout the world and recalled that, in the course of a trip he had made in 1941, he had found uranium in amounts that might be of interest to us in the gold mines of the Rand, in South Africa. A further investigation confirmed the presence of uranium, but not of sufficient richness for our needs. These findings were at considerable variance with Bain's estimate of what they should have shown.

After reviewing the entire situation with Guarin, Bain went home to Amherst from New York on a Sunday. While he was there, he took from his private collection a sample of the Rand gold-bearing rock, placed it on a photographic plate, and was delighted to find from the exposure very definite proof that the ore did emit beta rays of an intensity that indicated uranium content far beyond anything that had previously been suspected.

This made us feel certain that we had uncovered great possibilities. But we had a great deal of trouble convincing others, who insisted that it was impossible that uranium could have been overlooked in the Ra d ore for so many years. I discussed the matter with Sir Charles Hambro and Sir James Chadwick and they agreed with me that we ould pursue our investigation of the Rand vigorously.

A new assay confirmed Bain's opinion of the ore's richness, and proved that the Rand was probably a major potential source of uranium. It also led directly to the adoption of Bain's views that all placer deposits should be carefully considered; thus many other areas throughout the world came to be regarded as possible sources of uranium.

The uranium-bearing rock in the Rand mines occurs with the gold in a thin stratum. Under the normal extractive procedures, the uranium was completely mixed with the other nongold-bearing rock and deposited on the waste piles. It appeared difficult and expensive to recover, and many thought it would be completely impracticable even to attempt it. Nevertheless, with the wholehearted approval of General Smuts, the Prime Minister of the Union of South Africa, we organized a special group, headed by Dr. A. M. Gaudin of MIT, to try to devise a method by which it could be done. Gaudin had previously worked for the MED in developing concentration methods for low-grade Congo ores in his MIT laboratories. Within a comparatively short period of time, he came up with a most ingenious and satisfactory solution to our problem. Later the South Africans were able to improve Gaudin's process so that the Rand ores could be processed in mass.

The economic effect of these discoveries on the Union of South Africa has been tremendous. In 1959, well over $150 million worth of uranium was exported. It has made possible the working of many gold mines which, without this valuable by-product, would not have been able to operate. It is difficult to estimate how much the $700 million worth of gold produced during that same year would have been reduced, but it would have been by a substantial amount if uranium had not been recovered from what previously had been discarded as waste.

Another of Bain's technological breakthroughs involved his belief that uranium should normally be present in the monazite sands which are the principal source of thorium. To prove this theory, an agent was sent to the Lindsay plant in Chicago. This company, a refiner and dealer in the rare-earth class of chemicals for many years, had some old bins which contained unprocessed samples of monazite sands from one of the major sources of this material—the state of Travancore on the western coast of India.

Analyses of these samples proved the soundness of Bain's reasoning. This discovery emphasized the importance of gaining all possible knowledge of the world's monazite resources, in which we were already interested because of their being the principal source of thorium.

Although we were not sure of the value of thorium to our work, we had become convinced at an early stage of the desirability of arranging for long-term rights for thorium from the world's principal deposits of monazite, located in Brazil, the Netherlands East Indies and the Indian state of Travancore. Since Travancore fell outside the territory under the jurisdiction of the Combined Development Trust, negotiations with that state were conducted by the British.

Other agreements covering the sale of monazite sands were signed with Brazil and the Netherlands. During the time I was connected with the project, the Netherlands agreement remained inoperative. The Brazil agreement was finally abrogated by the Brazilian Government because of our failure to make use of its provisions, owing to the unexpectedly large production of uranium and consequent lack of interest in thorium.

Incidentally, it was at the start of negotiations with Brazil that the State Department was brought into the atomic energy picture for the first time. In a conference with President Roosevelt before his departure for Yalta, I suggested that it would be desirable to have Secretary Stettinius himself handle the initial talks with President Vargas, and that I would like to give him enough information about the project to enable him to do so. The President agreed and informed Stettinius almost immediately. Later I talked with him and arranged to have an officer accompany him to Brazil.

It was at this same conference that Mr. Roosevelt informed me that if the European war was not over before we had our first bombs he wanted us to be ready to drop them on Germany.

I should add that we also attempted to obtain first refusal rights on Sweden's uranium ores, and were completely unsuccessful. As far as I could see, this was largely because of the Swedish Government's concern about the possible reaction of the Russians. The Swedish negotiators also claimed that on constitutional grounds alone their government could not make any secret agreement without informing their Riksdag or at least their Foreign Ministry.

CHAPTER 13

MILITARY INTELLIGENCE: ALSOS I—ITALY

From January, 1939, until American troops finally entered Germany and we took into custody a number of the senior German scientists, we faced the definite possibility that Germany would produce a nuclear weapon before we could. For that reason it was absolutely essential for us to remain as fully informed as possible on German progress in this field.

When the Manhattan District was formed, American intelligence efforts on atomic energy matters were carried out separately, as a part of their general activities, by the Army and Navy intelligence agencies (G-2 and ONI), and by the OSS. In addition, a number of other intelligence agencies scattered among the various government departments were gathering scraps and bits of information within the enemy nations that might be useful in adding to the atomic picture.

In the fall of 1943, General Marshall asked me, through Styer, whether there was any reason why I could not take over all foreign intelligence in our area of interest. Apparently, he felt that the existing agencies were not well co-ordinated; and that, as a result, there were many gaps not being covered. Moreover, it was probable, he thought, that these agencies would not always recognize the particular importance to us of some of the information they might receive, since, for security reasons, we had to limit the number of outsiders to whom we needed to explain the kind of information we wanted.

In the course of my conversation with Styer, I was surprised to learn that there was considerably more friction between the various intelligence agencies than I had previously suspected. There seemed to be no alternative to my taking this added responsibility. As was customary, nothing was put in writing. We simply agreed that General Marshall would inform Major General Strong, then Army G-2, of the change and that I would take care of notifying OSS and ONI.

The first step was to discuss the problem personally with Strong, and then through Admiral Purnell, to establish proper contacts with ONI. Immediately thereafter I paid a call on Major General William Donovan, the head of OSS, and at the same time saw his Executive Officer, Colonel G. E. Buxton. Donovan designated Lieutenant Colonel Howard Dix to look out for our interests and to ensure that all atomic information collected by OSS would be forwarded promptly to the Intelligence Section of my office.

I was astounded to learn how thoroughly unsatisfactory the relationships were between G-2 and OSS. As I was leaving at the close of our discussion, Donovan remarked that I was the first general officer who had ever come to see him in his office. He appeared to be quite touched by this and insisted on personally escorting me out of the building and sending me back to my office in his own car, even going so far as to insist on holding the door for me while I got in. Buxton told me afterward that OSS would have supported us fully in any case, but that my call ensured the utmost in special treatment for the MED. I never tried to find out the reasons for the unfortunate relationships that had grown up among the various intelligence agencies, but we always enjoyed splended co-operation from every one of them.

My experience here was a graphic demonstration of the importance of extending common courtesy to those with whom you expect to conduct important business. There is no substitute for it. Going out of our way to establish initial contacts with other organizations and individuals through calls by senior personnel, instead of by letter or telephone, was common practice in the Manhattan Project.

The new intelligence mission of the MED was clear: We had to learn as soon as we could what the Germans might be able to do if they exerted every possible effort to produce an atomic weapon. Throughout the project, there was universal respect for the quality of German science. This was a feeling I shared wholeheartedly. Our scientific people were acutely conscious that European scientists had discovered the principle of fission, and that our enemies were continually harping on their proposed use of secret weapons. Although this was sometimes hard on our nerves, it did keep us from ever becoming overconfident of the superiority of American-British efforts in the field of nuclear physics. Unless and until we had positive knowledge to the contrary, we had to assume that the most competent German scientists and engineers were working on an atomic program with the full support of their government and with the full capacity of German industry at their disposal. Any other assumption would have been unsound and dangerous.

It made no difference whether we thought, as we did when our own work made us realize the enormous difficulties involved, that they probably would not be successful. The fact is that they did possess the necessary capabilities, particularly if they generally ignored safety considerations; and this I was certain they would do. Our chief danger was that they might come up with relatively simple solutions to the problems we were finding so difficult.

We did not make any appreciable effort during the war to secure information on atomic developments in Japan. First, and most important, there was not even the remotest possibility that Japan had enough uranium or uranium ore to produce the necessary materials for a nuclear weapon. Also the industrial effort that would be required far exceeded what Japan was capable of. Then, too, discussions with our atomic physicists at Berkeley, who knew the leading Japanese atomic physicists personally, led us to the conclusion that their qualified people were altogether too few in number for them to produce an effective weapon in the foreseeable future. Finally, it would have been extremely difficult for us to secure and to get out of Japan any information of the type we needed. I hoped that if any sizable program was started, we would get wind of it from one of the various intelligence-collecting agencies with which we maintained liaison. In that event, we would have immediately done everything we could to interfere with their operations.

Positive support for our reasoning that the Germans were vitally interested in atomic energy had come from Norway, where before the war, in the town of Rjukan, about seventy-five miles west of Oslo, the Norwegians had constructed a complex of hydroelectric and electrochemical plants. When the Nazis occupied the country in 1940, they had required the operators of the Rjukan works to enter into contracts to produce heavy water which was to be shipped to Berlin for experimental use in the development of atomic energy. In September of 1942 we had estimated that approximately 120 kilograms of heavy water were being delivered to the Nazis each month under the terms of this contract.

At my instigation, Strong, with the approval of General H. H. Arnold and Major General T. T. Handy,[1] had brought this matter to the personal attention of General Eisenhower, and suggested that the Rjukan plants be either bombed or sabotaged.

The first attempt to put these works out of commission involved the use of guerrilla forces. Some five months after my request, three Norwegians, especially trained in sabotage techniques, and wearing British uniforms, parachuted into Norway, where they were met by local guerrillas. After nearly a week of hard cross-country skiing, they arrived at Rjukan and attacked the factories there on February 27, 1943.

The first reports on this action were most encouraging. A news dispatch from Oslo, which was relayed to Stockholm, stated that damage was "not extensive except at the place where the attempt was made and there the devastation was total." Subsequent reports from Sweden were even more encouraging, calling this "one of the most important and successful undertakings the Allied saboteurs have carried out as yet during the war."

These same Swedish newspapers caused me some headaches when they went on to speculate at considerable length about the importance of heavy water, pointing out that "many scientists have pinned their hopes of producing the 'secret weapon' upon heavy water, namely an explosive of hitherto unheard-of-violence." These items were picked up by the London papers and finally, on April 4, 1943, New York readers were greeted by such headlines as "Nazi 'Heavy Water'

[1] Head of Operations Division, War Department, General Staff.

Looms as Weapon." Immediately, Dr. Harold Urey, who had discovered heavy water, was deluged with calls from reporters wanting more information. He neatly sidestepped all such inquiries with the statement that "So far as I know, heavy water's uses are confined solely to experimental biology. I have never heard of an industrial application for heavy water, and know of no way it can be used for explosives."

Meanwhile, the British were hard at work assessing the damage done to the Rjukan works in the February raid. Their first estimates indicated that heavy-water production had been set back by about two years. We had different information, but our suspicions were not confirmed until we learned definitely that the plant had resumed partial operations in April. Yet doubt can be contagious and, under our gentle prodding, Sir John Dill soon felt himself compelled to inform General Marshall that a more realistic appraisal of the damage indicated that the plant could be completely restored in about twelve months. After some discussion of launching another commando raid— a full-scale one this time—General Marshall, at my behest, proposed to Sir John Dill that, instead, the plants be made a first priority bombing objective. This proposal led ultimately to a massive air attack on Rjukan in November of 1943. Although this mission in itself was not particularly destructive, it apparently led the Germans to believe that more attacks would follow. This belief, together with the problem of constant sabotage by workers in the plants, and probably a lack of appreciation at high government levels of the possible value of the product, caused the Nazis to give up their attempts to repair the damage done by the saboteurs in February. All apparatus, catalyzers and concentrates used in the production of heavy water were ordered shipped to Berlin. Norwegian guerrillas interfered with every step of the transfer, successfully destroying much valuable equipment and even going so far as to sink the ferry which carried a large part of the heavy water.

Even before the MED took over the responsibility for atomic intelligence in the fall of 1943, I had had a number of discussions with Strong about the desirability of exploiting sources of information that would become available to us as the American Fifth Army advanced up the Italian peninsula. I thought that we might be able to learn something of German progress in the field of atomic energy, and thus estimate more realistically how much time we had left to complete our own project; possibly we might obtain some useful technical information. In this I had the concurrence and hearty support of Bush.

Some staff officers in G-2 and certain scientists in the OSRD were pursuing similar approaches to the solution of their own particular problems. The result of their efforts and mine was a decision to make an organized attempt to tap Italian sources of scientific and technical information.

After consultation with Bush and with me, and with our wholehearted agreement, Strong, in September, 1943, put the proposal to General Marshall. His memorandum of that date said, in part:

> While the major portion of the enemy's secret scientific developments is being conducted in Germany, it is very likely that much valuable information can be obtained thereon by interviewing prominent Italian scientists in Italy. . . . The scope of inquiry should cover all principal scientific military developments and the investigations should be conducted in a manner to gain knowledge of enemy progress without disclosing our interest in any particular field. The personnel who undertake this work must be scientifically qualified in every respect. . . . It is proposed to send at the proper time to allied occupied Italy a small group of civilian scientists assisted by the necessary military personnel to conduct these investigations. Scientific personnel will be selected by Brigadier General L. R. Groves with the approval of Doctor Bush and military personnel will be assigned by the Assistant Chief of Staff, G-2, from personnel available to him. . . .

He went on to indicate that "This group would form the nucleus for similar activity in other enemy and enemy occupied countries when circumstances permit."

Thus G-2 and I, and the Navy, which later asked to be represented on the mission, became partners in obtaining intelligence from an active theater overseas.

The Manhattan Project always carefully avoided drawing undue attention to its work and to its people. Code names for our projects were deliberately innocuous. Imagine my horror, then, when I learned that G-2 had given the scientific intelligence mission to Italy the name of "Alsos," which one of my more scholarly colleagues promptly informed me was the Greek word for "groves."[2] My first inclination was to have the mission renamed, but I decided that to change it now would only draw attention to it.

Alsos differed in many respects from other intelligence units then existing in either the American or British forces. Its purpose was to supplement, but not to overlap or duplicate, the other agencies, and we emphasized always that our people should make the utmost use of the resources of the already established units. This practice enabled us to eliminate surplus personnel and administrative confusion, and proved to be especially effective during the preliminary planning phases of selecting intelligence targets. It also secured the whole-hearted co-operation from G-2 organizations in the field.

Though Strong's memorandum made no mention of atomic energy, everyone concerned in the higher levels recognized that the mission's primary purpose was to obtain intelligence of atomic developments in Italy and Germany. Nevertheless it was logical to expect that, in the course of its work, the mission would also come upon data about other enemy projects; accordingly, it was directed to exploit to the fullest sources in a number of fields of technical interest. For this reason, after discussions with Bush, I recommended that the mission report directly to Strong, who would relay its findings to the appropriate agency. This was the procedure followed throughout the mission's existence.[3] My purpose in this was not altogether altruistic; for my principal concern, I admit, was to draw attention away from the mission's interest in atomic matters. So strongly did I feel about this that in November, 1944, I wrote to my personal representative with the mission:

[2] Actually it is "grove."
[3] Reports to me on atomic matters always came in sealed envelopes through G-2, without being opened.

The impression has been created that ALSOS has been acting solely for us. This is injuring both ALSOS and ourselves. ALSOS has a definite mission in many fields, one of which concerns us. Any idea on the part of those in authority that ALSOS is completely monopolized for our purposes must be corrected.

The original detachment forming the mission was to consist of thirteen military personnel, including interpreters, and not more than six scientists, either civilian or military. Its make-up was considerably different from that of other intelligence units. It included people who were capable of extracting through interrogation and observation detailed scientific information on atomic energy. It also contained people who were generally familiar with the research programs and interests of both the United States and Great Britain and, insofar as possible, of our enemies. The members of the mission had to have a general knowledge of enemy equipment and they had to be prepared to seek out not only military laboratories and technical personnel, but civilian scientists, technicians and facilities as well.

The objectives of Alsos in Italy were: "To obtain advance information regarding scientific developments in progress in enemy research and development establishments which are directed towards new weapons of war or new tactics . . . and to secure all important persons, laboratories, and scientific information immediately upon their becoming available to our own forces before their dispersal or destruction." Its efforts would enable us to select bomber targets, develop countermeasures against new weapons, organize counterpropaganda, plan our strategy and direct our own war research projects.

The mission was to conduct investigations in occupied Italy only, and was to advance with, or closely behind, our military forces to Rome, proceeding further north whenever it felt that important targets could be found. Unfortunately, our plans were based on the then-expected rate of advance to Rome. This rate was not achieved and, consequently, there was a considerable period of time during which the mission was unable to operate with full effectiveness.

Alsos was commanded from the beginning by Lieutenant Colonel Boris T. Pash. I had first met Pash in San Francisco, where he was working on a security matter involving the project at the time when such affairs were still handled in G-2. His thorough competence and great drive had made a lasting impression upon me. His unit originally consisted of an executive officer, four interpreters, four CIC agents and four scientists, Major William Allis (War Department), Lieutenant Commander Bruce S. Old (Navy Department), and Dr. James B. Fisk and Dr. John R. Johnson (both of the OSRD).

From information available to us in November, 1943, we considered it desirable to begin scientific investigation in Italy at the very earliest opportunity. Discussions with Vice Admiral Minissini of the Italian Navy had given us no grounds to believe that the Italians were working on an atomic weapon; nevertheless, we had evidence that all possible information on this subject was not reaching the MED through our established channels. On November 10, therefore, I urged G-2 to request the theater commander to authorize Alsos to conduct investigations immediately in that part of Italy which was then under American control.

The mission assembled in Algiers on December 14. After visiting the several commanders in whose areas they would be operating, the group departed for Naples, where they established contact with the Fifth Army Intelligence Section and the Italian Civil Government. They spent the next month and a half interviewing Italians at Naples, Taranto and Brindisi who might know something about the research efforts in Germany and unoccupied Italy. It soon became apparent that the most worth-while sources of Italian information were in Rome. Two plans were drawn up to exploit these sources. One would attach the Alsos mission to the troops of the Fifth Army, to enter Rome immediately after the city fell; the other would secure and bring back certain important scientists from Rome and northern Italy before Rome was captured. Neither of these plans could be carried out immediately, however; and since the Allied progress northward was slow and efforts to obtain information from Italians behind the enemy lines were generally unsuccessful, the mission gradually became inactive. By March 3, 1944, all its members had returned to the United States.

In spite of their disappointment over the inability to enter Rome, this first Alsos mission was most successful. Indeed, its accomplishments so far exceeded what we had considered possible that its conclusions were generally discounted, principally because its findings were essentially negative. While it discovered several items of immediate interest to both the Army and the Navy, any one of which would have justified its existence, the best that it could arrive at in our particular field was: "Almost all the evidence gained points toward no particular experimental activity by the Germans on explosives based on nuclear energy."

On the basis of this report, I felt justified in concluding that the German atomic energy effort was not so intensive as ours. Alsos' confirmation of the fact that control of the Germans' work still remained in the hands of research scientists indicated that they probably had not yet got to the point we had reached by the summer of 1942, when primary responsibility for our program passed from the OSRD to the Army.

However, we could not be absolutely sure that we were not being misled by the general lack of positive information. In its report at the close of its activities in Italy, the mission recommended that "steps be taken to obtain scientific intelligence in new theatres of military operation as they are developed," and that "careful consideration be given to the potential value of conducting scientific intelligence operations in the wake of the anticipated invasion of Western Europe." I concurred in these recommendations, as did Major General Clayton L. Bissell, who by then had succeeded Strong as G-2.

In December, 1943, just about the time the Alsos group was landing in Italy, I had sent one of my officers, Major R. R. Furman, to England to confer with officials in the British Government about the possibility of establishing a Manhattan liaison office in London and engaging in a joint Anglo-American intelligence effort. The idea was well received by the British and in January, 1944, we selected Captain Horace K. Calvert[4] to head the liaison office.

[4] Then head of MED security activities; previously with the Investigations Branch of Army G-2.

He was not picked for this important assignment because of his technical background, for he was an oilman-lawyer in peacetime. However, since he had had a thorough training in intelligence procedures and an extensive background in the project, we felt that he would be well qualified to recognize any danger spots in the German picture. Before he left, I gave him brief and quite general instructions. He was to gather all possible information on the various atomic energy efforts under way in Europe, particularly those being carried on by the Germans; to make use as far as possible of existing American and British channels; to keep his intelligence estimate up to date at all times and to report to us in Washington everything that he considered to be of importance. He was also expected to establish close and friendly relations with the Englishmen and Americans with whom we might have to deal from time to time, both in London and, as the situation developed, on the Continent.

We provided him with a letter of introduction from General Strong, to Colonel George B. Conrad, G-2 of ETOUSA.[5] Immediately after arriving in London, he reported to Conrad to present his credentials. I had always advised our officers to avoid any unnecessary demonstration of the usual military formalities whenever they felt that it might irritate our civilian scientists. Apparently Calvert had been well indoctrinated, for, somewhat flustered at reporting to a senior officer, instead of saluting, he just put out his hand and said, "Good morning. I am Major Calvert." Even now, some eighteen years later, he has not forgotten the horror that swept over him when he realized that he had failed to salute. He knew that Conrad and I had been roommates at West Point, and this only added to his embarrassment. Fortunately, Conrad accepted Calvert's greeting in the spirit in which it was offered and set him at ease by shaking hands and showing him to a chair. Calvert then told him of the basic purpose of the MED and of his own mission. Out of their conversation came arrangements for Calvert to have a desk in Colonel Conrad's office where he could go over all the raw intelligence data as they came through.

At the start, it was impossible to set up any criteria for the kind of information in which we might be interested; later, as Calvert

[5] European Theater of Operation, U.S.A.

became familiar with the various types of intelligence, he was able to eliminate certain classes of data as being of no interest.

After establishing himself in intelligence circles, Calvert called on Mr. Winant, our Ambassador to England, to whom he gave such general information as he felt to be necessary and appropriate. Again, he was furnished with a desk and the utmost support. Soon he was given the title of Assistant Military Attaché.

Calvert acquired a third desk in the British Atomic Energy Office, which carried the cover name of Tube Alloys. Here he maintained close liaison with the head of the office, Michael Perrin; his assistant, David Gattiker; and Lieutenant Commander Eric Welsh (of the British Intelligence); all of whom had a lively sympathy for his efforts.

Messages to and from Calvert passed through the American Embassy in London and the War Department. Cable messages were always sent in top secret code and, as an extra precaution, many of the key words were previously coded into the names of states, cities or other common words. For example, "New York" might be used for "uranium," "Indiana" for "plutonium," "Nevada" for "British Intelligence," and so on, the code being changed from time to time.

After a short time, Calvert was joined by another officer, Captain George C. Davis, three WAC's and two counterintelligence agents. Furman, in my office, was Calvert's Washington contact man. I used Furman primarily for special projects such as this one. His actions were always prompt and to the point.

In making his initial appraisal of the German atomic picture, Calvert knew it would take a combination of three requisite factors to make a bomb. Those were: (1) a sufficient number of top nuclear scientists and technical assistants; (2) the basic fuel for a bomb—uranium, and possibly thorium, probably combined with uranium; and (3) laboratories to develop it and industrial means to make it.

He started working on the fuel problem first, for we were sure of Germany's scientific and industrial ability to do the job. Thorium seemed out of the question, since it is mined chiefly in Brazil and India and, because of embargoes, Germany had been unable to import any since the war began, and had had only insignificant stocks on hand before the war. The basic fuel was thought to be uranium. Considering our own firsthand knowledge of the enormous industrial effort required to produce U-235, we were confident that we would have seen evidences of any such program had one existed. It seemed more likely that they would use plutonium. That they had enough to launch an atomic program seemed to be within the realm of possibility, for we knew there had been a large stockpile of refined uranium ore at Oolen, Belgium, a few miles outside Brussels, which originally had been the property of Union Minière.

The only other possible supply of uranium was the mines at Joachimsthal, Czechoslovakia, which was not a particularly significant source. Most of this ore was shipped to a uranium plant outside Berlin, the Auer-Gesellschaft. British Intelligence kept in touch with the activities of these mines, and in July, 1944, Calvert's group started periodic aerial surveillance over the entire mining area, studying the pictures in detail for new shafts and aboveground activity. Tailing piles from each mine were microscopically measured from one reconnaissance to the next. By knowing the general grade of the ore and measuring the piles, we could determine with some degree of accuracy the mine's daily production. There were no signs of extraordinary activity.

It would have been imperative for Hitler to enlist the aid of all his top scientists. Allied Intelligence had established that many of them were working on the "V" weapon; particularly at Peenemünde, but to our knowledge no nuclear physicists had been reported there. Calvert started a search for some fifty German nuclear scientists. He knew that there must be many young scientists who had come up since Hitler's rise to power of whom we had no knowledge; however, if we could locate a few of the top people, they should lead us to the rest. All the present and back issues of the German physics journals were scrutinized. Foreign-born nuclear scientists in the United States, like Enrico Fermi, O. R. Frisch and Niels Bohr, as well as anti-Nazi professors and scientists in Switzerland, Sweden and other neutral countries, were questioned in detail to obtain any past or present information they might have on the whereabouts of the German scientists. The names of all German scientists were placed on watch lists with American and British intelligence agencies which were daily scanning German newspapers that had been smuggled out. Before long we had recent addresses for a majority of the scientists in whom we were interested.

The third main category of pre-D-Day investigation, laboratories and industrial plants, was studied in much the same way. Lists were compiled of all of the precious metal refineries, the physics laboratories, the handlers of uranium and thorium, manufacturers of centrifugal and reciprocating pumps, power plants and other such installations as were known to exist in the Axis countries. These were placed on a master list from which they were not removed until we had positive information that they were not engaged in, or supplying, an atomic program. All plants where work of an unknown nature was being conducted were checked through aerial reconnaissance, the underground, OSS and all the numerous intelligence agencies.

By hard work and constant effort, Calvert was ready by the time the second Alsos mission reached Europe on the heels of the invading armies with a good list of the first intelligence targets, dossiers on all the top German scientists, where they worked and where they lived, the location of the laboratories, workshops and storage points of interest.

CHAPTER 14

A SERIOUS MILITARY PROBLEM

I think that even today few people are aware of one of the big risks that was taken by the United States during the war. There was no alternative; we had to take it. But it gave the handful who knew about it some bad hours. It grew out of the possibility that the Germans might use radioactive material to block the cross-channel attack of the Allied forces.

To cast back for a moment, it had begun to seem possible to us early in 1943 that the Germans could have progressed to the point where they might be able to use atomic bombs against us, or, more likely, against England. Although this possibility seemed extremely remote to me, a number of the senior scientists in the project disagreed. One even went so far as to urge that I should warn the American people in an official broadcast that the United States might be hit by an atomic bomb. Naturally, I was opposed to doing any such thing. What I thought more likely was that the Germans would use an ordinary explosive bomb containing radioactive material. If we were unable to neutralize the effects of such a weapon promptly, a major panic could easily sweep through the Allied countries.

However, as the plans for the invasion of Europe began to take form, we considered very seriously indeed the possibility that the Germans might lay down some kind of radioactive barrier along the invasion routes. We could not calculate with any certainty the likelihood of their doing this, for we were truly in the dark then about their progress in atomic development. It had always seemed to most of us that their best prospects lay in the use of plutonium, which would demand a much smaller industrial effort as well as considerably less

199

in the way of time, critical equipment and materials than any other method—provided they were willing to ignore safety precautions. This I felt the Germans would do, for considering what we already knew of their treatment of their Jewish minority, we could only assume they would not hesitate to expose these same citizens to excessive radiation. Hitler and his ardent supporters, we felt, would consider this a proper use for an "inferior" group, quite apart from the saving in effort and materials and time. Moreover, we knew that in the course of developing the plutonium process the Germans were certain to discover that tremendous quantities of highly radioactive fission products would be produced in their reactors. It would be perfectly natural for them to think of using these to lay down a barrier through which ground troops could not pass without disastrous results.

At the request of the Military Policy Committee, a three-man group, Conant, Compton and Urey, assisted by other project members, had made a study of radioactive poisoning; and on the basis of their report we had ordered a supply of portable Geiger counters[1] and were training a number of our personnel to use them.

It was an ugly prospect, and as D-Day in Europe approached, I had to make up my mind what action I should recommend with respect to the possibility that the Germans might use radioactive poisons. In reaching a decision, I sought the opinions of those I thought most capable of giving advice, but after their advice was all in, I was no better off than I had been in the beginning. This was a time when I was most appreciative of the sound counsel of Conant. He was never one to become unduly excited over the wild conjectures that the fertile minds of some of our people could produce, almost faster, it sometimes seemed, than we could disprove them—even to our own satisfaction, if not to theirs.

In any event, a decision had to be made. Would Eisenhower's troops encounter radioactive poisons? Should he be warned of the possibility? To me, the answers were "No" and "Yes." I saw General

[1] These were developed and manufactured by the Victoreen Instrument Company. A number of them were sent to England for use if needed. Others were placed in various locations in the United States.

Marshall on March 23, and recommended that we send an officer to England to warn General Eisenhower of what he might be up against. At the same time, in accordance with my usual custom,[2] I handed him a memorandum:

WAR DEPARTMENT
Office of the Chief of Engineers
Washington

22 March 1944

MEMORANDUM TO THE CHIEF OF STAFF

1. Radioactive materials are extremely effective contaminating agents; are known to the Germans; can be produced by them and could be employed as a military weapon. These materials could be used without prior warning in combating an Allied invasion of the Western European Coast.

2. It is the opinion of those most familiar with the potentialities of these materials that they are not apt to be used, but a serious situation would occur should any units of an invading Army be subjected to the terrifying effects of radioactive materials.

3. It is recommended that a letter similar to the draft enclosed be dispatched to General Eisenhower.

L. R. GROVES
Major General, C.E

Incl:
     Draft

The draft of the letter to General Eisenhower read as follows:

22 March 1944

General Dwight D. Eisenhower
Office of the Supreme Commander
Allied Expeditionary Forces
London, England

DEAR GENERAL EISENHOWER:

In order that your headquarters may be fully advised of certain materials which might be used against your Armies in a landing operation,

---

[2] I had adopted this practice to save his time. He did not retain these memoranda but handed them back to me after he had finished discussing them.

I have directed Major A. V. Peterson, who will be in England on temporary duty in the near future, to report to your office and to acquaint you, or such officers of your staff as you may designate, with the problems involved. The matter is of the highest order of secrecy.

Faithfully yours,

Chief of Staff

The use of officer messengers on highly secret missions, carrying no papers or very limited ones, was common practice in the Manhattan Project. Before their departure, I always instructed them carefully, going into the background and reasons for their missions so that they could answer any pertinent questions they might be asked and thus be of maximum assistance to the people with whom they talked. Written messages would have been extremely complex and confusing to anyone not thoroughly acquainted with our scientific progress in the atomic field. Worst of all, the reader might have had doubts about exactly what was meant. Then, too, there was always the problem of security. Every written message increased the chances of disclosing information to outsiders.

Peterson's mission was to inform Eisenhower that it was possible for the enemy to use radioactive poisons, but he was to emphasize our belief that they would not be used and that the invasion plans should be made accordingly. After he saw Eisenhower, he had longer discussions with Lieutenant General Walter Bedell Smith, the Chief of Staff, and other members of the SHAEF and ETOUSA staffs.

When he was asked if he could remain to help in preparing the necessary precautionary measures, although he was supposed to return at once, he said he could, and informed me by cable of his decision. During this period he talked with members of the Chemical Warfare Service, which was to use the Geiger counters, with the Signal Corps, which was to maintain and repair them, and with the Medical Corps, which had to be informed of the possibilities with which they might be confronted.

The Chief Surgeon, Major General Paul R. Hawley, issued two

cover orders designed to insure that GHQ would be promptly alerted if the Germans did resort to radioactive warfare, but worded in such a way as to disguise the real nature of the danger. One order said that trouble had been experienced with fogging (which always results when film is exposed to radiation) on certain photographic and X-ray films and that if any such trouble was noted by troops in the field, an immediate report should be made, citing lot numbers, so that defective film could be withdrawn from use. The order is given below in full:

RESTRICTED

OFFICE OF THE CHIEF SURGEON
EUROPEAN THEATER OF OPERATIONS
APO 871

3 May 1944

ADMINISTRATIVE MEMO NO. 60

SUBJECT: Report on Fogging or Blackening of Photographs or X-ray Film.

To: Surgeon, FUSAG,
            Surgeons, FUSA and TUSA,
            Surgeon, USSTAF,
            Surgeons, All Base Sections,
            Commanding Officers, All Hospitals.

1. Several instances of the finding of fogged and blackened photographic and X-ray film have been reported from the field.

2. A survey is being made by this office to determine the sources of supply and causes of these effects.

3. Until further notice, prompt report will be made directly to the Chief Surgeon, ETO, of the finding, by any medical agency in the European Theater of Operations, of photographic or X-ray film fogged or blackened without apparent cause. Report will include the following information:

a. Conditions under which fogging or blackening occurred: time, date, location, in cameras, sassettes or in storage.

b. In the case of roll film, whether fogging or blackening has occurred completely throughout the roll, or whether it has occurred at the beginning, middle or end of the roll. Also, whether it has occurred at the borders (unexposed portions) as well as at the center.

      c. In the case of cut film, whether fogging or blackening has occurred at the borders (unexposed portions) as well as at the center of the film.

By order of the Chief Surgeon:

J. H. McNINCH
Colonel, Medical Corps
Executive Officer

The other order, which follows, was aimed at securing immediate reports in case any symptoms of radioactive exposure were discovered among the troops.

RESTRICTED

OFFICE OF THE CHIEF SURGEON
EUROPEAN THEATER OF OPERATIONS
APO 871

3 May 1944

ADMINISTRATIVE MEMO NO. 58

SUBJECT: Report of an Epidemic Disease.

TO: Surgeon, FUSAG,
           Surgeons, FUSA and TUSA,
           Surgeon, USSTAF,
           Surgeons, All Base Sections,
           Commanding Officers, All Hospitals.

      1. A few cases of a mild disease of unknown etiology have been reported. While the symptomatology of this disease has not yet been well established, a few signs and symptoms seem to be relatively constant. The more important of these are:

      a. *Fatigue:* This varies between slight, in the milder cases, to profound, in the severer cases, and may even reach a state of prostration.

      b. *Nausea:* This may be slight, or in severer cases there may be active vomiting.

   c. *Leukopenia* is the most constant and reliable sign. It appears early and is in proportion to the severity of the attack, reaching a level of less than 1,000 cells per cu. mm. in the severest cases.

   d. *Erythema* may be observed in the severer cases. It is of a diffuse type and may be more pronounced upon the lower extremities.

   2. *Epidemiology:* Sporadic cases are very rare, cases tending to occur in smaller or larger groups. The admission of only one suspected case should cast considerable doubt upon the diagnosis, whereas the admission of several cases with similar symptoms should arouse suspicion of the presence of this disease.

   3. *Report:* In order that the epidemiology of this disease may be investigated thoroughly, prompt reports will be made directly to the Chief Surgeon, of the suspected presence of this disease in any medical agency in ETOUSA.

   By order of the Chief Surgeon:

J. H. McNINCH
Colonel, Medical Corps
Executive Officer

To avoid any possibility of a misunderstanding, General Eisenhower very wisely wrote the following letter to General Marshall:

Supreme Headquarters
ALLIED EXPEDITIONARY FORCE
Office of the Supreme Commander

11 May 1944

DEAR GENERAL:

   I have had a careful analysis made of the project which you instructed Major Arthur V. Peterson to explain to me. Since the Combined Chiefs of Staff have not brought this information officially to my notice, I have assumed that they consider, on the present available intelligence, that the enemy will not implement this project. Owing to the importance of maintaining secrecy to avoid a possible scare, I have passed this information to a very limited number of persons; moreover, I have not taken those precautionary steps which would be necessary adequately to counter enemy action of this nature.

The action taken by this Headquarters has been as follows:

a. Admiral Stark, General Spaatz, General Lee, and a very limited number of their staffs have been briefed on the project. No U.S. or British Commander participating in OVERLORD has been briefed.
b. Special equipment, U.S. and British, for use on the project has been earmarked in the United Kingdom for dispatch to the Continent at very short notice.
c. Channels have been established for the further supply of equipment and the provision of technical assistance if such are needed.
d. Medical channels have been informed as to the symptoms which would occur in these circumstances. This information has been sent out under suitable "cover," and I attach a copy of the letter lest you should require precise details in this respect.

I am writing in similar terms to General Ismay for the information of the British Chiefs of Staff.

Sincerely,

/s/ Dwight D. Eisenhower

General George C. Marshall
The Chief of Staff
The War Department
Washington, D. C.

As far as I know, General Marshall never told either the Combined Chiefs of Staff or the Joint Chiefs of Staff of the problem. From his discussion with Peterson, General Eisenhower evidently gained a very clear understanding of our views. When his letter arrived, I knew that our advice had been fully accepted and that nothing more could be done except to pray that we had not made a mistake. In making my recommendations, I was well aware I had assumed enormous responsibilities in connection with the invasion, and consequently, I was more than a bit relieved when the Allied troops made good their landing without any report of radioactive interference.

CHAPTER 15

MILITARY INTELLIGENCE: ALSOS II—FRANCE

At the same time that we were seriously considering the possibility that the Germans would prepare an impenetrable radioactive defense against our landing troops, we also had to plan confidently on a successful invasion. On the latter assumption, we again began to explore ways to exploit the intelligence opportunities that the invasion would open up. Dr. Bush and I urged Major General Clayton L. Bissell, now G-2, to establish a mission similar to the one that had produced such excellent results in Italy. Our recommendations fell into the grist mill of the General Staff and became the subject of a number of staff papers which contributed nothing of value to anyone's knowledge of the situation. At first I went along with the staff's efforts to develop a universal system which could be applied to our specific needs, but finally, by the end of March, 1944, I felt time was running out and I insisted that Bissell bring the matter of a scientific intelligence mission to the personal attention of General Marshall without further delay. The next day, G-2 proposed to the Chief of Staff "that a scientific intelligence mission be organized in the Military Intelligence Division, with the assistance of General Groves and Dr. Bush, and that it be sent into various portions of active theaters at suitable times. The military and civilian scientific personnel will be selected by General Groves and Dr. Bush, and the intelligence and administrative personnel by Assistant Chief of Staff, G-2." This proposal was approved by General Joseph T. McNarney for the Chief of Staff on April 4.

Colonel Pash was again designated chief of the mission. Bush and I agreed on Dr. Samuel A. Goudsmit, an extremely capable atomic physicist, and one not involved with the MED, for the scientific chief. An advisory committee to guide the work of the mission was established, consisting of representatives of the Director of Naval Intelligence, the Director of the OSRD, the Commanding General of Army Service Forces, and the Assistant Chief of Staff, G-2. This committee concentrated on the nonatomic areas.

To make absolutely sure that his operations would be fully supported, Pash was provided with a letter to General Eisenhower, signed by the Secretary of War, saying: "I consider it [the mission] to be of the highest importance to the war effort. . . . Your assistance is essential, and I hope you will give Colonel Pash every facility and assistance at your disposal which will be necessary and helpful in the successful operation of this mission."

Pash visited England in mid-May to make arrangements for establishing the mission on the Continent, and to open an Alsos office in London. Because of his letter, the mission obtained full support from General Bedell Smith; Brigadier General E. L. Sibert, G-2 of First U.S. Army Group; and Brigadier General T. J. Betts, G-2 of SHAEF; as well as from Conrad and the members of his section, who had been well prepared for Alsos by Calvert. The understanding co-operation given to Alsos by all these agencies throughout the European campaign made the difference between failure and the brilliant success it achieved. With their assistance, Pash was able to establish special channels for Alsos reports and targets. He also succeeded in negotiating out of existence the many problems that might have arisen in connection with seizing intelligence targets in liberated friendly territory.

In June, at the height of his preliminary work in London, Pash was interrupted by an order from Washington to investigate a most disturbing report from Italy. We had just learned that our efforts to bring a number of Italian scientists from German-occupied Rome had failed. In spite of the reports of success from an OSS agent to the effect that he had been in touch with these scientists, Pash now discovered that in fact no contact had ever been made. Fortunately, Pash was able to conclude after a severe interrogation of the agent that no lasting damage had been done. Pash ordered him to leave Rome at once, and never to talk about the operation. Needless to say, he was never used on our work again.

Pash's trip to Italy was not in vain, however. June 4, 1944, was notable as the day on which Rome finally fell to the American Fifth Army. With his customary initiative and energy, Pash left Naples immediately to secure the original Alsos targets in Rome. Most of these men were connected with the physics laboratory of the University of Rome, which was placed off-limits to Allied troops. The following day he took over the National Council of Research.

The pressure of business in London did not permit Pash to remain in Italy any longer, so he turned over the mission's activities there to Agent Bailey, who had accompanied him on this trip, and returned to England. We took steps promptly to reconstitute Alsos in Italy and, on June 17, Major R. C. Ham, Pash's Executive Officer, left London to establish the Mediterranean Section of Alsos. I had already arranged for Major Furman and Dr. John R. Johnson to join him in Rome to see what information might be developed there.

They arrived in Rome on June 19, and immediately started investigations that extended over the next six days. Drs. Wick, Amaldi and Giordani, all of the University of Rome, were questioned about their research activities. Their replies confirmed earlier indications that Italian scientists had had very few opportunities to visit Germany before the Italian armistice in July of 1943, and that practically no interchange had taken place since then. Italian scientific research and development had been generally disorganized and was almost militant in its resistance to the Fascist state—militant, at any event, after the armistice.

Both Wick and Amaldi had served in the Italian Army and since the surrender had been hiding in Rome. During the war they had engaged in theoretical research principally on isotope separation, neutron, infrared and cosmic rays. They had no direct information about German research in the field of nuclear fission, for they had never been asked to do any work with or for the Germans. They claimed not to understand the significance of heavy water, and they were not aware of any new activity at the Joachimsthal uranium mines in Czechoslovakia.

Wick had made a trip into Germany during June and July of 1942, and had seen and talked at some length with a number of German physicists at that time and, together with Amaldi, had been shown some of the correspondence between various German scientists; thus they were able to supply us with some useful information. They were most co-operative, and what they gave us was the basis for the compilation of brief accounts of the activities and locations of a number of Germans who were of outstanding interest to the MED. Although later investigations in Germany proved that some of the information obtained in Rome was not wholly accurate, in the main it was well worth the trouble we had gone to in collecting it.

The second phase of the Alsos mission, however, in France and Germany, was where our efforts really paid off, for here we were finally successful in obtaining positive rather than negative atomic information. In addition Alsos obtained important scientific information on matters, such as long-range rockets, outside its main preoccupation. This is a long story which cannot be told here, but G-2 was concerned about a number of scientific intelligence problems. In general, scientific intelligence had been a relatively neglected field in all the intelligence agencies, and the co-operation between G-2 and MED, with the advice and help of the OSRD, in Alsos was a pioneering achievement of significance.

Staffing of Alsos progressed gradually. By July 26, it had three operations officers and eleven scientists, most of whom were commissioned officers. Pash asked for some CIC detachments to increase the mission's capabilities and, after much co-ordination with many headquarters, these agents were eventually assigned to him. At the same time, Goudsmit was pressing for additional scientific people to handle the ever-growing load that was being placed upon Alsos as its reputation in the European Theater grew. We got them with the help of the OSRD, and by August 31, the mission had grown to seven operations officers and thirty-three scientists.

Throughout the European campaign, as far as atomic efforts were concerned, Alsos members had the tremendous advantage of knowing where they were going and whom and what they were seeking. When they landed on the Continent, they had in hand the fruits of Calvert's labors, in the form of a comprehensive list of intelligence "targets"—the names of key individuals, where they worked and where they lived; and the location of the laboratories, workshops and storage points, and other items of interest to us. At the head of the list was the famous French atomic scientist, Frédéric Joliot-Curie (later High Commissioner of Atomic Energy for France), and his equally famous wife, Irène Curie, the daughter of Madame Curie, discoverer of radium.

On August 9, 1944, advanced elements of the Alsos mission landed in France and entered Rennes. In going through the laboratories of the university there, they discovered a number of catalogues and other papers that provided information pointing to possible future targets.

Pash's first efforts in France were unproductive. Joining the Eighth Army Corps with one CIC agent, he tried to reach and search Joliot's summer home, as well as the homes of his colleagues, Pierre Auger and Francis Perrin, near l'Arcouest. Although the surrounding area was heavily mined, Agent Beatson led the way into Perrin's house on August 11, only to find that it had been completely stripped of furniture and personal effects and provided no information of any value. In trying to reach Joliot's house, the party came under small-arms fire from snipers and prudently abandoned the effort. When resistance finally ceased on the twelfth, the house was searched, but without any worth-while results.

Alsos' work on the Continent began in earnest on August 23 when Pash, Calvert and two CIC agents joined the leading elements of the Twelfth Army Group moving toward Paris. Learning that the entry into Paris would probably be from the south, the group joined the 102nd Cavalry Group, but when that unit was held up at Palaiseau, they moved cross-country and joined the 2nd French Armored Division, which had been chosen to lead the liberating forces into Paris.

They reached Joliot's house in the suburbs of Paris on the twenty-fourth. Servants there informed them that the professor was in Paris, probably at his laboratory. So, without further ado, they telephoned the laboratory and, finding Joliot away at the time, told one of his assistants that they would like to see Joliot, they hoped, in a day or two.

On August 25, they reached Paris, at the Porte d'Orléans, ahead of the French troops, and waited there for about half an hour until General LeClerc arrived with his armored division. The General led the triumphal entry, at 8:55 that morning, but tucked into the column, directly behind the first tank was an American jeep containing the first representatives of the U.S. Army: Pash, Calvert and two other Alsos agents. The sniper fire became a bit unpleasant, particularly as the open jeep was a much more attractive target than the well-buttoned-up tanks, so Pash's group left the column. They soon returned, however, finding that an exposed jeep was better off in a tank column than out of it. Later, toward evening, they broke off again, and succeeded in reaching Joliot's laboratory. There, on the steps of the university, they found Joliot and some of his staff, all wearing FFI arm bands. That evening they celebrated the liberation with Joliot by drinking some champagne he had reserved for the occasion. The American soldier's staff of life, the K ration, served as the hors d'oeuvres. In keeping with the scientific surroundings, the champagne was drunk from laboratory beakers.

In the course of their conversation with Joliot, the names of two of his former colleagues came up: Hans von Halban, born an Austrian in Leipzig and later naturalized as a French citizen, and Lew Kowarski. Both men had left France for England in June of 1940 and had been working in the British Tube Alloys Project in Canada. Joliot immediately surmised that there was some connection between them, Pash and Calvert, and the uranium problem. They did not openly tell him at first what they wanted of him. However, after an hour's conversation, Joliot willingly told them just what they wanted to hear: that it was his sincere belief that the Germans had made very little progress on uranium and they were not remotely close to making an atomic bomb. He said he had refused to perform any war work for the Nazis and had forbidden them to use his laboratories for such purposes. However, after the occupation commenced, he said he did allow two German scientists to move into his laboratory to continue academic work on nuclear physics. He added that he talked with them frequently and clandestinely checked their work at night after the laboratory was closed, thus keeping constant surveillance on their activity. How true this all was we never knew.

Arrangements were made immediately for Joliot to be interviewed in Paris by Goudsmit on August 27, and two days later, Calvert flew back to London with him. There interrogations were conducted for several days by Michael Perrin, assistant director of Tube Alloys, and others of the British staff, as well as by Goudsmit. At first Joliot, who claimed to have been actively engaged in the French underground resistance movement, seemed quite willing to discuss the scientific activities in his laboratory. Although he added little to the knowledge we already possessed, he did clarify a few doubtful items. The College of France, which was Joliot's laboratory, owned a cyclotron, and a number of German scientists of interest to the MED had spent varying lengths of time there operating it. Among them was Professor Erich Schumann, who headed the German Army Research conducted by the Ordnance Department and who, during the war, served as the personal adviser on scientific research to Field Marshal Wilhelm Keitel. Schumann was credited with initiating work on the German uranium project, although by the end of 1942 his responsibilities had been transferred to the Reich's Research Council. Another visitor to Joliot's laboratory was Dr. Kurt Diebner, who in 1939 had served as Schumann's right-hand man and who had continued nuclear research under the Reich's Research Council. Then there was Professor Walther Bothe, an outstanding German nuclear experimentalist in the physics laboratory of the Kaiser Wilhelm Institute for Medical Research. Dr. Abraham Essau, who until early 1944 was in charge of physics under the German Ministry of Education in the Reich's Research Council, had made a number of visits to Paris. Essau had been president of the Ministry's Bureau of Standards until January, 1944, when he was replaced as Plenipotentiary for Nuclear Physics by Walther Gerlach. There was also

Dr. Wolfgang Gertner, an able German scientist who, before the war, had been associated with Ernest Lawrence in the United States. Gertner was an outstanding German authority on cyclotron operations. Joliot's other visitors had included Dr. Erich Bagge, a member of the Kaiser Wilhelm Institute, who specialized in isotope separation, and Dr. Werner Maurer, an experimental physicist engaged in nuclear research.

Joliot consistently maintained that he had acquiesced in the Germans' use of the cyclotron with the distinct understanding that its use would not be of direct military assistance to their war effort. There was no independent evidence that this condition was made. There may have been a promise made to him by some of the German scientists, or they may have said that there appeared to be no military possibilities that could result from the use of the cyclotron, but I never found any real proof of Joliot's contention. Certainly, his subsequent behavior—and I shall come to that shortly—gave us room for doubt.

Bothe appeared to have been more or less in charge of the German scientists who visited the Paris laboratory. Apparently he had been rather high-handed in his treatment of Joliot, who showed no love for him. Joliot expressed the opinion that Bothe knew a great deal about our work in the United States.

We considered Joliot to be a competent judge in this matter, for he was in contact with von Halban and Kowarski, who, as I have said, were associated with the Canadian National Research Council, working on Canada's part in our joint project. We understood that von Halban had written his mother, who lived in Switzerland, that his Canadian work was the same as his prewar activity. That information had been relayed to Joliot. By this and similar means, he had been able to assemble considerable knowledge of our efforts—knowledge that he tried with little success to explain away.

Conversations with Joliot did reinforce our more recent surmises that the Germans were not so far along in the development of an atomic weapon as we had originally feared. Nevertheless, we felt he was evasive about his contacts, and we placed little faith in his statements. In any event, if what he said were true, his contacts in Germany and with German scientists were not too strong. He said he had only meager knowledge of the steps taken by the Germans in the field of nuclear fission, and we had no reason to doubt his contention.

And so the Alsos investigators were left once more without any positive information on enemy progress in nuclear research and development. On the other hand, we fully appreciated that the Germans must be embarked upon some sort of program, or they would not have found it expedient to use Joliot's laboratories.

In mid-September, after Paris was securely in Allied hands, Alsos headquarters was set up there, the mission's London business being turned over to Calvert. At the same time, through the link-up of the American Third and Seventh Armies, it was possible to reduce Alsos activities in the Mediterranean area and to reinforce their efforts in northern France and the Lowlands.

The mission was organized at this time into a number of groups of officers and CIC agents, who were to secure vital targets as they were seized by the assault elements of the advancing Allied armies. The good working relationships with the several high headquarters in the theater, which Pash and Calvert had established before the start of the campaign, ensured the mission whatever combat support it might need to reach its targets. And, as always, Calvert's group in London was invaluable in supplying it with the data it needed in order to operate—names, places, installations. Long before the invasion of France we had begun to get the reliable information that inevitably develops when intelligence officers keep plodding through the drudgery of examining thousands upon thousands of uninteresting reports.

The most difficult problem that Calvert's intelligence group had to tackle was to find out where Hitler was hiding his atomic scientists. They knew, as everyone did, that before the war the Kaiser Wilhelm Institute in Berlin had been a focal point for all atomic physicists and atomic research, not only in Germany, but in all of Europe. It was there that Otto Hahn and Fritz Strassman had carried out their startling experiments. It was also the home of Max Planck, the internationally famous atomic scientist.

As the war drew on, however, and the bombing of Berlin was stepped up, we had learned from both aerial reconnaissance and a Berlin scientist, who got word to us through the Norwegian underground, that research on uranium had been moved, presumably to a safer location, but where he did not know. Until that time, our intelligence had come in on a fairly regular basis, but then it virtually ceased. We were confronted with the problem of finding out where the Kaiser Wilhelm group had moved and what they were up to.

The first information had trickled through in the summer of 1943. It seemed so innocuous that we did not appreciate its full import until much later. It was in the form of a report from an ungraded[1] Swiss informant, received by the British Secret Intelligence, stating that a certain Swiss scientist, who was allegedly pro-Nazi, was aiding in the development of an explosive a thousand times more powerful than TNT. His experiments and research were being conducted in the greatest of secrecy in an unused spinning mill in Bisingen, Germany. Inasmuch as Allied Intelligence was receiving hundreds of reports of this nature daily, and coupled with the absence from this one of any telltale words or phrases, such as uranium, atoms, heavy water, cyclotrons or the like, Calvert catalogued this item but did not attach immediate importance to it.

Next, in the fall of 1943, American censorship had intercepted a letter from a prisoner of war in which he mentioned that he was working in a "research laboratory numbered 'D.'" The letter was postmarked Hechingen, Germany, which is three miles north of Bisingen, in the Black Forest region of Germany, where many secret German projects had been moved. But again the report was so scanty that one could hardly assume that Germany's atomic research was being carried on in these outwardly sleepy little villages.

It was not until the spring of 1944 that Calvert received his first solid information. Then the OSS reported from Berne, Switzerland, that a Swiss scientist and professor had said that Dr. Werner Heisen-

[1] A term used to indicate "unknown reliability," in military intelligence.

berg, an internationally famous nuclear physicist and one of Germany's top atomic scientists—if not the top—was living near Hechingen. We knew from other intelligence that Heisenberg was working on the uranium problem. With this new bit of information, Calvert knew that he had found the hiding place of Hitler's top atomic scientists. From that moment on, his group worked feverishly to learn all they could about the Bisingen-Hechingen area. Almost simultaneously with the Berne report came a message from one of Britain's most reliable agents in Berlin that other top-ranking atomic scientists had been seen in this same area.

Calvert's next big problem was to try to penetrate the area. To do that he would have to get somebody who knew it extremely well. British Intelligence located a vicar living in England who before the war had been Vicar of Bisingen. He was able to pinpoint and identify buildings and factories for us. He also pointed out buildings that had housed spinning mills. At the same time Calvert sent a very reliable and able OSS agent, Moe Berg, the former catcher of the Washington Senators and Boston Red Sox, and a master of seven foreign languages, into Switzerland to prepare for a surreptitious entry into the Hechingen-Bisingen area. While Berg was in Switzerland, he picked up additional information and, passing himself off as a Swiss student, even attended a lecture given by Heisenberg, who had been granted permission to travel outside Germany to deliver this one speech. When I heard of Calvert's plan for Berg to go into the Hechingen-Bisingen area, I immediately stopped it, realizing that if he were captured, the Nazis might be able to extract far more information about our project than we could ever hope to obtain if he were successful. Instead, we continued to send agents into Sweden and Switzerland to see what they could find, and, at the same time, intensified our search for prisoners of war who had recently been in the area.

Starting in July, Calvert put the Bisingen-Hechingen area under constant air-photo surveillance. The pilots who flew these missions were never told of the nature of the suspected targets, lest they be interrogated in the event of a crash landing. At first our aerial reconnaissance produced nothing new. Then in the fall of 1944, we had our biggest scare to date. After one aerial sortie it was observed that near the town of Bisingen a number of slave labor camps had been erected with incredible speed. Ground had been broken and a complex of industrial sites had mushroomed within a period of two weeks. Railroad spurs had been constructed; mountains of materials had been moved in; power lines had been erected; and there was every indication that something was being built that commanded the utmost priority. Aerial interpreters, intelligence officers, our own technicians and scientists were all baffled after studying the photographs. Nobody could offer any sensible explanation of this new construction. All we knew was that throughout the past year we had been getting reports that this area was housing Germany's top atomic scientists. The only thing upon which we could all agree was that whatever the construction was, it was unique. Naturally the first question that came to our minds was whether this was the start of Germany's "Oak Ridge." If it was, we did not want to bomb it immediately, since that would only drive the project underground and we would run the risk of not finding it again in time. Yet we could not let construction progress too far, particularly since this was just at the time when it was thought that the Germans might withdraw to the Black Forest and make it a redoubt area. Fortunately our anxiety was short-lived, and the fear of a German atomic plant was dissipated almost as quickly as it had arisen, when some British mining experts recognized that what we had been observing so closely was nothing more than a new form of shale-oil-cracking plant.

When Alsos moved northward into the Lowland countries in September, Pash entered Brussels with the advanced elements of the mission. There he set about securing Union Minière's offices and records. From these, he soon determined that all our previous information about the disposition of the Belgian ore had been correct.

Mr. Gaston André, in charge of uranium at Union Minière's main office, gave us much valuable information about the movement of uranium from Belgium during the German occupation. Before the war, when Union Minière was the world's leading supplier of uranium and radium, a number of German firms had purchased uranium products for normal peacetime uses, as well as for retrading purposes. The shipments involved in such transactions normally consisted of less than a ton per month of assorted refined materials, but since June, 1940, orders from a number of German companies had increased spectacularly.[2]

A preliminary study conducted by Union Minière indicated that a quantity of material was still in Belgium. Part of it was ready for shipment, but probably had not yet been removed. When I learned of this, I immediately sent Furman back to Europe with instructions to locate and secure the material. He and Pash conferred with General Bedell Smith, who arranged for the British 21st Army Group to support Alsos in its recovery operations, without revealing to the British the name or purpose of the material being sought. The area where they expected the ore to be was then in the front lines of the British sector and under light sniper fire. Pash and two of his agents hunted for it from September 19 to 25 before they finally found it. The captured ores amounted to sixty-eight tons, which were placed under joint American and British control and removed from Belgium to the United States by way of England.

Information obtained in Belgium led to further investigations in Eindhoven, near Antwerp, where we learned that in May of 1940, nine cars containing approximately seventy-two tons of uranium ores

[2] Auer-Gesellschaft, a well-known German chemical firm, which had not been one of Union Minière's customers before the war, suddenly became an outstanding consumer of uranium products, and had bought about 60 tons of refined material from them during that period. The next large German shipment was in November, 1941, and consisted of about 9 tons of uranium products to the Deutsche Gold und Silber Scheideanstalt (Degussa), an Auer subsidiary. Degussa had been a prewar user of small amounts of uranium for making ceramic colorings. It had an outstanding reputation in the field of metal refinement.

During June of 1942, unusually large amounts of uranium products had been shipped to Roges GMBH, a wartime trading agency which was considered to be directly connected with the German Ministry of Trade and Finance. The uranium products ordered by Roges consisted of about 115 tons of assorted refined and partially refined materials. In addition to these, they obtained 610 tons of crude materials, 17 tons of uranium alloys and about 110 tons of impure rejects from the refining process. In January and May of 1943, they obtained 60 tons and 80 tons of additional refined products.

had been shipped out to Le Havre, France, ahead of the German invasion. Apparently, the Germans had seized two of the nine carloads at Le Havre, while the remainder were rerouted to Bordeaux. I instructed Alsos to obtain clearance from Supreme Headquarters, and then to locate this material and secure as much of it as possible. Pash and Calvert concentrated at first on an area in the vicinity of Périgueux, France, and finally in early October expanded their search to include much of southwestern and southern France. They were greatly hindered in their search by the presence of several thousand German troops, who had been cut off south of the Loire River by the Seventh Army. Eventually, they found thirty tons of the missing ore in Toulouse, but the remaining forty-two tons eluded us.

Calvert had by then determined where almost all of the Union Minière ore in Germany was located and asked permission to make plans to go behind the German lines to get it. I denied his request, for I thought that any such attempt would be doomed to failure, and, what was more important, it would alert the Germans to the fact that we considered the ore to be of such value that we would take great risks to obtain it.

Throughout this stage of the campaign, one Alsos team had been visiting stockades to question prisoners of war, giving special attention to men who had served with labor battalions in Germany. They turned up a number of good leads. One took them, in November, 1944, to the abandoned office of a Paris company that handled rare earths. It had been taken over by Auer-Gesellschaft, the well-known German chemical firm, and was run by a Dr. Ihee whose trail we had crossed earlier in Brussels. During his frequent absences from the Paris office, his business was conducted by his representative, a Dr. Jansen, and his private secretary, Ilse Hermanns. Among the few items of intelligence found in the Paris firm's office was a list of registered mail, which indicated that one of the last outgoing letters was addressed to Miss Hermanns at Eupen. This town was in American hands by then and the investigation was pressed until both Miss Hermanns and Jansen were found. Although we got very little useful information from Ilse Hermanns, documents found among Jansen's effects indicated that both had recently visited Ihee at Oranienburg near Berlin, and that Jansen had also recently visited his mother at Hechingen. Since we had previously learned of thorium deliveries to Oranienburg, and considering our vital concern with Hechingen, our interest was greatly aroused. Yet Jansen, it turned out, had very little information about either place. He told us that Ihee was in charge of the rare earths department of Auer, and that his main office was at Oranienburg, but he appeared to have only a superficial knowledge of what Auer was producing. He said that Ihee had visited Paris about once every six weeks and in the meanwhile traveled extensively in southern France, for what purpose he did not know, though he did speak of a search for monazite. As for Hechingen, Jansen knew only that it was located within a zone that had been restricted for military reasons; but otherwise he knew of no activity there.

Jansen's information, though unremarkable, did focus our attention on Ihee, whom we eventually caught up with. It also strengthened our suspicions regarding Oranienburg and Hechingen.

As the American armies approached the city of Strasbourg, we made careful plans. The main items of interest to us there were the personnel and facilities at the University of Strasbourg. The Nazi authorities had always treated this as an entirely German institution; it was staffed throughout by Germans and the faculty was working on at least a part-time basis on German war projects. Pash maintained close liaison with the Strasbourg T-Force Command, a unit of the Sixth Army Group, to ensure that the Alsos advanced party would be among the first to enter the city and to make certain that the T-Force knew what the Alsos objectives were.

On November 25, 1944, the advance party from Alsos joined the T-Force in Strasbourg, took over the university laboratories and the offices and residences of all target personnel, and by the end of November had located, interned and placed under guard seven physicists and chemists, all of whom were German citizens.

The scientific members of Alsos began a detailed study of all captured objectives as soon as the military situation permitted. They obtained various leads and items of information concerning medical research, aircraft and naval matters. In the field of our particular interest, they found four faculty members whose backgrounds and occupations warranted their separation from the other internees, for possible transfer to the United States. Subsequent questioning, however, failed to show that any of them had engaged in direct research on nuclear weapons and their answers provided very little worth-while information.

One of the men in whom we were particularly interested was Professor C. F. von Weizsäcker. We had thought he was with the University but though we found his house, he appeared to have fled before the entry of the Allied forces. In contrast to the meager information obtained from the German scientists themselves, the documents and personal correspondence found in their offices, laboratories and home files provided us with much intelligence of real value.

While this information was in itself unclassified, it was possible, through the notes of meetings, fragments of computations, descriptions of experiments and vague hints in personal correspondence, to assemble a revealing picture of the German nuclear research program.

This operation at Strasbourg was by far the most successful that Alsos had conducted up to that time. The information gained there indicated quite definitely that Hitler had been apprised in 1942 of the possibilities of a nuclear weapon. Nevertheless, all evidence from Strasbourg clearly pointed to the fact that, as of the latter part of 1944, the enemy's efforts to develop a bomb were still in the experimental stages, and greatly increased our belief that there was little probability of any sudden nuclear surprise from Germany.

I still wanted further confirmation, however, for Strasbourg also established that the enemy was definitely engaged in a program of nuclear energy research and indicated the places in Germany where that work was going on. We were also fairly sure of the personnel and locations involved in the industrial effort supporting this program. After all these indications had been analyzed by both the MED and the OSRD, Alsos was in possession of a dependable guide for its subsequent operations in the German homeland.

The German Ardennes offensive (the Battle of the Bulge) caused us some real consternation when, for a brief period, it appeared that the Allied forces might have to evacuate Strasbourg temporarily. This created a very serious problem for us, for it inevitably would have compromised Alsos for future operations within Germany. Fortunately, evacuation did not become necessary.

CHAPTER 16

THE PROBLEM OF THE FRENCH SCIENTISTS

After Joliot's initial interrogations in England by Alsos agents, he returned to London on a number of visits and met occasionally with Sir John Anderson, who, by that time, had become Chancellor of the Exchequer. Sometime in November of 1944, Joliot discussed atomic energy and France's position in this field with General de Gaulle. Shortly after this we learned with dismay how much knowledge Joliot had managed to acquire of our own atomic effort.

The circumstances that made this possible go back to 1939, when a group of French scientists, working under Joliot's leadership, had patented a number of inventions that they claimed would provide means for controlling the energy of the uranium atom. They assigned their rights in these patents to the Centre Nationale de la Recherche Scientifique, an agency of the French Government.

One of Joliot's assistants in this work was Hans von Halban. In June of 1940, when France was collapsing under the German onslaught, von Halban had left for England, taking with him the entire French supply of heavy water, a number of scientific papers, and a verbal commission from Joliot to act for the Centre in attempting to obtain the best possible terms to protect future French interests in the atomic field. He apparently engaged in prolonged negotiations with the British to this end, for it was not until September, 1942, that he finally entered into a formal agreement with representatives of the Committee of the Privy Council for Scientific and Industrial Research and the Imperial Trust for the Encouragement of Scientific and Industrial Research.

Under the terms of the agreement, von Halban and Kowarski assigned to the British their own rights in the French patents and promised to ask Joliot to try to persuade the Centre to assign its collective rights as well. In return, the British would offer to the Centre certain rights in a series of other patents in this same field.

At the same time, the British employed von Halban and three of his associates from the Centre, eventually, as I have said, assigning them to the laboratories of the Tube Alloys Project in Montreal. By 1944, a number of other Centre scientists had left France to join the Free French Provisional Government in Algiers. The French working in the Montreal laboratories maintained contact with their former colleagues in North Africa and, through them, with their former leader, Joliot, who remained in Paris throughout the German occupation.

In August, 1943, President Roosevelt and Prime Minister Churchill signed the Quebec Agreement, one clause of which specifically stated: "that we will not either of us communicate any information about Tube Alloys to third parties except by mutual consent." Sir John Anderson was the British representative in drafting this agreement. As far as I could ever learn, neither he nor any other British representative mentioned the arrangement that was already in effect with the French through von Halban.

We first learned of the British-French arrangements when Sir John Anderson mentioned it in discussing with Ambassador Winant his desire to give certain information to the French. Mr. Winant requested an *aide-mémoire* on the whole matter. As soon as he received it he transmitted it immediately to Washington. Those of us who had been closely associated with the British throughout our joint venture were amazed by this first indication of their obligations to a third party. It was clear that the Chancellor felt it necessary to placate Joliot and also that the British felt themselves obligated to disclose to Joliot certain information concerning the progress that we had achieved in atomic research. We then made every possible effort to learn more about the British-French agreement, the exact conditions of which were still completely unknown to us. In the meantime, on the basis that Anderson was thoroughly cognizant of the purpose of the Quebec Agreement, having helped to prepare its text, I continued to object strongly to his stand.

Earlier, in November of 1944, the British had asked that von

Halban be permitted to visit London. I had consented, provided that
he would not be allowed to go on to the Continent, and the British
had agreed. However, immediately upon von Halban's arrival in London, the question of his seeing Joliot was raised. Sir John felt that to
prevent their meeting would raise serious political controversy. He
went on to express his opinion that Joliot was interested only in
the commercial aspects of atomic energy, that he already knew most
of what we had done, which was not true at that time, and that von
Halban was an honorable man who could be trusted. Winant, who
had been brought into the matter without warning and consequently
was not fully informed, was persuaded by the Chancellor that the
status quo with the French must be maintained at all costs and that
von Halban was an appropriate agent for this task. Under the circumstances, Winant's decision was understandable; he erred only in
neglecting to secure approval from Secretary Stimson, who was the
representative of the President in all matters pertaining to atomic
energy.

Thus, on December 3, when Bundy and I telephoned the Ambassador to express our concern lest von Halban might be permitted to
visit the Continent, we were surprised and distressed to learn that he
had been in Paris with Joliot since November 24.

After Mr. Winant had given in to Sir John and consented to von
Halban's departure, events had moved at a swift pace. Winant agreed
that von Halban could report to Joliot on his negotiations with the
British that had culminated in the 1942 agreement, that he should
attempt to persuade Joliot not to press France's demands for admission to partnership in the atomic project at that time, and that he
should ascertain the status of his patents under the new French Government. The British then furnished von Halban with a written agenda
—a "barest outline"—of some of the courses we were pursuing and
sent him on his way to Paris.

Upon his return to London, von Halban was closely questioned by
my agents about his discussions with Joliot and it became obvious, as
we had expected, that he had not held the conversation within the
bounds of any "barest outline." Vital information relating to our

research had been disclosed—information that had been developed by Americans with American money, and that had been given to the British only in accordance with interchange agreements subsidiary to the Quebec Agreement. It confirmed facts that Joliot might have suspected, but which he otherwise could not have known. This information had always been scrupulously regarded as top secret.

At the long conference Secretary Stimson and I had with the President on December 31, 1944, we discussed the French problem in detail. Mr. Roosevelt remarked that obviously Ambassador Winant had been deceived, and stated categorically that he himself had no knowledge of any British-French agreement involving atomic energy.

When we finally obtained a copy of their 1942 agreement with von Halban we could see that the British had obtained nothing more than the hope that von Halban might succeed in convincing the Centre to assign its rights. The French patents in question, which were on file in Australia, we felt would be ineffective in the United States.

At about the same time we learned that the British considered most of the French scientists in Canada to be representatives of the French Committee for National Liberation. They were not primarily British Civil Servants, as we had understood, but had been employed under an agreement that could be terminated whenever "their scientific position in the French Government Service" made it desirable. We discovered, in fact, that one of them, Gueron, had been on the French Government's payroll throughout his service in Montreal.

It should be re-emphasized that the American Government had no knowledge of the British obligations prior to this time. Dr. Bush did recall that the British had once asked him to use his influence with the Patent Office to have the French patents accepted. He had flatly refused, however, and knew nothing more of the matter except for what Sir John had written to him on August 5, 1942: "I am glad to say that we have now concluded an agreement with Halban and Kowarski whereby we have acquired all their rights in their invention. We have also taken positive steps, which we believe will be successful, to acquire the rights of the other French inventors associated with these inventions." No other mention was made here or elsewhere of the French Government or any of its agencies.

In December, 1942, Conant had learned from Michael Perrin, of the British Tube Alloys Project in Washington, that the British were dealing with von Halban on his patents, but were encountering difficulties growing out of what Perrin called von Halban's grand ideas on the future of France. Again, there had been no indication of any commitments to the French Government.

I cannot help but feel that the British should have disclosed to us, prior to the Quebec Agreement, any arrangements that they had previously entered into with third parties. Had they done so, we could have arrived at an understanding concerning the obligations that had been assumed by the United Kingdom, and we could have adopted a joint policy relating to any future problems that might grow out of such agreements. Had this been done, our difficulties at the end of 1944 would have been avoided.

Having effected a breach in the Quebec Agreement, Joliot proceeded to exploit it. He met with the Chancellor in February, 1945, and made it clear to Sir John that, while France had no immediate desire to press the issue, if she were not eventually admitted to full collaboration with the United States and Britain in the project, she would have to turn to Russia.

Thus, France acquired a bargaining power out of all proportion to anything to which her early patents entitled her. She was enabled to play power politics with our accomplishments and to bring, or threaten to bring, Russia into the picture. The United States was forced to sit quietly by while a large measure of the military security that we had gone to such pains to maintain was endangered and prematurely compromised by the actions of other governments over which we had no control.

In May of 1945, the French Government instructed Joliot to begin work on an atomic energy project. Joliot turned to his colleague, Pierre Auger, who had been working in the Montreal laboratories. Anticipating our concern, the British hastened to assure us that Auger would not participate in the actual work, but would limit his activities strictly to putting the French back on the right line if they made any serious errors. While Dr. Chadwick and I were both confident of Auger's integrity, we realized that naturally his greatest loyalty was to his own country.

We continued to watch developments in France closely, but by this time the end of our labors was in sight and our approach to the security of such information as was involved would soon be radically altered. Nevertheless, one of the first aspects of the project that we brought to President Truman's attention after he assumed office was the breach in the Quebec Agreement which grew out of the British-French liaison.

My sole source of satisfaction in this affair came from a remark made by Joliot to an employee of the United States Embassy in Paris: while the British had always been most cordial to him and had given him much information, he said, he got virtually nothing from the Americans he encountered.

CHAPTER 17

MILITARY INTELLIGENCE:
ALSOS III—GERMANY

Alsos began its operations in Germany on February 24, 1945, near Aachen. During the early phases of the Allied thrust into Germany, its work was directed primarily at targets of concern to agencies other than the Manhattan Project. It was quite interesting—academically, at least—to watch the pattern of the total German scientific effort emerge. I have always considered Goudsmit's opinion much to the point: "On the whole, we gained the definite impression that German scientists did not support their country in the war effort. The principal thing was to obtain money from the government for their own researches, pretending that they might be of value to the war effort. One genuine selling point which they used extensively was that pure research in Germany in many fields was far behind the United States."

Although most of our objectives in Germany lay in the French zone of advance, one that was particularly important to us—the Auergesellschaft Works in Oranienburg, about fifteen miles north of Berlin—lay in what was to be the Russian zone. The information that Alsos had uncovered in Strasbourg had confirmed our earlier suspicions that the plant was engaged in the manufacture of thorium and uranium metals which were to be used in the production of atomic energy and hence probably for the manufacture of an atomic bomb. Since there was not even the remotest possibility that Alsos could seize the works I recommended to General Marshall that the plant be destroyed by air attack.

When he approved, I sent Major F. J. Smith, of my office, to explain the mission to General Carl Spaatz, who was then in command of our Strategic Air Forces in Europe. Spaatz co-operated wholeheartedly and, in a period of about thirty minutes during the afternoon of March 15, 612 Flying Fortresses of the Eighth Air Force dropped 1,506 tons of high explosives and 178 tons of incendiary bombs on the target. Poststrike analysis indicated that all parts of the plant that were aboveground had been completely destroyed. Our purpose in attacking Oranienburg was screened from Russians and Germans alike by a simultaneous and equally heavy attack upon the small town of Zossen, where the German Army's headquarters were situated. I have since learned that as an entirely unexpected bonus the Zossen raid incapacitated General Guderian, then Chief of the German General Staff.

Members of Alsos entered Heidelberg in March and occupied the laboratories in which we were interested. Among the scientists taken into custody were Walther Bothe and Richard Kuhn, the Kaiser Wilhelm Institute Director for Medicine, Wolfgang Gertner and Beckner. From them we learned that Otto Hahn had been evacuated to Tailfingen, a small town south of Stuttgart, near Hechingen, and that two other scientists, Werner Heisenberg and Max von Laue, were at Hechingen, and that the experimental uranium pile at Berlin Dahlem had been removed to Haigerloch, another small town near Hechingen. They reported a shortage of heavy water, explaining that their only source of it had been in Norway. The pieces of the puzzle were beginning to fall into place at last.

Bothe disclosed that the total German effort on atomic physics had consisted of himself and three helpers: Heisenberg with ten men; Dopel in Leipzig, assisted by his wife; Kirchner in Garmisch with possibly two assistants; and Stetter in Vienna with four or five others. Hahn, he said, was engaged in work on chemical problems.

The Heidelberg group told us that Gerlach's approval was required before any physicist could obtain the means for scientific work. If he wanted the highest priority rating, called DE, he had to have the additional approval of Albert Speer, Minister of Armaments and Munitions.

Later, Bothe expressed his belief that the separation of uranium isotopes by thermal diffusion was impossible and indicated that the only work on isotope separation being done in Germany involved the centrifugal method. He added that this work was under the direction of Dr. Harteck. Bothe said he knew of no element higher than 93, although he recognized that since element 93 was a beta emitter, 94 must exist. He repeatedly expressed his opinion that the uranium pile as a source of energy was decades away and that the use of uranium as an explosive was altogether impracticable. He claimed not to know of any theoretical or experimental work being done in Germany on the military applications of atomic fission, but he agreed that such work could be under way without his knowledge.

After repeated questioning about the military value of the cyclotron, Bothe admitted that it had been regarded as a means for obtaining radioactive material for bombs. He said that all secret documents relating to his work had been burned in accordance with his government's instructions. The files of his institute were examined carefully and later his home was searched; but to no avail. Some of his personal letters, however, did cast doubts upon his assertion that he knew nothing of the work being done at Bisingen and Sigmaringen. From other sources, the interrogators learned that Bothe had returned a considerable quantity of uranium to Degussa after he had no further use for it.

Kuhn was present throughout Bothe's interrogation. When it was over, he called one of the Alsos men aside and told him about the technical and scientific library of the German Chemical Society, of which he was the custodian. He claimed that it was the best of its kind in the world and included accounts of most of the German chemical activities in the war. To avoid the risks of heavy bombing, the library had been concealed in a number of caves and eventually was moved to a salt mine. Quite evidently, Kuhn preferred to have it taken over by the Americans rather than by the Russians. Unfortunately, it was behind the Russian lines.

Wolfgang Gertner generally confirmed the information we had received from Bothe. He had been separated from Bothe ever since

Heidelberg had fallen and there was no evidence of collusion in their answers. Certainly there was no reason to believe that Gertner knew the details of Bothe's testimony. Gertner said that he had worked with Joliot in Paris, from September of 1940 to July of 1943. He and Joliot, who had been close friends, had discussed the possibility of an atomic bomb and they had agreed that its development was not feasible. He said that his work in Paris had been confined largely to pure scientific research without any specific military applications. After leaving Paris, he had joined Bothe at Heidelberg, where he said he still engaged in pure scientific work. He had reached the conclusion that it would be impossible to develop an atomic bomb because of the difficulties involved in separating isotopes. He further believed that, of all the separation methods, the centrifuge process offered the best prospects of success, but the low production rates that had been achieved by that method appeared to rule it out.

Gertner and Harteck, who was concerned with centrifuge separation, were quite friendly with each other. Gertner believed in the future possibilities of the uranium pile as a source of energy. He confirmed our previous information that the Germans' experimental pile was not self-sustaining and that it had been moved from Berlin Dahlem to Haigerloch, where it was under the control of Heisenberg's group in Hechingen.

At about this time a major problem arose in Washington. The division of Germany into three zones of occupation had been arranged at Yalta. Later, when it was decided to establish a fourth zone to be occupied by the French, the readjustment of the American zone's boundaries was handled by a committee of representatives of the State Department and of the Joint Chiefs of Staff. All the information that had been developed by Alsos indicated that the principal German work on atomic energy was being conducted in the general area Freiburg-Stuttgart-Ulm-Friedrichshafen, a large part of which would be turned over to the French. Hechingen lay near the center of this area and was in the French Army's zone of advance, far removed from the zone of any American unit.

As I saw it, there could be no question but that American troops must be the first to arrive at this vital installation, for it was of the utmost importance to the United States that we control the entire area that contained the German atomic energy activities. General Marshall agreed. However, the State Department representatives who were working on the boundaries refused to go along without a full explanation of our reasons. This I would not give. I presented the matter to Secretary Stimson, who, on April 5, decided that any further attempt to get the State Department to adopt our views would not be feasible. Consequently, I was forced to initiate some drastic measures to accomplish our purpose. One of these became known as Operation Harborage.

According to this plan, American troops would have to get into and hold the area long enough for us to capture the people we wanted, question them, seize and remove their records, and obliterate all remaining facilities, for my recent experiences with Joliot had convinced me that nothing that might be of interest to the Russians should ever be allowed to fall into French hands. Having reached this conclusion, I discussed the matter with Secretary Stimson and General Marshall together. After I had outlined briefly what I wanted, and we had considered the possible value of the information we might gather, we all turned to the big wall map in the Secretary's office. To my great embarrassment, I was unable to find Hechingen on this map, and both the Secretary and the Chief of Staff were equally unsuccessful. Finally, Mr. Stimson summoned his aide, Colonel William H. Kyle, who succeeded eventually in locating our target at the bottom of the map, not more than two feet above the floor. If a photographer had been present at that time when the four of us were almost on our hands and knees, gazing intently at this point barely off the floor, he might well have caught one of World War II's more interesting photographs.

A short discussion followed, during which General Marshall asked me how I would ensure the capture of our objectives. I suggested that the necessary American troops, possibly as much as a reinforced corps, should cut diagonally across the advancing French front. Marshall agreed and sent for Major General J. E. Hull, Head of the Operations Division, War Department, General Staff, telling him to issue instructions to General Eisenhower that would take care of our requirements.

Although the details of the operation were worked out directly between Operations Division and the European Theater, I was asked to furnish Eisenhower's headquarters with certain essential information. General Marshall made it clear to me that, in so doing, I and my representatives were always to point out that, while the War Department approved this action, and, indeed, considered it highly important, my requests should not be construed as orders to be carried out at all costs.

At this point, the State Department representatives who were engaged in revising the zones of occupation apparently began to have some misgivings, for I was approached on April 7 to see whether I did not think it advisable to keep all or part of Baden in the American zone. Two days earlier, I would have said yes, but Mr. Stimson had made his decision not to press the matter and I did not wish to reopen it, particularly since the proposed revision would still leave the Hechingen area in the French zone.

Lansdale went to Europe to make the necessary arrangements for Harborage. He reported directly to General Bedell Smith, at Rheims, who, on April 10, called a conference to discuss the operation. Major General H. R. Bull, SHAEF G-3, felt that a corps, consisting of one airborne and two armored divisions, should be attached to the Sixth Army Group for this action. There was apparently a difference of opinion about how heavily the Germans were defending Hechingen and the powerful force proposed was calculated to overcome the worst possible opposition. General Smith finally decided, however, that any major offensive on this front would be contrary to General Eisenhower's policy of maintaining the Sixth Army Group on the defensive while attacking in the north. He did agree, nevertheless, to draw up the necessary plans for the attack, in case it later became possible to launch it. Upon learning of this development, in order to avoid any possible misunderstanding, I cabled Lansdale to remind him once more that our wishes were not to be construed as orders from the War Department.

During this entire period, Brigadier General Eugene L. Harrison,

G-2 of the Sixth Army Group, went out of his way to help Pash in all the military operations in that Group's sector.

Since the war, I have had occasion to discuss Operation Harborage and other Alsos operations with a number of the officers who were involved. In the course of these discussions, I have made it a point to tell them how much I always appreciated the co-operation given my representatives throughout the European Theater, when the only justification that they had for their apparently outlandish requests were simple memoranda addressed "To Whom It May Concern," signed by either Secretary Stimson, General Marshall, or in a few cases by Colonel Frank McCarthy, the Secretary of the General Staff, and stating that their mission was of the utmost importance and that the Secretary of War would appreciate any assistance that could be rendered. Invariably, I have been told that it was not a case of kindness on the part of anyone in the European Theater, for these letters were most unusual and they realized that the matters involved must be of paramount importance. But over and above this, I have always felt great pride and pleasure upon hearing from these same commanders that while my officers were far from high-ranking, they were obviously of such ability and so convinced of the importance of their mission and the strength of their backing that they would have accomplished their missions no matter what obstacles stood in their way.

Ever since Calvert had found out in the fall of 1944 where the Union Minière ore was probably hidden, we had been particularly interested in a salt mine near Stassfurt, bearing the formidable title of Wirtschaftliche Forschungs Gesellschaft (WIFO). We believed that the largest portion of the Belgian uranium ore had been stored there. Early in 1945, I transferred Major J. C. Bullock from the Manhattan Project to G-2 so that he could serve with the Alsos mission for the express purpose of recovering it.

.However, when the time finally came, Alsos was already so heavily committed that we had to set up a special unit for this particular task. Using Lansdale as its head, whom I had sent over to make certain that Operation Harborage would move smoothly, we hastily formed an American-British group consisting of Lansdale, Calvert, some CIC agents borrowed from the Alsos mission, Sir Charles Hambro, Michael Perrin and David Gattiker. Sir Charles, who had been associated with us for some time in Washington, was in his colonel's uniform of his World War I unit, the Coldstream Guards. Perrin and Gattiker, although normally civilians, also wore British uniforms.

While waiting for Operation Harborage to get under way, Lansdale, Pash and Sir Charles had met with Brigadier General E. L. Sibert, G-2 of the Twelfth Army Group, to discuss means of getting into Stassfurt, which then lay between the American and Russian armies. Although Sibert was somewhat hesitant about doing anything that might disturb the Russians, his hesitation was quickly overcome by General Bradley's terse opinion: "To hell with the Russians."

Accordingly, Lansdale's group joined the 83rd Division, in whose zone Stassfurt lay. The WIFO plant was seized quickly and without incident. It was in a terrible condition from repeated bombings, but fortunately the manager had stayed on the job. Hidden in his house was an inventory of the plant's property, which showed the whereabouts of the missing ore. Approximately eleven hundred tons of it were soon found stored in barrels under open sheds above-ground.

Most of the barrels were either broken or rotten, and it was obvious that the ore would have to be repacked before it could be moved. Complicating the problem was the fact that there were still many German units in the area. Fortune smiled upon Lansdale's group again when the CIC agents found a barrel factory close at hand. The owner of the plant, who was also the local burgomaster, was soon prevailed upon to round up a sufficient number of laborers and to resume operations. During the next two weeks, with Agent Schriver in charge, and while under intermittent enemy fire, this factory turned out about twenty thousand fruit barrels.

Lansdale, in the meantime, had gone back to SHAEF, where he saw General Smith, and procured the services of a truck company. Trucks were in great demand at this period and the men, all Negroes with one white lieutenant, were exhausted from lack of sleep. They were further handicapped by being far from their normal maintenance bases. Nevertheless, they performed splendidly, and with the use of forced labor to repack and load the ore, the entire tonnage was removed during three days and nights to an airport hangar at Hildesheim, near Hanover, well behind the Allied lines. A small amount of the ore was lost en route because of the number of truck ditchings caused by the extremely rough roads.

The route was well marked by the usual arrows, bearing the name "Calvert." This led to a difference of opinion among the troops who happened to see anything of the operation. From the "Calvert" most Americans guessed that it involved whiskey. Observing the ore's hue and noting that it was escorted by Hambro, a member of a well-known London banking family, many of the British were convinced that it was gold.

From Hanover, a considerable tonnage was moved by air to England. There was too much, however, to carry all of it in this manner, so arrangements were made to move the remainder by rail to Antwerp about two hundred miles away, and thence by ship to England. The precautions for insuring its delivery proved inadequate and somewhere along the line, probably in a switching yard, three cars disappeared, but after an intensive search, Agent Schriver found them, much to our relief.

From England, the ore was sent over to the United States. Before the WIFO operation began, I had assured Union Minière that, if they had not previously received payment from some other source, we would pay for it at the current rate for uranium ore; provided that higher authority in our government did not consider it captured booty, in which case payment could not be made. I also assured the British that, subject to the same proviso, the ore would be credited to our joint account with the Combined Development Trust.

At last we were absolutely sure of all that we had set out to learn. Accordingly, on April 23, I handed the following memorandum to General Marshall:

In 1940 the German Army in Belgium confiscated and removed to Germany about 1200 tons of uranium ore. So long as this material remained hidden under 'he control of the enemy, we could not be sure but that he might be ,,eparing to use atomic weapons.

Yesterday I was notified by cable that personnel of my office had located this material near Stassfurt, Germany, and that it was now being removed to a safe place outside of Germany where it would be under the complete control of American and British authorities.

The capture of this material, which was the bulk of uranium supplies available in Europe, would seem to remove definitely any possibility of the Germans making any use of an atomic bomb in this war.

During this period we were still rounding up the few remaining German scientists and laboratory facilities and material. As always, Pash led the pack. The capture at Nordheim of Professor Osenberg, Chief of the Planning Staff in the German National Research Council, together with all records of the German research agencies, was particularly useful, for the documents listed the ultimate destinations of all evacuated research personnel and laboratories.

The continuing Alsos interrogations in Heidelberg were also proving most informative. It was becoming apparent that there were two groups in Germany working on the uranium pile, the first under Diebner at Frankfurt and the second under Heisenberg. Heisenberg's group had been started in 1939 as a co-operative project of the most important physicists in Germany, with headquarters at the Institute of Physics in Berlin. There had been a certain amount of competition between the two groups, and quarrels over who would get materials continued even after all research had been officially consolidated under Gerlach. In Gertner's opinion, the work done under Diebner was not so good as that over which Heisenberg had supervision.

Having pretty well exhausted its Heidelberg sources, Alsos next turned its attention to the Frankfurt area, where the uranium metal required by the German project had been produced. It found there that the degrees of purity achieved were not particularly high.

Following closely behind the advancing American front, on April 12, Alsos moved in and seized Diebner's laboratory and offices, which were located in an old schoolhouse. Pash's people found, however, that the majority of the scientists, together with most of their documents, materials and equipment, had been evacuated on April 8, to carry on their work elsewhere. Nevertheless they picked up some uranium oxide, various pieces of equipment, an extensive physics laboratory and many files. From these last it appeared that Germany's military interest had been aroused in early 1940 by the experiments of Hahn and Strassman. It had been suggested then that uranium could be used to form an explosive, as well as to serve as a source of energy. Work to this end had been started by Heisenberg's group in Berlin, using uranium ore from Joachimsthal, which had been transformed into powdered U-238. This attempt at making a pile, however, was unsuccessful, primarily, I believe, because of the clumsiness of the experimental equipment. Heisenberg's group continued experiments with their apparatus until about the end of 1941, always with negative results. In spite of their failures, Heisenberg and von Weizsäcker calculated that by making a number of modifications to their equipment a self-sustaining pile could be built. The work was transferred to Leipzig, where, in 1942, a pile gave positive results, but was not self-sustaining. This led to the initiation late in 1942 of the so-called large-scale experiments at Berlin Dahlem. Finally, late in 1944, an exponential pile was constructed in Berlin. This, however, was what might be termed purely academic scientific experimentation.

By April 14, the military situation in the north had changed radically. It was decided then that the Allied armies would hold up short of Berlin and would devote themselves to cleaning up their flanks. General Eisenhower therefore approved Operation Harborage as it had originally been visualized. While no specific time was set for the attack, General Smith assured us that it would take place within two weeks and possibly sooner, if it appeared that the French were advancing faster than expected, which they did not do.

German resistance began to disintegrate so quickly at this point that, in spite of its lack of armored strength, the French Army started to move rapidly toward Hechingen. On April 21, the French crossed the line at which they had been ordered to halt, apparently for the purpose of getting to Sigmaringen, where the survivors of the Vichy Government were located. Whatever their intent, we could not sit idly by while they moved into this vital area.

Something had to be done, and, as usual, Pash did it. He asked for help, and General Harrison gave him operational control of the 1279th Engineer Combat Battalion. With this force he seized Haigerloch on April 23 and immediately began dismantling the laboratory. Its major feature was the exponential pile, which had been brought there from Berlin in February and concealed in a tunnel under a high cliff. The Alsos detachment was greatly assisted by the arrival of a number of British scientists under the leadership of Sir Charles Hambro, and was able to complete its operations in Haigerloch before the French reached there.

In the meantime, Pash, with one company of the 1279th Engineers, moved on to Hechingen, which he captured on April 24. Efforts to take this town the night before had been strongly resisted, but the final attempt was virtually unopposed. Pash seized a large atomic physics laboratory and a number of the leading German physicists, including von Weizsäcker and Wirtz.

The next morning he moved into Tailfingen, where they took over a large chemistry laboratory and captured Otto Hahn and Max von Laue. At Stadtilm, Alsos had found signed receipts for all the secret reports and documents that had been sent to the various scientists. But as the men were picked up, one by one, they all announced blandly that everything had been destroyed. Hahn, however, answered promptly, "I have them right here."

The capture of Hahn was simple. A German on the street, when questioned, pointed the way to an old school building which contained his laboratory. After the school was surrounded by troops, F. A. C. Wardenburg and James Lane, both chemical engineers from du Pont and two of our Alsos scientific personnel, walked in and asked for Hahn. They were shown into his laboratory and started their interrogations. "It was just like a business call on a customer," was their apt description. By now French Moroccan troops were in the area, yet the mission still had not found the German stores of heavy water and uranium oxide that had been used in the Haigerloch pile. Fortunately, the French were few in number, and the many German units scattered throughout the countryside kept them fully occupied while Alsos was getting its job done.

Skillful questioning of the German scientists by Goudsmit and his associates finally disclosed the hiding place of the heavy water and uranium and, on April 26, the heavy water was removed from the cellar of an old mill near Haigerloch and sent back to Paris. About one and a half tons of small metallic uranium cubes were dug up from a plowed field just outside the town. These, too, were quickly dispatched to Paris. Both water and uranium were then shipped to the U.S., to be disposed of by the Combined Development Trust.

On the twenty-seventh, the German scientists were taken to Heidelberg for further questioning, and later removed to Rheims. As they were in the act of leaving, von Weizsäcker suddenly blurted out the information necessary to locate the still missing records of the German research programs. They were sealed in a metal drum, which had been deposited in the cesspool in back of von Weizsäcker's house.

By the end of April, Alsos was heavily engaged in mopping-up activities. Most of the material we wanted had been secured. A few important scientists—notably Heisenberg—still eluded us. But, generally, our principal concern at this point was to keep information and atomic scientists from falling into the hands of the Russians.

Our primary objective now, therefore, was to get hold of Heisenberg, who had left Hechingen about two weeks before Pash's arrival. He and two other members of the Hechingen group were known to be either in Munich or at Urfeld in the Bavarian Alps. Major Ham took charge of the search in Munich, while Pash moved on to Urfeld.

Ham located Gerlach without any difficulty, on May 1, in the physics laboratory of the University of Munich. After an intensive search, Diebner was picked up in Schöngeising, about twenty miles southwest of Munich. On May 3, both scientists, together with their papers, were taken to Alsos Forward Headquarters in Heidelberg.

Ernest O. Lawrence, above, director of the Radiation Laboratory at the University of California, at the panel of a cyclotron; Arthur H. Compton, left, director of the Metallurgical Laboratory, University of Chicago; Enrico Fermi, below left, who built the first atomic reactor at Chicago; and, below right, Albert Einstein with Harold C. Urey, who directed the research at the SAM Laboratories at Columbia University.

Typical Hanford scene (top) before construction; above, a completed reactor at Hanford in operation; left, construction workers at Hanford; and opposite page, top, the Hanford construction camp, which housed 45,000 workers. Project offices of U.S. Corps of Engineers and du Pont are in foreground.

(*du Pont Photo Library*)

Right, some of the thousands of people who built Oak Ridge; below, view in 1942 of the Clinton Engineer Works area, where the city of Oak Ridge was built (*MED photos*); and opposite page, below, the gaseous diffusion plant at Oak Ridge.   (*USAEC photo*)

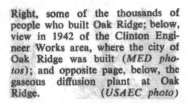

Top, the control dugout at Alamogordo, the point from which the bomb was actually fired. Above, the carrier for Jumbo. When loaded, it was drawn by two heavy tractors. Left, Jumbo after the blast. Horses also helped to usher in the atomic age. Below, some of the mounted guards at Los Alamos.

(MED photos)

A scientific discussion in the author's office. Left to right, Sir James Chadwick, the senior British scientific adviser; General Groves; and Richard C. Tolman, one of the latter's two senior American scientific advisers. *(U.S. Army photo)*

Left to right, James B. Conant, Chairman of the National Defense Research Committee, the other senior scientific adviser to General Groves; Vannevar Bush, Chairman of the Office of Scientific Research and Development; and General Groves, Commanding General, Manhattan Engineer District. *(MED photo)*

The air crew of the Hiroshima bombing. Standing, left to right, Lt. Col. John Porter, ground maintenance officer; Capt. Theodore J. Van Kirk, navigator; Maj. Thomas W. Ferebee, bombardier; Col. Paul W. Tibbets, pilot; Capt. Robert A. Lewis, co-pilot; and Lt. Jacob Beser, radar countermeasure officer. Front row, left to right, Sgt. Joseph Stiborik, radar operator; Sgt. George R. Caron, tail gunner; Cpl. Richard H. Nelson, radio operator; Sgt. Robert H. Shumard, assistant flight engineer; and T/Sgt. Wyatt E. Duzenberry, flight engineer. Not shown: the weaponeer, Capt. W. S. Parsons, USN, and his assistant, Lt. M. R. Jeppson.

(*Official U.S. Air Force photo*)

Ten minutes before the take-off for Nagasaki. Left to right, Col. Paul W. Tibbets; Rear Adm. W. R. E. Purnell; Capt. William S. Parsons (with cap); and, in his flight equipment, Comdr. F. L. Ashworth, the weaponeer.

(*MED photo*)

The air crew of the Nagasaki bombing. Front row, left to right, S/Sgt. Edward Buckley, radar operator; M/Sgt. John Kuharek, flight engineer; Sgt. Raymond Gallagher, assistant flight engineer; S/Sgt. Albert Dehart, tail gunner; and Sgt. Abe Spitzer. Standing, left to right, Capt. Kermit K. Beahan, bombardier; Capt. James F. Van Pelt, Jr., navigator; Capt. Charles D. Albury, co-pilot on this flight (normally airplane commander); Lt. Fred J. Olivi, third pilot (normally co-pilot); and Maj. Charles W. Sweeney, pilot. Not shown: the weaponeer, Comdr. F. L. Ashworth, USN, and his assistant, Lt. P. M. Barnes.

(*Official U.S. Air Force photo*)

The now familiar smoke column over Nagasaki (opposite page) just after the bombing on August 9, 1945. Note wing of escort plane in lower right corner.

(*Official U.S. Air Force photo*)

The author with J. Robert Oppenheimer, who headed the Los Alamos Laboratory.

(*MED photo*)

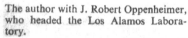
Secretary of War Robert Patterson awarding the Distinguished Service Medal to Brig. Gen. Kenneth D. Nichols, the District Engineer, to whom much of the success of the project was due. The MED shoulder patch was authorized a few days after the Hiroshima bombing; it was worn by several thousand military personnel assigned to the project.

(*MED photo*)

In the meantime, Pash was about twenty miles ahead of the advanced elements of the Seventh Army as they pushed through the Bavarian Alps. Finding the bridge into Urfeld destroyed, his group of eight dismounted from their vehicles and joined a ten-man reconnaissance patrol, entering Urfeld in the late afternoon of May 2. About an hour later, a German unit tried to pass through the town and a hot fire fight broke out. Shortly afterward, the patrol withdrew, leaving Pash in control of the town but alone. I could never straighten out in my own mind just what happened that night. But the following will give a general picture.

At sundown Pash was told that a German general wished to see him. The general was admitted and immediately surrendered his entire division to the very surprised chief of the Alsos mission. Pash, thinking fast, replied that it was getting late and he did not wish to bother his own general, who was right behind him with a large force, with any formalities that night; so the Germans would have to wait until morning to have their surrender officially accepted. No sooner had the first general departed than another German commander, this time of seven or eight hundred men, arrived and the procedure was repeated. By now, Pash was thoroughly alarmed for the safety of his infinitesimal force, and, as soon as it became thoroughly dark, he withdrew quietly toward the American lines. Shortly before this he had found Heisenberg, but with the difficult situation confronting him he felt it wiser to leave him in his home for the time being. During the night, he was able to obtain the support of an infantry battalion. He returned at dawn on May 3 and picked up Heisenberg, who was waiting in an office with his bag already packed. When Pash entered, Heisenberg greeted him with: "I have been expecting you." Heisenberg was immediately evacuated to Heidelberg.

Pash's last effort typified the boldness with which he carried out every one of his operations, and clearly demonstrated his ability to stick to his objective, which, in this case, had been to catch Heisenberg. Heisenberg was one of the world's leading physicists and, at the time of the German break-up, he was worth more to us than ten divisions of Germans. Had he fallen into Russian hands, he would have proven invaluable to them. As it was, he has always remained on the side of the West. Judging from other actions taken at that time, we seemed to be almost alone in our appreciation of the potential value of German scientists to the Russians.

With Heisenberg in our custody, all his colleagues began to talk freely. We soon learned that, although Gerlach had been in administrative charge of the work, he had only a superficial knowledge of its technical details. Diebner was not particularly co-operative and seemed to be rather antagonistic toward Heisenberg. Gerlach and Heisenberg were on cordial terms and appeared to consider Diebner an inferior scientist. Heisenberg was, outwardly at least, actively anti-Nazi, but was nevertheless strongly nationalistic. None of them seemed to know very much about the Allies' nuclear fission efforts; Gerlach spoke several times of the poor quality of German technical intelligence. There was no German counterpart of Alsos.

After Hamburg fell, Harteck was captured. He had written a letter on April 24, 1939, in which he advised the War Ministry that:

We take the liberty of calling to your attention the newest development in nuclear physics which in our opinion will perhaps make it possible to produce an explosive which is many orders of magnitude more effective than the present one . . . it is obvious that if the possibility of energy production outlined above can be realized, which certainly is within the realm of possibility, that country which first makes use of it, has an unsurpassable advantage over the others.

However, Harteck, like the other German scientists, seemed to have come to feel that, while there was some hope of producing energy from a uranium pile, it was unlikely, if not entirely impossible, that a workable weapon could be developed.

The various possibilities open to the Germans were never systematically and completely investigated. This was because their work was seriously deficient in over-all direction, unity of purpose and co-ordination between the participating agencies. Originally, there had been a number of more or less competing groups, one under

Army Ordnance, another under the Kaiser Wilhelm Institute of Physics, and still another under the Postal Department. There was continual bickering, as might be expected, over supplies and material, and surprisingly enough, in the light of most American scientists' pleas for freedom from the restrictions of compartmentalization, there was a generally nonco-operative attitude regarding the exchange of information between the various groups. Many German scientists worked alone on their individual projects and did not seem to feel any compulsion to work for the national interest. The basic reason for this was probably the generally accepted belief that the development of a nuclear weapon was not possible.

In any case, the development of nuclear energy in Germany never got beyond the laboratory stage, and even there the principal consideration was its use for power rather than for explosives. Other scientific objectives seemed to be more important and received greater governmental attention and support.

The status of the German effort at the close of the war in Europe was reminiscent of the early phases of our project in the United States, when committees were appointed only to be superseded by other committees. At times it seemed as though more thought had to be devoted to organization than to solving the problems under study.

The German National Research Council had been under the Ministry of Education until 1943, when it began to function as a central agency for research work with possible military applications. About a year after its formation, it was placed directly under Göring. It assigned research projects to universities and individuals, allocated funds, established priorities and handled scientific personnel.

In the fall of 1944, Himmler's Security Service Organization apparently became interested in the atomic project and formed a War Research Pool, which remained under Göring to avoid duplication and useless work. Himmler's people did not seem to be entirely satisfied with progress under the National Research Council, however, and they subsequently proposed a plan to remove all obstacles to the project and obtain maximum results. Although this plan was sound, it came too late.

The Germans carried out their centrifuge research at Celle. This work had started at the University of Hamburg under Dr. Harteck, but had to be relocated several times because of Allied bombings. It finally moved into the spinning mill near Hechingen, where it was seized by Alsos. The equipment was very small in comparison to our own, but it seemed to be operating satisfactorily.

The German scientists whom we had assembled at Rheims—Hahn, von Laue, von Weizsäcker, Wertz, Bagge and Korshing—seemed fairly content with their lot and, having given their parole not to leave the house in which they were billeted, settled down quickly to their new routine. But on May 7, they were moved to Versailles, where they were subjected to treatment that I considered entirely unacceptable in that they were not segregated from other prisoners of war. The scientists themselves were most indignant at being considered war criminals and repeatedly asked to see Joliot. A couple of days later this group was joined by Heisenberg and Diebner, and discontent continued to grow until we found it necessary to do something about it. The principal trouble stemmed from a SHAEF order that prohibited preferential treatment for any German national.

I succeeded, on May 11, in having our guests moved to Le Vésinet, where the situation quickly began to improve. The professors were well pleased with conditions there in spite of the dismaying plumbing and electrical problems with which they were confronted. As a matter of fact, they kept themselves occupied and amused for a time in repairing and restoring these essential services. However, we soon began to run into trouble with the troop units in the areas, as well as with the French local inhabitants who accused our officer-in-charge of harboring Marshal Pétain. To counteract this, we put out the story that the group consisted of active anti-Nazis whom we were holding in protective custody.

Nevertheless, it was obvious that the scientists soon would have to move again, and, on the fourth of June they were taken to Huy in Belgium. There, our troubles began all over again because the local commander insisted upon issuing POW rations to the professors. By this time they were becoming quite upset by the lack of contact with their families and with other scientists that had been promised them when we took them into custody. Fortunately, their house at Huy contained a broken piano and they kept busy for a while repairing it. Trouble began in earnest, however, when they read in their newspapers of the extension of the Russian zone in occupied Germany to include Dr. Diebner's home in Thuringia. This was a serious matter from my point of view, too, since Frau Diebner had worked with her husband throughout the war on everything that he had done. After a frantic search, we were relieved to learn that she had moved to Neustadt in the American zone.

Dr. Gerlach joined the troupe at Huy on the fourteenth. By now they were all definitely unhappy. It was during this period that Heisenberg indicated that the documents that the Alsos mission had captured were misleading, because the German scientists had progressed much further in their work than they had indicated in their reports, and because their knowledge had advanced materially since we had brought them together in custody and made it possible for them to exchange information.

On July 3, the group was sent by plane to England. Upon arriving there, they were taken to a country estate called Farm Hall, which had been thoroughly prepared for their arrival and custody. I was most amused to read, therefore, the transcript of one of their early conversations:[1]

DIEBNER: I wonder whether there are microphones installed here?
HEISENBERG: Microphones installed? (Laughing) Oh, no, they're not as cute as all that. I don't think they know the real Gestapo methods; they're a bit old-fashioned in that respect.

Immediately the group began working hard to build up a legal case against their detention. This was a period of much conjecture for them. Some thought the British would not let them have any

[1] All reported conversations were translated from the spoken German.

contact with British scientists, because, Heisenberg suggested, the British Government were frightened about some of the professors. "They say," Heisenberg continued, " 'if we tell Dirac or Blackett where they are, they will report it immediately to their Russian friend Kapitza, and Comrade Stalin will come and say: "What about the Berlin University professors? They belong in Berlin," and it is possible that the Big Three will decide it at Potsdam and then Churchhill will come back and say: "Off you go" and the whole group is to return to Berlin and then we'll be in the soup.' "

Suspicions of each other's motives cropped up during this period and there were frequent displays of temper at the manner in which some of the professors had voluntarily surrendered their secrets. At the same time, the group was beset by doubt and a sense of insecurity. They were greatly concerned about the continued lack of news from their families and there was considerable speculation about what we were doing in the field of atomic energy. Said Bagge: "I am convinced they [the Anglo-Americans] have used these last three months mainly to imitate our experiments."

After V-E Day, a number of searches for specific information and materials were conducted in various parts of Germany. Alsos sent groups to Berlin and Salzburg, but, by that time, I was no longer too much concerned with their work, beyond insuring that no information remained that might eventually fall into Russian hands. These operations only confirmed what we already knew and it was quite clear that there was nothing in Europe of further interest to us.

By the end of the war, the value of Alsos had been demonstrated in so many ways that there was a strong current of opinion in favor of continuing the organization, suitably transformed, as a permanent instrument, under G-2, for the collection of scientific intelligence information. As to whether or not this was feasible, there were arguments in both directions. Subsequent history has certainly proved the need for prompt and accurate information on the progress of potential enemies in such fields as rocket fuels, satellite guidance, thermonuclear reactions, seismology and biological weapons. Again this is a broad problem, outside the present story. The fact is that Alsos, after considerable discussion and against the wishes of the lower levels of G-2, was broken up shortly after the end of the war. The 114 men and women who were with the mission on V-E Day (28 officers, 43 enlisted men, 19 scientists, 5 civilian employees and 19 CIC agents) were gradually reduced by attrition until, on October 15, 1945, the "MED Scientific Intelligence (Alsos) Mission" was officially disbanded.

PART II

CHAPTER 18

TRAINING THE AIR UNIT

There has never been an improvement in weapons comparable in degree and in sudden impact to the atomic bomb. In the case of other developments, such as explosives, the airplane, the tank, long-range artillery, armor-clad warships, submarines, and even rifles, it took years, if not decades and centuries, after their first use for their revolutionary influence upon warfare to be felt. In the case of the atomic bomb it took only a few hours.

In employing it in battle, the Army Air Force would obviously play a major role, provided, of course, that the bomb could be airlifted. Consequently, I called on General H. H. Arnold in the spring of 1944 and discussed the whole situation with him, including our prospects for success and the date upon which the bomb would probably be ready. He had previously been told about the project in very general terms; now, I wanted to fill him in more completely and to make certain of his support.

In calling upon him at this time, I was, of course, assuming that our work would be successful. This was normal procedure. Always we assumed success long before there was any real basis for the assumption; in no other way could we telescope the time required for the over-all project. We could never afford the luxury of awaiting proof of one step before proceeding with the next. Just as we nearly finished building and equipping the gaseous diffusion plant before we knew that we could make its most vital part, the barrier, so here we began to prepare for combat operations a year before we knew that we could produce an atomic explosion.

We were also still uncertain about the final size and weight of the implosion bomb, although we were quite certain of these factors for the gun type. However, after discussing this problem with Oppenheimer and his senior associates at Los Alamos, I had decided it would be reasonable to plan on using the B-29 as the carrier plane for both types, though we might have to make certain modifications, all of which appeared feasible, in the plane's bomb bay and bomb-handling equipment. Nevertheless, we all recognized that problems then unforeseen might make the use of the B-29 impossible.

When I told Arnold there was a chance that we might not be able to fit the bomb into the B-29, no matter how hard we tried, he asked me what I would do then. I said that if the B-29 could not be used, we would have to consider the use of a British plane, the Lancaster, which I was sure the Prime Minister would be glad to make available to us.

This brought from him the characteristic reply that I had hoped and expected to hear: that he wanted an American plane to deliver our bomb, and that the Air Force would make every effort to ensure that we had a B-29 capable of doing the job. Because the use of a British plane would have caused us many difficulties and delays, I, too, was most anxious to use the B-29 if it could be done. Fortunately, as time went on, we were able to make changes in the design of both types of bombs, so that it became possible to fit them into a specially modified B-29.

Arnold and I agreed that the Air Force would have three major responsibilities: First, it must be able to provide planes able to carry the bombs. These planes would have to have sufficient weight-carrying capacity, adequate bomb bay dimensions, large enough bomb doors and the necessary range. Second, a highly competent tactical unit must be organized, equipped and trained by the time we were able to produce enough fissionable material for a bomb. Third, the bomb must be delivered without fail upon the target.

In addition we would need Air Force assistance in the ballistic tests of the bombs, and special air transportation to move materials and equipment, particularly in the final stages of our preparations for the actual bomb drop.

Both to avoid bothering Arnold with the many details of our

joint effort and because he was frequently absent from the country for extended periods, I asked him to appoint an officer with whom we could deal. He designated Major General Oliver P. Echols as our liaison, and emphasized to him that the Air Force must do everything necessary to ensure the success of the mission. Echols named Colonel Roscoe C. Wilson to serve as his alternate, and it was with him that we conducted most of our business. He was a most fortunate choice, for his personality and professional competence ensured the smooth co-operation essential to our success. Through his efforts, the necessary air support was always provided by the subordinate Air Force commands, if not willingly, at least without delay.

While I can say the same of every other Air Force officer with whom I had any dealings in the project, I have always felt particularly grateful to Wilson, for he had to bear the brunt of all our many minor problems with the Air Force as well as a major responsibility for a number of our principal activities. I am sure that he must have had many difficult moments with his Air Force colleagues, as he denied them, for security reasons, information they considered essential to understand the reasons for his requests.

On the more difficult problems, our discussions were often with both Echols and Wilson. Only on very major problems did I feel that it was necessary to see Arnold; however, as with the Secretary of War and General Marshall, I always saw him without delay whenever I felt it was desirable.

In August of 1944, I notified Wilson that the Air Force should initiate planning along the following general lines:

1. During the period September 1, 1944, to February 1, 1945, we would provide them, for training purposes, with several hundred[1] high-explosive bombs having ballistic characteristics similar to those of the Fat Man (implosion-type) atomic bomb. At the time, I suggested that these practice bombs might prove to be effective weapons in themselves, and if so, they might be made available for operational use.

2. The work on the Thin Man (gun-assembly) bomb was

[1] This number was not reached.

essentially completed, but it would be some time before we would have enough U-235 to load it. We expected the first Thin Man to be ready in June of 1945.

3. The Fat Man was still under development, but there was some hope that the first of this model could be delivered as early as January of 1945.

These dates were the earliest by which the Air Force would have to be ready.[2]

On the basis of this information, Wilson developed a general plan for Air Force support to the project:

1. One very heavy bomber squadron with special units attached would be assigned to our support at the earliest practicable date.

2. Personnel for these units would be selected as soon as possible.

3. The units would be assembled at a base in the Southwest for special training.

4. Modification of aircraft would begin in time to deliver three planes prior to September 30, 1944, with enough more to be delivered before January 1, 1945, to provide a total of fourteen modified B-29's.

5. Flight testing of the Fat Man would continue through the period September 30, 1944, to January 1, 1945. Training would be conducted using the high-explosive training bomb, with particular emphasis upon ground and air techniques for handling atomic bombs.

6. Specialists from the Manhattan Project and the Air Force would be present at all times to supervise the technical aspects of training.

7. During the training period all equipment would be thoroughly tested and ballistic data for the bomb would be assembled.

8. At the end of 1944, a field party would visit the area from which the bomb would be delivered against the enemy, to make all necessary arrangements.

The effects of General Arnold's wholehearted co-operation became very evident when elements of the Air Force appeared reluctant to

[2] In order to avoid any possible unnecessary delay in the use of the bomb, the dates I gave Wilson were in advance of my actual expectations.

furnish the necessary number of B-29's. Their reaction was quite understandable, for these planes were in such short supply that it was impossible to give the crews that were to operate them overseas even the desired minimum of training.

After exploring the situation, Wilson told me that the only way we could get the planes was for me to go see Arnold myself. Even then, in his judgment, we would not get them until shortly before the actual operation.

General Arnold did not hesitate. He fell in with my request almost without discussion, without any evidence of disinclination, and without any suggestion that I might get along with a smaller number. This was typical of all my experiences in dealing with him. He fully realized the importance of the project and never expressed the slightest doubt of our ultimate success.

As I left Arnold's office to walk down to Wilson's to tell him how I had made out, he met me in the hall and said that Arnold had just telephoned him. When he added that he was completely surprised and quite impressed by Arnold's prompt agreement, I replied that I was not the least bit surprised; it was just what I had expected.

Not only did Arnold's action in this case provide us with the planes we so urgently needed, but it indelibly impressed upon all his staff that MED requests were to be granted without argument. This stood us in good stead until a few months before our actual operations against the Japanese began, when Parsons reported that he did not think our planes were in the best operating condition, and said they should be replaced by new ones. Investigation showed that he was right, but again Wilson felt it would be impossible to do much about it because of the tremendous demands for B-29's in the Pacific Theater.

When I appealed to Arnold, however, his response again was quick and emphatic. He said that in view of the vast national effort that had gone into the Manhattan Project, no slip-up on the part of the Air Force was going to be responsible for a failure. He then asked me how many new planes I needed, and I replied that as a minimum I would need one to carry the bomb. While several more to carry instruments and make observations would be desirable, they were not essential. I made it clear, however, that no matter what else might suffer, we must have one plane that would be in absolutely perfect working condition when we were ready for the final take-off. Immediately Arnold said that he would order fourteen new planes for us, and fourteen more to be placed in reserve to meet emergency needs. He repeated that no matter what else might go wrong, no one would ever be able to say that the Air Force did not do its utmost to support the Manhattan Project. In this, he was entirely correct.

The new lot of B-29's was delivered to Wendover Field, Utah, during the spring of 1945. These planes had fuel injection engines, electrically controlled reversible propellers, and were generally much better than their predecessors, particularly from the standpoint of ruggedness.

General Arnold and I had agreed at our first meeting that for the job at hand we should create an Air Force unit that would be self-sustaining; for that reason, Arnold decided to organize it as a composite group, built around a heavy bombardment squadron.

Our first problem was the selection of its commanding officer. In our discussions it was understood from the start that the responsibility for organizing, manning, equipping and training the unit would be Arnold's; nevertheless, this in no way relieved me of my responsibility to be certain that the group was properly prepared for its mission. Arnold, of course, was as vitally concerned as I was with obtaining for it people of the highest quality, and he could make the selection better than I. I never had the slightest doubt that he would choose as commanding officer the best person available, and in Colonel Paul W. Tibbets he did. Tibbets had been the Operations Officer of the 97th Bombardment Group in the North African and European Theater of Operations, where he had flown the usual number of combat missions, and had then been returned to the United States. Since his return, he had been engaged in testing the B-29 and in formulating the instructions for its use in combat. He was a superb pilot of heavy planes, with years of military flying experience, and was probably as familiar with the B-29 as anyone in the service.

In selecting the other officers of the group a serious mistake was made—one for which I have since regretted my own lack of foresight.

Insofar as possible the group's officers should have been men who might reasonably be expected to remain in the regular service after the war. We should have recognized the importance of this but, as far as I know, nobody did. Although this mistake made no difference in the accomplishment of our immediate goal, in the postwar years it has been most unfortunate that we have not had in the regular service as many men as possible who were experienced in the use of atomic bombs in actual war. Indeed, sixteen years after Hiroshima and Nagasaki, only four—Tibbets and Ashworth, a weaponeer, and the two bombardiers, Ferebee and Beahan—remain on active duty.

In September, 1944, the 393rd Heavy Bombardment Squadron was detached from the 504th Bombardment Group to form the nucleus of the new unit—the 509th Composite Group.[3] The 393rd was picked because of the fine reputation it had gained during its training. The squadron was about to go overseas with its parent organization when the new orders were received, and the unit soon found itself at Wendover Field, Utah, on the Nevada border, about a hundred miles from Salt Lake City.

We chose Wendover as the home station for a number of reasons. It was close enough to Los Alamos by air for proper liaison. It was only a short distance by air from the Salton Sea area, where many of our ballistic tests were to be conducted with dummy bombs. It was located in a sparsely populated area, well removed from centers of population, which simplified security problems. The existing facilities at Wendover were sufficient for our needs, so no new construction would be required. Finally, we could take it over immediately without unduly upsetting any other Air Force operations.

The total authorized strength of the 509th Group was set at 225 officers and 1,542 enlisted men. It was a completely self-contained unit, including besides its Heavy Bombardment Squadron, a Troop Carrier Squadron and all other essential supporting units.

Later, the First Ordnance Squadron Special, Aviation, was activated. This unit consisted of skilled welders, machinists and explosives

[3] The 509th Composite Group was formally activated on December 17, 1944, at Wendover Field. It was assigned to the 315th Bombardment Wing of the Second Air Force. Headquarters of the latter was at Colorado Springs, Colorado.

workers. They had been hand-picked with considerable care from the entire Army Air Force, with a few specialists coming from the ground forces.

In September, after it became clear that we would use both a gun-assembly bomb (the Little Boy) and an implosion bomb (the Fat Man),[4] we decided to freeze the external shapes of the three models then existent—one Little Boy and two Fat Men. This was to permit completion of the necessary modifications to the B-29 so that the training of the 509th would not be delayed. The first planes were ready in October and were delivered to Colonel Tibbets at the Wendover Army Air Base, which went under the code name of "Kingman" and sometimes "W-47." The ballistic tests were begun that same month and were continued until August 8, 1945, by which time each of the two models finally adopted had been dropped in combat.

Throughout its operations at Wendover the 509th was assisted by a special base unit. This included a reserve pool of ordnance personnel and flight test specialists to help in conducting the extensive tests of the components of the bombs and of the full-sized dummy bombs.

In December, the 393rd Squadron was sent to Batista Field, Cuba, for two months of special training in long over-water flights. The program also included training in high-altitude visual and radar bombing. On these practice missions, formation flights were habitually avoided and the crews grew accustomed to operating singly. This was not because we anticipated sending unescorted single planes against Japan, but rather because we were not sure that the escort planes would accompany the bomb-carrying plane all the way from its take-off to the target and back; therefore, we wanted it to be fully capable of independent navigation.

This special training proved very fortunate indeed when later General Curtis LeMay adopted the plan of using a single plane for each bombing mission. This placed all navigational responsibility on

---

[4] The Little Boy was known variously as the Thin Man, the Skinny One and a number of similar names; the Fat Man was also called the Big Boy, the Round Man, the Big Fellow and the like.

the navigator of that plane and was completely at variance with the standard Air Force navigational procedures. Normally, bomber missions were flown by massive formations, with the lead plane carrying a thoroughly competent navigator. The navigators in the other planes were not as a rule nearly so experienced. Fortunately, because of our insistence upon a high state of navigational training throughout the 393rd Squadron, the 509th Group was not caught short when LeMay's proposal was put into effect.

After the 393rd Bombing Squadron returned to Wendover from the Caribbean, its training continued, and the fliers gained much valuable experience in the course of the ballistic testing of dummy bombs similar in dimensions and weight to the atomic bombs that were eventually used. At first the dummies were inert; later some were filled with normal high explosives. They were never, of course, loaded with any fissionable material. Most of our ballistic testing was conducted at a range in the Salton Sea area. Out of these tests came the information we needed to aim the final bombs accurately.

At the same time, under the supervision of Commander F. L. Ashworth, Parsons' assistant at Los Alamos, a long series of tests of the three bomb models was being conducted in order to obtain ballistic data and to determine the best procedures for dropping the bomb. The tests also provided valuable experience in designing and assembling some of the subunits of the weapons.

The Ballistics Group of the Los Alamos Ordnance Division did the research on the problem of aircraft safety in delivery. This group was concerned with such matters as the shock pressure that the B-29 could safely withstand, the flight maneuver that would carry the plane the greatest distance away from the burst in the least time, and the use of special shock-bracing for the crew.

Throughout the fall and winter of 1944-45, the Delivery Group at Los Alamos, which later would bear the primary responsibility for developing the facilities and equipment for assembling the atomic bomb at the overseas base, continued its program of design and production of mock bombs. During this period the final Fat Man design was adopted, and we were able to discontinue our work on the other Fat Man model. Numerous tests were run on bombs that were generally complete, except for the fissionable material.

As our time schedule became tighter, component designs were necessarily altered as little as possible, though many changes were necessary because of earlier wrong guesses. Emphasis during this period was on correcting faults as they became apparent in the tests. As always, the Little Boy was far ahead of the Fat Man from the point of view of design and development, since the group working on the Little Boy had known almost since the beginning the course to be followed.

A new division, called Alberta, was set up at Los Alamos in March of 1945 to take charge of everything involved in the preparation and delivery of the bomb. The members of Alberta's Weapons Committee became concerned with the need to start work on a new design for the Fat Man that would be based on our most recent knowledge. They realized, however, that this program could not be allowed to interfere with their main task, which was to get the existing model ready as quickly as possible. The job of redesigning the Fat Man, from a sound engineering point of view, was barely organized by the end of the war. Today a great deal of attention would doubtless be given to this problem and possibly many would argue that the improvements should be completed before the bomb was actually used in combat. I overruled any such idea and insisted that there be no delay in the use of the existing models, which were satisfactory, though they did not have the explosive power that the redesigned Fat Man would have had.

CHAPTER 19

CHOOSING THE TARGET

Throughout the period when we were planning our atomic bombing operations against Japan, American strategy was based upon the assumption that an invasion of the Japanese homeland was essential to ending the war in the Pacific.

Under the strategic concept approved by the Joint Chiefs in July of 1944, Kyushu was to be invaded on October 1, 1945, with the final assault on Tokyo following in December of that year. This plan was not dependent in any way upon Russian co-operation.

After the Yalta Conference, however, a debate sprang up over whether it would not be better to encircle Japan and defeat her by attrition than to defeat her by direct attack. Both General MacArthur and Admiral Nimitz, when asked for their opinion, voted for a direct assault. Supported by these highly respected judgments, the Joint Staff reiterated its previous preference for a strategy of invasion, reporting in April, 1945, that:

In view of all factors, we should follow the strategy of early invasion and our course should be:

a. Apply full and unremitting pressure against Japan by strategic bombing and carrier raids in order to reduce war-making capacity and to demoralize the country, in preparation for invasion.

b. Tighten blockade by means of air and sea patrols, and of air striking force and light naval forces to include blocking passages between Korea and Kyushu and routes through the Yellow Sea.

c. Conduct only such contributory operations as are essential to establish the conditions prerequisite to invasion.

d. Invade Japan at the earliest practicable date.

e. Occupy such areas in the industrial complex of Japan as are neces-

sary to bring about unconditional surrender and to establish absolute
military control.

It was estimated that a force of 36 divisions—1,532,000 men in
all—would be required for the final assault, and it was recognized
that casualties would be heavy.

The Joint Chiefs of Staff approved the concept and on May 25 the
directive for the Kyushu invasion was issued to General MacArthur,
Admiral Nimitz and General Arnold. The target date for the invasion
of Kyushu was now November 1, 1945.

The record of a meeting between President Truman and the Joint
Chiefs on June 18, just before the American delegation left for Potsdam, reads:

THE PRESIDENT said he considered the Kyushu plan all right from
the military standpoint and, so far as he was concerned, the Joint Chiefs
of Staff could go ahead with it; that we can do this operation and then
decide as to the final action later.

In such a climate, no one who held a position of responsibility in the
Manhattan Project could doubt that we were trying to perfect a
weapon that, however repugnant it might be to us as human beings,
could nonetheless save untold numbers of American lives.

However, one very important question arose: should the United
States delay any contemplated military action in the expectation that
an effective atomic bomb would be produced as scheduled? To any
experienced soldier it was obvious that, once an advantage had been
gained over an enemy as dangerous as Japan, no respite should be
given. If the bomb had been scheduled for delivery in early November,
a few days after the scheduled date of the Kyushu invasion, I would
have advised a delay in the landing operation. I expressed this point
of view in conversations with Secretary Stimson and Harvey Bundy,
but I also told them and General Marshall that I would consider it a
serious mistake to postpone any feasible military operation in the
expectation that the bomb would be ready as a substitute at some
later date.

There has been much discussion since the war about the decision to use the atomic bomb against Japan. Decisions of this nature must always be made by only one man, and, in this case, the burden fell upon President Truman. Under the terms of the Quebec Agreement, the concurrence of Prime Minister Churchill was necessary; nevertheless, the initial decision and the primary responsibility were Mr. Truman's. As far as I was concerned, his decision was one of noninterference—basically, a decision not to upset the existing plans.

When we first began to develop atomic energy, the United States was in no way committed to employ atomic weapons against any other power. With the activation of the Manhattan Project, however, the situation began to change. Our work was extremely costly, both in money and in its interference with the rest of the war effort. As time went on, and as we poured more and more money and effort into the project, the government became increasingly committed to the ultimate use of the bomb, and while it has often been said that we undertook development of this terrible weapon so that Hitler would not get it first, the fact remains that the original decision to make the project an all-out effort was based upon using it to end the war. As Mr. Stimson succinctly put it, the Manhattan Project existed "to bring the war to a successful end more quickly than otherwise would be the case and thus to save American lives."

Certainly, there was no question in my mind, or, as far as I was ever aware, in the mind of either President Roosevelt or President Truman or any other responsible person, but that we were developing a weapon to be employed against the enemies of the United States. The first serious mention of the possibility that the atomic bomb might not be used came after V-E Day, when Under Secretary of War Patterson asked me whether the surrender in Europe might not alter our plans for dropping the bomb on Japan.

I said that I could see no reason why the decision taken by President Roosevelt when he approved the tremendous effort involved in the Manhattan Project should be changed for that reason, since the surrender of Germany had in no way lessened Japan's activities against the United States. A little later some of the scientists began to express doubts about the desirability of using the bomb against Japan. A number of these men had come to the United States to escape racial persecution under the Hitler regime. To them, Hitler was the supreme enemy and, once he had been destroyed, they apparently found themselves unable to generate the same degree of enthusiasm for destroying Japan's military power.

At this same time a debate arose about how the bomb should be employed. Should we conduct a demonstration of its power for all the world to see, and then deliver an ultimatum to Japan, or should we use it without warning? It was always difficult for me to understand how anyone could ignore the importance of the effect on the Japanese people and their government of the overwhelming surprise of the bomb. To achieve surprise was one of the reasons we had tried so hard to maintain our security.

President Truman knew of these diverse and conflicting opinions. He must have engaged in some real soul-searching before reaching his final decision. In my opinion, his resolve to continue with the original plan will always stand as an act of unsurpassed courage and wisdom—courage because, for the first time in the history of the United States, the President personally determined the course of a major military strategical and tactical operation for which he could be considered directly responsible; and wisdom because history, if any thought is given to the value of American lives, has conclusively proven that his decision was correct.

At about this time, the spring of 1945, another job was dropped into our laps at the MED. The first inkling I had of this added responsibility came in the course of a conversation with General Marshall. We had been discussing the progress of the work and, having mentioned our anticipated readiness date, I suggested that the time was fast approaching when we should begin to make plans for the bombing operation itself, even though we still had no assurance that the bomb would be effective. I asked him to designate some officer in the Operations Planning Division (OPD) of the General Staff with whom I could get in touch so that planning could be started. After a moment's hesitation, General Marshall replied: "I don't like to bring too many people into this matter. Is there any reason why you can't take this over and do it yourself?" My "No, sir, I will" concluded the conversation, which constituted the only directive that I ever received or needed.

General Marshall's position on this matter came as a complete surprise to me. I could easily understand, in fact I favored, restricting the knowledge of our work to the smallest possible number of people. I realized too that he might be questioning whether it was wise to bring into the operational planning officers who might not be able to understand the technical problems involved. But I had never imagined that he would want to keep the execution phases of our project entirely apart from the OPD.

I immediately informed Arnold of this development, and together we went over the general problems facing us. Our most pressing job was to select the bomb targets. This would be my responsibility.

For many months before this, target criteria had been discussed over and over again with and among the members of the Military Policy Committee. They were finally established only after thorough discussions with Oppenheimer and his senior advisers at Los Alamos, and particularly with Dr. John von Neumann. Before approving them, I went over them with my deputy, General Farrell, and then with Brigadier General Lauris Norstad,[1] to be sure that nothing had been overlooked.

I had set as the governing factor that the targets chosen should be places the bombing of which would most adversely affect the will of the Japanese people to continue the war. Beyond that, they should be military in nature, consisting either of important headquarters or troop concentrations, or centers of production of military equipment and supplies. To enable us to assess accurately the effects of the bomb, the targets should not have been previously damaged by air raids. It was also desirable that the first target be of such size that the damage would be confined within it, so that we could more definitely determine the power of the bomb.

[1] Chief of Staff, Army Strategic Air Force.

In April, Farrell, Parsons and Major J. H. Derry of my office met in Washington and later at Los Alamos with three senior men[2] from the Operations Analysis Group of Arnold's office, so that the Air Force representatives could become familiar with technical aspects of the bomb. Norstad directed his people to find out all they could about the performance and capabilities of the specially modified B-29; to establish the cruising data for the plane; and to determine the best way to fly the loaded plane at various altitudes.

The next step was to set up a special committee to recommend specific targets. Three of the members came from Arnold's office: Colonel William P. Fisher, Dr. J. C. Stearns and D. M. Dennison; the others were from the MED: Farrell, von Neumann, R. B. Wilson and William G. Penney, a member of the British team at Los Alamos. This group met for the first time on May 2 in Washington. I opened the meeting by pointing out the importance of the committee's task, the need for the highest degree of secrecy, and the number of targets I thought we should have (four, initially). I emphasized General Marshall's opinion that the ports on the west coast of Japan should not be ignored as possible targets, since they were vital to the Japanese communications with the Asiatic mainland.

General Norstad told them that the facilities of the 20th Air Force would be made available to any degree necessary, through Stearns and Fisher, to provide related data, operational analyses, maps and any other data or information that might be needed.

Norstad and I then departed, leaving Farrell in direct charge of the work from that point forward. I was kept constantly informed on the committee's progress by Farrell and the others from the MED, particularly von Neumann, with whom I frequently discussed the many scientific and technical problems with which we were faced.

Certain fundamental guidelines were established at this first meeting, among them the probable maximum range of the loaded B-29; the need for visual bombing; the general weather conditions desired over the targets; the expected blast and damage; and the need to have three targets available for each attack.

[2] F. R. Collbohm, W. T. Dickinson and D. M. Dennison.

One of the most difficult problems was attempting to estimate, or even guess, the probable explosive force of the weapon. An accurate guess was of vital importance because the closer we came to being right, the more effective the bomb would be. The optimum height of burst was entirely governed by the explosive force. If the altitude of burst we used was below or too high above this optimum, the area of effective damage would be reduced; and it was possible, if it was much too high, that all we would produce would be a spectacular pyrotechnical display which would do virtually no damage at all. We calculated that if the bomb was detonated at 40 per cent below optimum altitude or 14 per cent above, there would be a reduction of 25 per cent in the area of the severe damage.

At first, considerations of possible fallout and of direct radiation definitely favored a burst at a maximum altitude. I had always insisted that casualties resulting from direct radiation and fallout be held to a minimum. After the Alamogordo test, when it became apparent that the burst could be many hundreds of feet above the ground, I became less concerned about radioactive fallout from too low a burst.

For the present, however, since we did not know the size of the explosion, our plans had to be based on conservative detonation heights. For the Little Boy, it was estimated that the explosive force would have a TNT equivalent ranging from 5,000 to 15,000 tons. For this weapon, the corresponding desirable height of detonation would vary then from 1,550 feet to 2,400 feet. For the Fat Man, it was thought that the magnitude of the explosion would range from 700 to 5,000 tons. This would require detonating heights between 700 feet and 1,500 feet. We could only hope that we could tighten up these estimates considerably, particularly for the Fat Man, after testing the implosion bomb. This element of doubt meant that we had to have fuses for four different height settings. By then most of us in the project were thoroughly inured to such uncertainties. Indeed, three days later, on May 14, Oppenheimer informed me that he and von Neumann had concluded after a thorough discussion that the probable explosive power of the Fat Man was still uncertain, and that the views of the Target Committee should be amended accordingly. They estimated that the maximum altitude for which we should be prepared to set fuses was about twice the minimum, and even for the Little Boy the minimum and probable altitude was only two-thirds of the maximum.

It was agreed that visual bombing was so important from the standpoint of hitting the target that we should be prepared to await good weather. This was based on an estimate that there was only a 2 per cent chance that we would have to wait over two weeks.

The committee recommended that we have spotter aircraft over each of the three alternative targets so that the final target could be selected in the last hour of flight. In case the delivery plane should reach the target and find visual bombing to be impossible, they thought it should return to its base with the bomb. The drop should be made with radar only if the plane could not otherwise return. Radar and navigational developments should be followed closely so that these conclusions could be altered, if desirable.

We all recognized that the plan to use visual bombing, with the possible long delays entailed, required that the bomb be so designed that it could be held for at least three weeks in a state of readiness that would permit its being dispatched on twelve hours' notice. This was not considered too great a problem at the time, and later we could ignore it as it became obvious that we would normally have at least forty-eight hours' notice of possible suitable weather.

It was generally agreed that if a plane in good condition had to return to base with the bomb, it would probably be able to make a normal landing. Frequent practice landings had been made with dummy bombs, some of them filled with high explosives. The committee advised that special training in landing with dummy units should be given to all plane crews who were to carry the bomb.

If the bomb had to be jettisoned, extreme care would have to be exercised. Under no circumstances should it be jettisoned near American-held territory. Prior to actual take-off, definite instructions would have to be furnished the weaponeer to guide him in case of trouble. Careful calculations and the experience gained by the Air Force in England from missions involving bombs as large as two thousand pounds indicated that there was no reason to fear for the safety of the bombing plane if its flight were properly controlled. Discussion of the possible radiological effects did indicate, however, that it would be unwise for any aircraft to be closer than two and a half miles to the burst. To provide protection against blast effects, a distance of five miles was advisable. Also, no plane should be permitted to fly through the radioactive clouds.

At its third meeting, the Target Committee was informed that General Arnold and I had concluded that control over the use of the weapon should reside, for the present, in Washington. This announcement was necessary because some of the Air Force people on the committee had displayed a total lack of comprehension of what was involved. They had assumed that the atomic bomb would be handled like any other new weapon; that when it was ready for combat use it would be turned over to the commander in the field, and though he might be given a list of recommended targets, he would have complete freedom of action in every respect.

I felt, and so did General Arnold, that this was too complicated and all-important a matter to be treated so casually; and, regardless of our own feelings, we doubted whether either Mr. Stimson or General Marshall would ever approve, though until now there had been no discussion of this vital point. I had always assumed that operations in the field would be closely controlled from Washington, probably by General Marshall himself, with Mr. Stimson fully aware of and approving the plans. Naturally I expected that the President also would share in the control, not so much by making original decisions as by approving or disapproving the plans made by the War Department. It was quite evident by now, however, that the operation would not be formally considered and acted upon by either the Joint Chiefs of Staff or the Combined Chiefs. One of the reasons for this was the need to maintain complete security. Equally important, though, was Admiral Leahy's disbelief in the weapon and its hoped-for effectiveness; this would have made action by the Joint Chiefs quite difficult.

When I had visited him in his office six or seven months before to show him a status report on the project, he told me of his long experience with explosives in the Navy and emphasized his belief that nothing extraordinary would come out of our work. He reminded me that no weapon developed during a war had ever been decisive in that war. Then he went on to say he was sorry that I was involved in the project as it would have been much better for me to have had a different and more usual assignment. His views were undoubtedly colored by his feeling, expressed as early as July, 1944, that Japan's early surrender was inevitable, and would be brought about by combined Naval and Air Force action. In any case, as he frankly admitted after the war, he never had any confidence in the practicability of the atomic bomb. I want to make it clear, however, that he and the other Joint Chiefs, as well as Field Marshal Wilson of the Combined Chiefs, knew of our general plans before the operation began, even though not informed officially of all the details.

Reinforcing my impression that control would not be passed to the field was General Arnold's strong desire to retain personal control over the Strategic Air Force, keeping Norstad, its Chief of Staff, in Washington where he could closely supervise his activities. Not until Spaatz was sent over to Guam just prior to the bombing did this situation change appreciably. As far as I know, however, there was never any doubt in the minds of those most concerned that control of the actual bombing would be retained in Washington.

The cities that the Target Committee finally selected, and which I approved without exception, were:

1. Kokura Arsenal, one of the largest munitions plants in Japan, which was engaged in the manufacture of a wide variety of weapons and other defense materials. The arsenal covered an area of about four thousand by two thousand feet and was contiguous to railway yards, machine shops and electric power plants.

2. Hiroshima, a major port of embarkation for the Japanese Army and a convoy assembly point for their Navy. The city, in which the local Army headquarters, with some twenty-five thousand troops, was situated, was mainly concentrated on four islands. The railway yards, Army storage depots and port of embarkation lay along the eastern

side of the city. A number of heavy industrial facilities were adjacent to the main metropolitan area.

3. Niigata, a port of growing importance on the Sea of Japan. It contained an aluminum reduction plant, and a very large ironworks, together with an important oil refinery and a tanker terminal.

4. Kyoto, an urban industrial area with a population of about one million inhabitants. It was the former capital of Japan, and many displaced persons and industries were moving into it as other areas were destroyed. Also, it was large enough to ensure that the damage from the bomb would run out within the city, which would give us a firm understanding of its destructive power.

With these selections in hand, I prepared a plan of operations for General Marshall, recommending his approval. This report was in my office when I went to see Secretary Stimson about another matter. In the course of our conversation, he asked me whether I had selected the targets yet. I told him that I had and that my report was ready for submission to General Marshall. I added that I hoped to see the General the next morning.

Mr. Stimson was not satisfied with this reply and said he wanted to see my report. I said that I would rather not show it to him without having first discussed it with General Marshall, since this was a military operational matter. He replied, "This is a question I am settling myself. Marshall is not making that decision." Then he told me to have the report brought over. I demurred, on the grounds that it would take some time. He said that he had all morning and that I should use his phone to get it over right away.

While we were waiting, he asked me about the targets. When I went over the list for him, he immediately objected to Kyoto and said he would not approve it. When I suggested that he might change his mind after he had read the description of Kyoto and our reasons for considering it to be a desirable target, he replied that he was sure that he would not.

The reason for his objection was that Kyoto was the ancient capital of Japan, a historical city, and one that was of great religious significance to the Japanese. He had visited it when he was Governor

General of the Philippines and had been very much impressed by its ancient culture.

I pointed out that it had a population of over a million; that any city of that size in Japan must be involved in a tremendous amount of war work even if there were but few large factories; and that the Japanese economy was to a great extent dependent on small shops, which in time of war turned out tremendous quantities of military items. To reinforce my argument, I read from the description of Kyoto, included in my report, which had now arrived. I pointed out also that Kyoto included 26,446,000 square feet of plant area that had been identified and 19,496,000 square feet of plant area as yet unidentified. The city's peacetime industries had all been converted to war purposes and were producing, among other items, machine tools, precision ordnance and aircraft parts, radio fire control and gun direction equipment. The industrial district occupied an area of one by three miles in the total built-up area of two and one-half by four miles.

Mr. Stimson was not satisfied, and without further ado walked over to the door of General Marshall's office and asked him to come in. Without telling him how he had got the report from me, the Secretary said that he disagreed with my recommendation of Kyoto as a target, and explained why. General Marshall read the target description of each of the four cities, but he did not express too positive an opinion, though he did not disagree with Mr. Stimson. It was my impression that he believed it did not make too much difference either way.

General Marshall never knew how he came to be caught unawares in this matter. He never indicated any displeasure about it to me and I doubt whether he gave it any thought. If he did, he probably guessed what had happened. Personally, I was very ill at ease about it and quite annoyed at the possibility that he might think that I was short-cutting him on what was definitely a subject for his consideration.

After some discussion, during which it was impossible for me discreetly to let General Marshall know how I had been trapped into by-passing him, the Secretary said that he stuck by his decision. In the course of our conversation he gradually developed the view that the decision should be governed by the historical position that the United States would occupy after the war. He felt very strongly that anything that would tend in any way to damage this position would be unfortunate.

On the other hand, I particularly wanted Kyoto as a target because, as I have said, it was large enough in area for us to gain complete knowledge of the effects of an atomic bomb. Hiroshima was not nearly so satisfactory in this respect. I also felt quite strongly, as had all the members of the Target Committee, that Kyoto was one of the most important military targets in Japan. Consequently, I continued on a number of occasions afterward to urge its inclusion, but Mr. Stimson was adamant. Even after he arrived in Potsdam, Harrison sent him a cable saying that I still felt it should be used as a target. The return cable stated that he still disapproved, and the next day he followed it with another which said that he had discussed the matter with President Truman, who concurred in his decision. There was no further talk about Kyoto after that.

Nothing is more illustrative of the relationship between Secretary Stimson and me than this episode. Never once did he express the slightest displeasure or annoyance over my repeated recommendations that Kyoto be returned to the list of targets. Nor did I ever feel that he wanted me to remain silent, once I had learned his views, on a matter of such great importance. I believe the affair was also typical of his attitude toward other senior officers.

Events have certainly borne out the wisdom of Mr. Stimson's decision. I think, however, he did not foresee that much of the criticism he so scrupulously sought to avoid would come from American citizens; certainly he never mentioned this possibility to me. After the sudden ending of the war I was very glad that I had been overruled and that, through Mr. Stimson's wisdom, the number of Japanese casualties had been greatly reduced.

When our target cities were first selected, an order was sent to the Army Air Force in Guam not to bomb them without special authority from the War Department. About six weeks or so after Mr. Stimson refused to approve Kyoto, I suddenly realized there was a danger that the Air Force might remove it from the list of proscribed cities. I spoke to Arnold, who promptly saw to it that it remained on the reserved list and that the Air Force Command on Okinawa was also notified. If we had not recommended Kyoto as an atomic target, it would not of course have been reserved and would most likely have been seriously damaged, if not destroyed, before the war ended.

CHAPTER 20

TINIAN

As our schedules became more definite in the latter part of 1944, we began to plan for the overseas operating base from which the B-29's would take off for Japan. This meant that we had to inform the theater commanders and arrange for their assistance.

This need became more urgent when I began to receive reports that Admiral Nimitz was making inquiries about what the 509th Group was going to do in his theater. He had learned, somehow, that this unit would be operating in the Central Pacific, although he had never been told of its purpose. In order to avert what might develop into an embarrassing situation, I arranged, through Admiral Purnell, to have Commander Ashworth go to Guam to tell Nimitz about our plans and make certain of essential Navy co-operation. He went armed with a letter from Admiral King, stressing the importance of Ashworth's message, and stating that only two officers were to be told of the purpose of our operations.

When Ashworth reported to Nimitz' headquarters, he first saw Vice Admiral Charles H. McMorris, who insisted on receiving the message himself, saying that he was the alter ego of Nimitz and could take care of everything. Ashworth replied firmly that his message was for Nimitz and was to be delivered to him alone.

After hearing what Ashworth had to say, Nimitz picked McMorris as the other officer to be informed. Later, at Nimitz' request, and with Admiral King's consent, Captain Tom B. Hill, his gunnery officer, was brought into the picture and served as the immediate liaison officer with us throughout the operation.

Shortly before Ashworth left to see Nimitz, I had obtained General

Marshall's approval to tell Lieutenant General M. F. Harmon, the Deputy Commander of the 20th Air Force in the Pacific, and two of his staff about our plans, for by this time we felt certain that our bombing operations would be conducted from his area. Arnold arranged for this to be taken care of by General Norstad, who visited the Pacific in January of 1945. While he was there he also saw Brigadier General H. S. Hansell, Jr., Commanding General of the 21st Bomber Command, to which the 509th Group was to be assigned, but told him only that the 509th was a special organization. Unfortunately, soon after this, Harmon and the two staff officers disappeared in flight en route from Guam to Washington. Lieutenant General Barney McK. Giles took over the area command on May 4, and I was able to brief him thoroughly before he left Washington.

Norstad had returned from his trip to the Pacific convinced that Guam should be the 509th's base of operations, primarily because of its fine deepwater harbor, with the additional attraction of its excellent maintenance facilities. Discussions with Ashworth and Tibbets, however, led me to believe that the one hundred miles of flying distance to the target that could be saved if we based on Tinian might well be an overriding consideration. The flight distance was extremely critical, because the plane carrying the bomb would be loaded to capacity and, in case of engine failure, might find itself in serious trouble.

Tibbets' experience in the Caribbean training had indicated that we could expect to lose approximately five out of every forty engines in a flight comparable to the one we would have to make to and from the target. We decided with Norstad's concurrence to use Tinian, regardless of the attractions of Guam; to provide for an intermediate emergency landing point, possibly on the Bonin or Volcano islands; and to make arrangements with the Navy to recover survivors and equipment in the event of a landing at sea.

The reasons for choosing Tinian instead of Guam were contained in Ashworth's report, who in February, 1945, after seeing Nimitz, had investigated both sites:

1. The construction of necessary facilities at Guam could not be completed before August, 1945.

2. Incoming shipments at Guam already far exceeded its port capacity.

3. Tokyo is approximately 1,450 air miles from Tinian—about one hundred miles less than it is from Guam.

4. An airfield was already in existence at Tinian which, with certain modifications, would be suitable for the 509th Group.

5. Adequate harbor facilities would be available at Tinian on about March 15, 1945.

6. Although Tinian was under the jurisdiction of the Army, the 6th Seabee Brigade was stationed there, and would be available for such work as the 509th Group might require.

Work on Tinian began at the end of February under the general supervision of the Navy. Before the first month had passed, it became apparent that we needed a personal representative on the island to ensure that there would be no delay in the preparations. For this I selected Colonel E. E. Kirkpatrick, who had just been assigned to my office. I had known him well when I had had charge of construction operations, and I had complete confidence in his ability to carry out smoothly any mission that involved dealing with other services.

In this instance his job was to see that all physical installations would be ready when they were needed. In some ways his assignment was an odd one in that it could be construed to indicate a lack of confidence in the support we would get from the authorities in the area. Actually, it was only another example of the extreme care we took to lessen the chances of failure. Nimitz told me after the war that it was the first time he had ever known the Army to send an ambassador to the Navy, adding that that was what he always considered Kirkpatrick to be. That thought had occurred to me, too, when I selected him. His two brothers were both graduates of the Naval Academy, a circumstance I felt might prove to be helpful. Actually it was not so great a factor in Kirkpatrick's success as his ability to get along with people and at the same time to achieve his objectives.

I am sure that, but for Kirkpatrick's presence, we would not have been ready, even with a wild last-minute rush, and that the delivery of the bomb either would have had to be postponed or made under much less favorable circumstances. A major source of delay would have been the congestion of Tinian's unloading facilities. Another would have been the local authorities' human tendency to take care of first things first, rather than to prepare for an uncertainty far in the future.

Before leaving for Tinian, Kirkpatrick had been instructed, as was usual, to prepare a code using names of states and cities to designate persons and things that would be repeatedly mentioned in messages. They would also be encoded in the usual secret code afterward, but this precaution was taken to protect security as the messages went through communication channels.

When his first message arrived in Washington it created considerable commotion in the message center for the code words were also the names of a number of important naval vessels known to be in the Pacific. Wild rumors spread rapidly. I immediately cabled him: "Destroy code. Send no messages. New code coming by courier."

All Kirkpatrick's messages came to me through Guam, except for his handwritten reports which he mailed directly as informal letters. Even in these, because of possible censorship, he had to be cautious. From their tone, however, I surmised that things might not be going so well as they should. Consequently, I cabled him that unless he felt it would be detrimental to the progress of the work he should return to Washington for a short conference. He came at once and explained the basic difficulty.

Because of the shipping jam, each vessel was required to wait its turn for unloading. This could mean a delay of as much as three months in the unloading of some of our vital equipment. It was a simple matter, through Purnell, to have Admiral King cable Nimitz that all of our material must be unloaded immediately upon its arrival at Tinian. This was quite upsetting to the normal operations on the island, but it was typical of the support that we unfailingly received from Admiral King, Admiral Purnell and the entire Navy at all times.

The very size of the military operations in the Guam area and the secrecy that shrouded our project made it impossible for the importance of our work to percolate down to all the various commanders and staffs with whom we became involved from time to time. However, as anyone familiar with military affairs will realize, orders signed Stimson or Marshall or King or Nimitz carried unprecedented weight during the war years. An example of that basic fact of military life was what happened when on one occasion Kirkpatrick needed more cement than he had. There was none on Tinian, but he had learned of some on another island. The naval officer who controlled it refused, as was quite natural, to turn it over to him unless he explained why he needed it. This, of course, Kirkpatrick could not do, but he saw Captain Hill, who sent a peremptory order over Nimitz' signature to the naval officer to turn over all the cement Kirkpatrick wanted at once and to ask no questions about need or purpose for there was no intention of telling him.

At the time we chose Tinian as our base, both Iwo Jima and Okinawa were still held by the Japanese. Neither island had been considered suitable for our purposes, in any event, for they were too close to the Japanese homeland for the safety of our assembly operations. In the spring of 1945, however, we decided to have an emergency bomb-handling facility installed on Iwo Jima, so that if the plane carrying the bomb had aircraft trouble, the bomb could be shifted to another plane. On April 6, the necessary instructions were issued by Admiral Nimitz' headquarters, with highest priority assigned to the work.

Kirkpatrick visited Iwo Jima in the middle of April and showed a copy of the order to the local authorities. It was agreed then that the job was to be completed by July 1. But when an officer courier delivered some MED equipment there on July 1, he found and reported that little work had been done and that the island engineer was now planning on completion on July 15. Kirkpatrick investigated promptly and found conditions even worse than reported. This time he was assured that August 1 would be the completion date. He thereupon took the matter up with Captain Hill and the situation was corrected just in time. This was another instance where having a responsible officer in the area checking on every detail paid off, for if Kirkpatrick had not been there, I doubt if the handling facilities at Iwo Jima would be ready yet.

I had begun to think about the make-up of the field crews for Tinian as early as June of 1944. These would be the men who would carry out the last-stage experiments and tests and be responsible for the final assembly of the bomb. At the time, I felt it would be wise to delay their selection, since the skills and experience needed might change with the design of the bomb, and there seemed to be so many people anxious to volunteer. Consequently, the thirty-seven-man team of twelve civilians, seventeen Army enlisted men, seven Naval officers and one Army officer was not chosen until early May, 1945.[1]

This group of specialists from Los Alamos formed part of what was known as the First Technical Service Detachment. Each civilian was required to wear a uniform and received an assimilated Army rank. Two of the scientists were commissioned as reserve officers in recognition of the duties and responsibilities they would have in connection with the bomb assembly. After they had been appointed, we found that this was not a happy solution for one of them so we placed him in an inactive status and sent him over as a civilian.

The job of the Los Alamos group on Tinian was to provide and test bomb components and also to supervise and inspect the actual assembly of the bombs. They were to inspect the bomb prior to take-off, test the completed unit, and co-ordinate the various project activities on the island. They were not responsible for providing any advice concerning the use of the weapon.

There was considerable furor over the exchange of information between Tinian and Los Alamos, but, though there were difficulties at each end, the loudest complaints came from Los Alamos, where the people did not receive all the information that they thought they would like to have. The reason for this was simply that very little information was being sent. Direct communications between Los Alamos and

[1] The key positions were manned as follows: In charge—Parsons; Scientific and Technical Deputy—Ramsey; Operations Officer and military alternate to Parsons—Ashworth; Fat Man Assembly Team—Warner; Little Boy Assembly Team—Birch; Fusing Team—Coll; Electrical Detonator Team—Lieutenant Commander Stevenson (a reserve officer and a scientist); Pit Team—Morrison and Baker; Observation Team—Alvarez and Waldman; Aircraft Ordnance Team—Dike. Special Consultants included Serber, Penney, and Captain Nolan, an Army medical officer.

Tinian were not permitted for security reasons and what messages did pass had to be relayed through Washington, which required the use of a fairly elaborate table of codes which had been prepared by Alberta. To get around this difficulty and particularly to assure the Los Alamos scientific people that the information to which they were entitled was not being held back, we arranged for Dr. J. H. Manley to come in from the laboratory to Washington. He remained there to pass on information as it was received from Tinian and Los Alamos. In the process he had to rewrite all messages to eliminate any chance of our codes being broken.

The 509th Group had begun to stage out of the United States at the end of April. Approximately eight hundred of its members sailed from Seattle in the early part of May. The advance air echelon reached Tinian around the twentieth, and by the middle of July, all elements of the group were present on the island. The shipment of their equipment progressed smoothly; and the men quickly settled down to the routine of life in the field and began the last phase of their training.

Prior to shipment overseas, all Army Air units were customarily inspected by an outside team of officers to make certain that all equipment was in perfect order. For the 509th, this inspection was well along toward completion when the Second Air Force, which exercised administrative control over the 509th, learned of it and peremptorily ordered it discontinued with the comment, "They can't inspect the 509th because they are ignorant of the mission."

Upon its arrival in the theater, the 509th Group was placed under the direct operational control of the 21st Bomber Command (now under Major General Curtis LeMay) of the 20th Air Force. When LeMay had come to Washington in June, we had met to discuss the delivery of the bomb. General Farrell, who was devoting most of his attention to planning the actual operation, was also present.

This was my first meeting with LeMay and I was highly impressed with him. It was very evident that he was a man of outstanding ability. Our discussion lasted about an hour, and we parted with everything understood and with complete confidence in each other. This feeling lasted throughout the operation and into the years since then.

I explained to him the anticipated outcome of our work, describing the probable power of the bombs, their expected delivery dates and probable production rates, and said that we fully expected to drop each bomb as soon as it was ready. I also went into the general organization and state of training of the 509th Group; the responsibilities of the supporting groups from Los Alamos; the factors governing the altitude from which the bomb would have to be dropped, which was approximately the maximum altitude of the B-29; the approximate weights of the two types of bomb; the targets that we had selected; and the type of instructions that would be issued to the field. I made it perfectly clear that the conduct of the operation would be entirely under his control, subject, of course, to any limitations that might be placed upon him by his instructions. Finally, I explained the roles of the two weaponeers, Parsons and Ashworth—the men who would actually arm the bomb—giving him a résumé of their particular qualifications.

LeMay asked a few very pertinent questions, and then announced that he would want to carry out the bombing operation using a single unescorted plane. In explaining his reasons for preferring this radical tactic, he pointed out that the Japanese were unlikely to pay any serious attention to a single plane flying at a high altitude, and would probably assume that it was on either a reconnaissance or a weather mission. I replied that I thought his plan was sound, but that this phase of the operation came under his responsibility. I added, however, that some arrangement should be made for the necessary observation planes to be present in the general area at the time the bomb was dropped.

After their arrival on Tinian all combat crews went through the seven days of regular schooling required in the theater before combat flying could be undertaken. During the next twenty days they flew practice missions over various targets, such as Truk.

Because they had been modified to carry the atomic bomb, the B-29's of the 509th Group could not easily carry standard conventional bombs. They could, however, deliver bombs having the same shape as the Fat Man, and such a bomb had been developed and produced to provide training and experience to the crews. Known as the Pumpkin, this bomb contained 5,500 pounds of explosives, and was designed for blast effect only, with a proximity fuse that would permit its use for an air burst. Although it was primarily a training device, we had always recognized that it could have tactical uses; now as part of the group's security cover, we let it leak out on Tinian that its mission was the delivery of Pumpkins in battle. We also hoped that analysis of the results obtained by the use of the Pumpkins might help us to refine the ballistic data for the real bomb.

The Pumpkins began to arrive at the end of June. Reaction to these bombs were mixed. The members of the 509th who, with a few exceptions, still did not know the real reason for their training, were somewhat disappointed that they had spent so much time in practicing to deliver this fairly modest weapon. On the other hand, some members of the other Air Force units based on Tinian, who likewise did not know what the 509th's real purpose was, became quite enthusiastic about the effectiveness of the Pumpkin's air bursts over enemy targets and set up a clamor to have more of them made available to their theater.

To familiarize the plane crews with the general areas of the targets and to ensure more certain navigation and target recognition, the cities selected for the Pumpkin missions were in the general vicinities of, but outside, the atomic targets. The bombings were carried out at the same high altitudes, and daylight visual bombing was specified; however, radar could be used if visual aiming proved impossible.

The flights were planned to pass over Iwo Jima in both directions. The altitudes to be used depended on the weather, with a limitation of ten thousand feet or less going to the target, and eighteen thousand feet or more returning, so that the local defenses would not be unnecessarily alerted. All planes were required to carry a maximum load of fuel.

Every effort was taken to make these flights as much like the main job as possible. For this, wide deviations from usual bombing techniques would be necessary because of the terrific power of the explosion and the high air burst that would occur instead of the usual impact or delayed explosion.

As I have explained, a high air burst was necessary for maximum results. It was also dictated by our desire to eliminate, if possible, or in any case to decrease, residual radioactivity on the ground below the burst; to decrease to a negligible degree any harmful fallout downwind; and to diminish to a minimum serious radioactive injuries to the population in the bombed area. We felt that the high burst would confine casualties for the most part to nonradioactive injuries; namely, those due directly and indirectly to the force of the unprecedented explosion.

To be well removed from the point of burst, the bombing plane would have to maneuver as no heavy bomber had ever had to maneuver before. As soon as the bomb was "away," the plane was to make a sharp diving turn to get as far as possible from the point of explosion. This was one of the reasons why the run was made at the then un- precedented altitude of some thirty thousand feet. The high altitude also greatly reduced the danger of gunfire from enemy airplanes, per- mitting the removal of the fuselage turrets and all other armament except for the tail guns. This weight reduction appreciably increased the plane's range and the height at which it could fly.

Studies made at Los Alamos had determined that with a bomb of twenty thousand tons of TNT equivalent, a B-29 plane ten miles away from the burst would be safe from destruction by a factor of two. Under these conditions, the aircraft, which had been designed to with- stand a force of four times gravity, would be subjected to a force equivalent to no more than two times gravity. It was calculated that by making a sharp diving turn, the sharpest possible consistent with safety, the B-29 could reach a point at least ten miles from the burst by the time the bomb exploded.

In late June, as the forces under General MacArthur and Admiral Nimitz approached within bombing range of the Japanese homeland, we suddenly realized that they had not been told about the ban on certain cities, for at the time it was imposed they had been too far away to make it necessary. This concern was soon removed, however, for when we brought the matter to the attention of the Joint Chiefs, they hastily reserved our targets from all air attack.

CHAPTER 21

ALAMOGORDO

We were fairly sure by now that we would be able to test the Fat Man, the implosion-type bomb, sometime around the middle of July. (At no time was there any idea of testing the gun-type bomb.) Planning for this operation, which carried the code name of Trinity, had begun back in the spring of 1944 when Oppenheimer and I decided that a test might be necessary to make certain that the complex theories behind the implosion bomb were correct, and that it was soundly designed, engineered, manufactured and assembled—in short, that it would work.

We thought then that we might want to explode the first bomb inside a container, so that if a nuclear explosion did not take place or if it was a very small one, we might be able to recover all or much of the precious plutonium. Also, we wanted to prevent its being scattered over a wide area and creating a health hazard that would make it necessary to guard the area against trespassers for many years.

Consequently we ordered from Babcock and Wilcox a heavy steel container, which because of its great size, weight and strength was promptly christened Jumbo.[1] To move it from the manufacturing plant in the East to New Mexico, it had to be loaded onto specially reinforced cars and carefully routed over the railroads. At the nearest railroad stop to the test site it was unloaded onto a specially built trailer with some thirty-six large wheels, and then driven overland about thirty miles to Alamogordo.

[1] Its dimensions were as follows: Inside diameter, 10 feet; straight length between the heads, 25 feet. It had an intershell 6 inches thick of solid steel plate. The shell, after the attachment of the two closure heads by welding, was banded with a number of layers of ⅜-inch-thick steel bands, so that the overall thickness of the main part of the shell was approximately 14 inches.

But by the time of the test we had decided we would not need to use Jumbo, for we had learned enough to be reasonably certain of a fair-sized nuclear explosion. Even if it were as low as 250 tons, as many of our scientists were predicting, the container would only create additional dangers.

It is interesting to speculate about what would have happened, with the actual explosion of almost twenty thousand tons, if we had used Jumbo. That the heat would have completely evaporated the entire steel casing is doubtful. If it did not, pieces of jagged steel would probably have been hurled for great distances.

The scientist in charge of the test was Dr. K. T. Bainbridge, who had the unusual qualification of being a physicist with undergraduate training in electrical engineering. He was quiet and competent and had the respect and liking of the over two hundred enlisted men later on duty at Alamogordo. His first step, with the assistance of Oppenheimer, Major W. A. Stevens, who was in charge of construction activities at Los Alamos, and Major Peer DeSilva, the head of security at Los Alamos, was to select a site.

I had ruled out using Los Alamos for the test on grounds of security and also because I doubted if the area could be expanded sufficiently. Later, we decided that we would need a site measuring approximately seventeen by twenty-four miles, that it should be in a generally non-populated area, and that it should be no further from Los Alamos than necessary. I added one special prohibition: that it should have no Indian population at all, for I wanted to avoid the impossible problems that would have been created by Secretary of the Interior Harold L. Ickes, who had jurisdiction over the Bureau of Indian Affairs. His curiosity and insatiable desire to have his own way in every detail would have caused difficulties and we already had too many.

After looking at several other sites, the committee finally settled on Alamogordo as being entirely satisfactory. It was on an air base, but was far removed from the airfield itself. Arrangements were promptly made with Major General U. G. Ent, under whose control the base came, for us to use the Alamogordo area.

We were not sure until the last moment exactly when we would be ready. The time depended entirely upon the delivery and successful assembly of the bomb's components. I felt that I had to be present personally, especially in case anything went wrong, and naturally Bush and Conant wanted to be there, too.

Because of our uncertainty about the date, the three of us decided to visit several of the MED's installations on the Pacific Coast during the days preceding the test. This would enable us to get to Alamogordo promptly if the date of the test were advanced. We went first to Hanford; from Hanford to San Francisco, where we visited the Berkeley Laboratory; from San Francisco to Inyokern; then to Pasadena; and from there to Albuquerque, where we switched from plane to automobile and drove to Alamogordo.

Air travel has improved considerably since those days. The field we used at Pasadena was very small, and our approach to it was impeded by some high-tension lines at the end of the strip. As he came in, our pilot found himself lined up on the taxiway and quite low. Instead of circling the field, he came in over the wires and then side-slipped, landing with a terrific bounce—both horizontal and vertical. Our landing brought everyone out of the small operations office, including one of my security officers who had missed the plane in San Francisco, and who was waiting to rejoin us in Pasadena. He remarked afterward that, if not the first, at least the second thought that flashed through his mind was: "How am I going to explain the accidental death of Bush, Conant and Groves, without publicity to the project and resulting breaches of security?"

We left the next morning from March Field in Riverside in order to be sure that the predicted Los Angeles fog would not interfere with our taking off. One of the members of our plane crew had been telling me for several days that he was looking forward to spending a night in Los Angeles so that he could see his mother. After we were airborne again, I asked him: "Did you see your mother all right?" He replied, "No, sir, I didn't dare go and see her because she has always worried about my flying, and I knew that after yesterday's landing I would not sound too convincing about its safety."

After arriving at the Alamogordo base camp on July 15, a brief review of the situation with Oppenheimer revealed that we might be in trouble. The bomb had been assembled and placed at the top of its hundred-foot-high steel tower, but the weather was distinctly unfavorable. There was an air of excitement at the camp that I did not like, for this was a time when calm deliberation was most essential. Oppenheimer was getting advice from all sides on what should and should not be done. After discussing matters with General Farrell, who had been there for several days, I concluded that the best thing I could do was to introduce as much of an atmosphere of calm as possible into the very tense situation.

The main problem was the weather. We had obtained the very best men that the armed forces had on long-range weather forecasting, and, for a considerable period, they had been making accurate long-range weather predictions for the test site. The only time they were not right was on the one day that counted. The weather that evening was quite blustery and misty, with some rain. Fortunately, the wind seemed to be in the right direction.

We were interested in the weather for a number of reasons: First and foremost, we wanted to avoid as much radioactive fallout[2] as possible, particularly over populated areas. This was a matter that had not received any attention until about six months earlier, when one of the Los Alamos scientists, Joseph Hirschfelder, had brought up the possibility that it might be a real problem. For this reason, we felt it would be desirable to explode the bomb when rain was unlikely, since rain would bring down excessive fallout over a smal area instead of permitting it to be widely distributed and therefore of little or no consequence. In reaching this decision we could no ignore the old reports that heavy battle cannonading had sometime brought on rain, even though no scientific basis was known for an such phenomenon.

Second, it was extremely important that the wind direction b

---

[2] Radioactive fallout is the falling to earth of particles of airborne matte which have been made radioactive through the effects of a nuclear explosio These are of varying danger to life, depending on how long they retain a hi degree of radioactivity; the time varies greatly with the different elements, fro seconds to many decades (for one isotope of cobalt). With the bomb explosi only one hundred feet off the ground, we expected that a great deal of mater from the tower and the ground surrounding it would be made radioactive a carried as small particles for great distances though the air.

NOW IT CAN BE TOLD

satisfactory, because we did not want the cloud, if one developed, to pass over any populated areas until its radioactive contents were thoroughly dissipated. It was essential that it not pass over any town too large to be evacuated. The city about which we were most concerned was Amarillo, some three hundred miles away, but there were others large enough to cause us worry. The wind direction had to be correct to within a few degrees.

Third, we wanted suitable flying weather so that we could have observation planes flying over the near-by areas; and finally, we wanted to avoid prior heavy rain or continuous dampness, which might ruin our electrical connections, both for firing the bomb, and for the various instruments.

Many of Oppenheimer's advisers at the base camp (and by 6 P.M. these included not only the senior scientists, but many in secondary positions) were urging that the test be postponed for at least twenty-four hours.[3] I felt that no sound decision could ever be reached amidst such confusion, so I took Oppenheimer into an office that had been set up for him in the base camp, where we could discuss matters quietly and calmly. The only other persons taking part in this conversation were some weather forecasters whom we called in. Since it was obvious that they were completely upset by the failure of the long-range predictions, I soon excused them. After that, it was necessary for me to make my own weather predictions—a field in which I had nothing more than very general knowledge.

I was extremely anxious to have the test carried off on schedule. One reason for this was that I knew the effect that a successful test would have on the issuance and wording of the Potsdam ultimatum. I knew also that every day's delay in the test might well mean the delay of a day in ending the war; not because we would not be ready with the bombs, for the production of fissionable material would

---

[3] I found out later that Bainbridge and his principal associates who were at the bomb tower did not want a postponement. The weather seemed better at that point. One of them, R. B. Wilson, insisted that if the shot was postponed, it would be days before it could be made, for the key personnel were completely worn out and would have to have a rest before starting the necessary checks again. I think he was right.

continue at full tilt anyway, but because a delay in issuing the Potsdam ultimatum could result in a delay in the Japanese reaction, with a further delay to the atomic attack on Japan. Obviously, a reasonable time had to be allowed for the Japanese to consider the ultimatum.

From a purely technical point of view, also, it was desirable to avoid a postponement, for the chances of short circuits and a misfire would increase appreciably with every hour that our connections were subjected to excessive moisture. To an even greater degree, though not of such vital importance to the test, the connections to and within our instruments, which naturally had not been so carefully put together as those for the bomb, would be subject to damage. I also recognized the concern of our security officers that every hour of delay would increase the possibility of someone's attempting to sabotage the tests. The strain had been great on all our people, and it was impossible to predict just when someone might give way under it. There was always the chance, too, that a trained saboteur might be present, either within or without our organization, awaiting his opportunity.

At the end of our discussion, Oppenheimer and I agreed that there was no need to postpone the test for a day, but that we might have to put it off for an hour or two.

It had originally been scheduled for 4 A.M. on July 16. This hour had been fixed with the thought that an explosion at that time would attract the least attention from casual observers in the surrounding area, since almost everyone would be asleep. We expected there would be a tremendous flash of light, but thought it would not be great enough to waken many people who were well removed from the burst. Then, too, we wanted the darkness for our photography.

Oppenheimer and I agreed to meet again at 1 A.M., and to review the situation then, with the understanding that we should be ready to set the bomb off on schedule if the weather had improved by that time. I urged Oppenheimer to go to bed and to get some sleep, or at least to take a rest, and I set the example by doing so myself. Oppenheimer did not accept my advice and remained awake,

I imagine constantly worrying. Bush, Conant and I were quartered in a tent that had not been set up very well and the canvas slapped constantly in the high wind. Bush and Conant told me afterward that they could not sleep at all, and did not understand how I could under such conditions.

About 1 A.M., Oppenheimer and I went over the situation again, and decided to leave the base camp, which was ten miles from the bomb, and go up to the control dugout, which was about five miles away. The only people at the control dugout were those whose duties required them to be there. I had arranged for Farrell to be at this point during the test, while I would be at the base camp, the nearest point to the explosion where we permitted anyone to be out in the open. Although there was an air of excitement at the dugout, there was a minimum of conflicting advice and opinions. This was because everyone there had something to do, checking and rechecking the equipment under their control.

As the hour approached, we had to postpone the test—first for an hour and then later for thirty minutes more—so that the explosion was actually three and one half hours behind the original schedule. While the weather did not improve appreciably, neither did it worsen. It was cloudy with light rain and high humidity; very few stars were visible. Every five or ten minutes, Oppenheimer and I would leave the dugout and go outside and discuss the weather. I was devoting myself during this period to shielding Oppenheimer from the excitement swirling about us, so that he could consider the situation as calmly as possible, for the decisions to be taken had to be governed largely by his appraisal of the technical factors involved.

Shortly before we determined the hour that the test would go ahead, we received word from Captain Parsons, who was at the Albuquerque Airfield ready to take off in one of the observation planes, that the Base Commander objected to planes' flying in the area because of weather conditions. I decided that we would hold the test whether the planes were allowed to take off or not. Actually they did take off, but because of the weather their observations were not of as great value as we had hoped. The main reason for having observation planes in the air was to enable Parsons to report later on the relative visual intensity of the explosion of the test bomb and that of the one we would drop on Japan.

Once the decision was made to go ahead, no additional orders were needed. At thirty minutes before the zero hour, the five men who had been guarding the bomb to make certain that no one tampered with it left their point of observation at the foot of the tower. They came back to the dugout by jeeps,[4] reaching there in ample time. Their instructions were very definite about what they should do in case of motor trouble, and since Kistiakowsky was one of the five, I knew that they would find a safe position even in the event of a complete breakdown. After all, in the thirty minutes allotted for their return they could walk several miles from the point of detonation, and I was sure that they would not walk slowly. They also had the key to a padlock guarding the firing circuit at the dugout. Nevertheless, if they had not come back in time, we would have had a real problem, particularly if we had no idea where they were. Fortunately, the problem did not arise. As they left the bomb tower, previously emplaced large lights were turned on as a marker for the observing planes. They also illuminated the tower so that it could be seen with field glasses from the vicinity of the dugout. This, we felt, would serve as a deterrent to any but the most courageous saboteur.

A little later, leaving Oppenheimer at the dugout, I returned to the base camp. There was quite a hubbub of excitement there, but not too much tension. Most of these people had spent many months and even years in preparation for this moment, but now they had no further responsibility for the conduct of any part of the actual test.

Our preparations here were simple. Everyone was told to lie face down on the ground, with his feet toward the blast, to close his eyes, and to cover his eyes with his hands as the countdown approached zero. As soon as they became aware of the flash they could turn over and sit or stand up, covering their eyes with the

---

[4] In a previous check test with a hundred tons of TNT, a jeep had broken down. This time, to be on the safe side, they were using several.

smoked glass with which each had been supplied. We thought that the
time necessary for this movement would be sufficient to eliminate any
danger of eye injury. As we approached the final minute, the quiet
grew more intense. I, myself, was on the ground between Bush and
Conant. As I lay there, in the final seconds, I thought only of what I
would do if, when the countdown got to zero, nothing happened.

I was spared this embarrassment, for the blast came promptly with
the zero count, at 5:30 A.M., on July 16, 1945.

My first impression was one of tremendous light, and then as I
turned, I saw the now familiar fireball. As Bush, Conant and I sat
on the ground looking at this phenomenon, the first reactions of
the three of us were expressed in a silent exchange of handclasps.
We all arose so that by the time the shock wave arrived we were
standing.

I was surprised by its comparative gentleness when it reached us
almost fifty seconds later. As I look back on it now, I realize that
the shock was very impressive, but the light had been so much greater
than any human had previously experienced or even than we had
anticipated that we did not shake off the experience quickly.

Unknown to me and I think to everyone, Fermi was prepared to
measure the blast by a very simple device. He had a handful of torn
paper scraps and, as it came time for the shock wave to approach,
I saw him dribbling them from his hand toward the ground. There
was no ground wind, so that when the shock wave hit it knocked
some of the scraps several feet away. Since he dropped them from a
fixed elevation from near his body which he had previously measured,
the only measurement he now needed was the horizontal distance that
they had traveled. He had already calculated in advance the force of
the blast for various distances. So, after measuring the distance on the
ground, he promptly announced the strength of the explosion. He
was remarkably close to the calculations that were made later from
the data accumulated by our complicated instruments.

I had become a bit annoyed with Fermi the evening before, when
he suddenly offered to take wagers from his fellow scientists on
whether or not the bomb would ignite the atmosphere, and if so,

whether it would merely destroy New Mexico or destroy the world. He had also said that after all it wouldn't make any difference whether the bomb went off or not because it would still have been a well worth-while scientific experiment. For if it did fail to go off, we would have proved that an atomic explosion was not possible. Afterward, I realized that his talk had served to smooth down the frayed nerves and ease the tension of the people at the base camp, and I have always thought that this was his conscious purpose. Certainly, he himself showed no signs of tension that I could see.

There was at least one member of the project who was completely unprepared for the events of that morning. He was an enlisted man—a cook, I believe. He had returned from pass during the evening and, as it was afterward reported to me, had possibly had a bit too much to drink. Somehow, the military police in searching the barracks before the explosion had missed him and he was lying in his bunk, half-asleep, when the explosion came. It blinded him temporarily, but after a few days he was perfectly all right. There were many stories afterward about his determination never to take another drink.

Another casualty of the explosion was Jumbo. At the time of the blast it was standing on end some five hundred yards away from the tower. The explosion knocked it into a slanting position, where it remained for years as a silent witness to the power of the infinitesimal atom.

I had planned to remain at Alamogordo for a number of hours after the explosion to make certain that there was no fallout problem. In order to make full use of the time, I planned to discuss and settle a number of matters involved in our operations against Japan with the members of the Los Alamos group, some of whom were due to leave almost immediately for Tinian. I had also counted on having a discussion with Oppenheimer on some other important points. These plans proved utterly impracticable, for no one who had witnessed the test was in a frame of mind to discuss anything. The reaction to success was simply too great. It was not only that we had achieved success with the bomb; but that everyone—scientists, military officers and engineers—realized that we had been personal participants in, and eyewitnesses to, a major milestone in the world's history and had a sobering appreciation of what the results of our work would be. While the phenomenon that we had just witnessed had been seriously discussed for years, it had always been thought of as a remote possibility—not as an actuality.

Shortly after the explosion, Farrell and Oppenheimer returned by jeep to the base camp, with a number of the others who had been at the dugout. When Farrell came up to me, his first words were, "The war is over." My reply was, "Yes, after we drop two bombs on Japan." I congratulated Oppenheimer quietly with "I am proud of all of you," and he replied with a simple "Thank you." We were both, I am sure, already thinking of the future and whether we could repeat our success soon and bring the war to an end.

I soon decided that the most useful thing I could do was to make certain that the other phases of the operation were going smoothly, and my first concern was with the steps that were being taken to ensure that no damage resulted from fallout. This was the responsibility of our Chief Medical Officer, Colonel Stafford Warren, who had made elaborate preparations for gathering data and for the conduct of our safety precautions.

My greatest concern was over radioactive fallout and the possibility that it might concentrate on a populated area or even an isolated ranch. We had a network of carefully instructed men equipped with Geiger counters who observed the moving cloud as long as they could see it and took repeated Geiger counter readings after it had passed. These were reported back to us at Alamogordo.

As it happened, all went well, but if there had been excessive fallout that would have made it necessary to move out any civilians, we had provided for the emergency. We had military trucks standing by. We were also fully prepared to have martial law declared over as large an area as might be necessary. Naturally, the likelihood that any such action would be required decreased as the distance from the explosion site increased. The first reports began coming in within about half an hour after the explosion and throughout the next critical three hours we had a good picture of the situation at all times.

When I went to Warren's headquarters in the base camp soon after the explosion, I was not pleased to discover that he had been so busy getting ready that he had gone without sleep for almost forty-eight hours. Although his decisions were sound and his instructions were clear, I was sure from listening to them as he talked over the telephone, that—quite understandably—his mind was not working so quickly as it normally did, by any means. Fortunately, we had at Alamogordo a Navy doctor who was familiar with our activities—Captain George Lyons, and I suggested that he spell Warren for a few hours to give him some rest. I was displeased, too, with myself, because I felt that I had fallen down in not making certain that Warren would be in first-class physical shape to handle the situation. However, as the reports came in, it was evident that we would probably have no trouble.

At about 11 A.M., a major security problem arose. We had stationed at the Associated Press office in Albuquerque an officer whose job was to keep any alarming dispatches on the explosion from going out. Shortly before eleven, the AP man told our officer that he could no longer hold back on the story; that, if there was nothing put out by the Army, he would have to put his own stories on the wire. This was relayed to me from Albuquerque.

We had prepared for an official release weeks before by having General R. B. Williams, who had replaced General Ent in command of the Air Forces in that area, and who also had administrative control over the Wendover Base and the 509th Group, write a letter to the Commanding Officer of the Alamogordo Air Base. This letter was delivered in person by Lieutenant W. A. Parish, Jr., from my office, a very smooth young Texas lawyer whose calm, unfailing courtesy and great firmness in carrying out his instructions had been my reason for picking him for this assignment. He had no difficulty whatever in adapting himself to a situation and getting the job done.

The letter said merely that the Commanding Officer would carry out any instructions that the Lieutenant might give him. Naturally, his first query when the letter was presented to him was: "What is it all about?" Parish replied: "I am sorry, Colonel, I cannot tell you." That this was not all pleasing to the Colonel is an understatement, according to the account I received from Parish. The Colonel next asked what the instructions would be and received the same reply. As it was reported to me, both General Williams and his ways of doing business, as well as ours, were most severely criticized. The Colonel then said that he did not wish to have anything to do with the matter, so he would turn Parish over to his Executive Officer, whom he told to carry out General Williams' directions.

My only means of communicating with him was by an open telephone circuit. He had been told the afternoon before that the test would probably take place during the night. Later he was told that all planes at Alamogordo Base should be held on the ground until further notice. Arrangements had also been made with the Civil Aeronautics Authority, as well as with the Air Force and Navy, to ensure that the entire area would be barred to all aircraft during the crucial hours. This was quite upsetting to the base, for it was there that regular B-29 crews received their final training before leaving for the Pacific Theater. Every unit commander wanted his crews to have as many hours in the air as possible, and time was always at a premium. All that the people on the base knew was that their training schedules were being upset for some unexplained reason.

During the early morning hours of the sixteenth, the Executive Officer and Parish stayed at the air control tower at the base. The Executive Officer was most co-operative, apparently fully appreciating the difficult position in which Parish found himself. There was never any question of "Why?" Simply: "What do you want me to do now?" Many men were already on the landing field when the explosion occurred, and not long after several thousand men were there preparing for take-offs. After a reasonable wait the aircraft were released and training resumed without too much harm having been done.

Parish had been provided with a release to be issued by the Commanding Officer at Alamogordo Base. Every word in this release was numbered, so that it was a simple matter to alter it without disclosing any secrets to an unauthorized listener-in. When the AP man at Albuquerque became insistent, I telephoned to Parish and made the necessary deletions and insertions in the release, and told him to have it given out at once. It read:

ALAMOGORDO, N.M., July 16

The commanding officer of the Alamogordo Army Air Base made the following statement today:

"Several inquiries have been received concerning a heavy explosion which occurred on the Alamogordo Air Base reservation this morning.

"A remotely located ammunition magazine containing a considerable amount of high explosives and pyrotechnics exploded.

"There was no loss of life or injury to anyone, and the property damage outside of the explosives magazine itself was negligible.

"Weather conditions affecting the content of gas shells exploded by the blast may make it desirable for the Army to evacuate temporarily a few civilians from their homes."

In the meantime, a great deal of excitement had swept the civilian community throughout New Mexico, and particularly in El Paso, Texas. With the usual freakiness of such explosions, the bomb had done little damage at the base camp, or anywhere else near by, but had cracked one or two plate-glass windows at Silver City, New Mexico, 180 miles away. The El Paso papers carried banner headlines of the explosion and the phenomena occurring in that area.

Through the close supervision and co-operation of the Office of Censorship under Byron Price and Jack Lockhart, no news of the explosion appeared in any Eastern paper, except for a few lines in the early morning edition of a Washington paper. On the Pacific Coast, however, it got on the radio, and spread up and down the Coast.

One disturbing element in our preparations for the press release, and the reason I had held it up for so long, was our uncertainty whether we would have to evacuate some of the civilian population. Because of this possibility, when I talked to Parish I inserted the reference to gas shells.

Our release did not fool everybody. Several days after I got back to Washington, Dr. R. M. Evans, of the du Pont Company, came to see me about some of the operating problems at Hanford. After we had finished and as he was leaving, he turned, his hand on the door-knob, and said, "Oh, by the way, General, everybody in du Pont sends you their congratulations." I quickly replied, "What are you talking about?" He answered, "It's the first time we ever heard of the Army's storing high explosives, pyrotechnics and chemicals in one magazine." He went on to add that the radio announcement on the Pacific Coast had been teletyped in to Wilmington from Hanford. My only response was: "That was a strange thing for the Army to do, wasn't it?"

Mrs. O'Leary, my secretary, had been told the day before the test to be in the office by 6:30 the next morning to receive a message. Because of the delay in the test, I could not telephone her until after 7:30 Washington time. I had left with her a special code sheet which I would use to pass on the results of the test either by tele-phone or by teletype. In addition, I made use of another code, of which only she and I had copies, so that I could safely talk to her over the telephone. In my message to her, I gave her the salient facts which should be reported by cable to the Secretary of War at Potsdam.

Before his departure for Potsdam, Secretary Stimson had set up a special channel of communication between himself and me when he designated George L. Harrison as his representative in Washington to cover atomic affairs. This arrangement not only made it possible for the Secretary and me to get in touch with each other quickly, but it also was a security safeguard. A cable, and particularly a number of them, between the Secretary and me would have aroused considerable curiosity in the Army's communications center, and thus would have increased our security hazards. As soon as Mrs. O'Leary finished decoding my message she went over to the Pentagon to Harrison's office. She showed him the message and helped him draft the first cables to Mr. Stimson.

When it became clear late in the afternoon that radioactive fallout was not going to be a problem, I started my plane trip back to

Washington with Bush, Conant, Lawrence and Tolman. They were still upset by what they had seen and could talk of little else. I learned later that the effects of the test on all who had witnessed it, particularly the scientists, were quite profound for a number of days. As for me, my thoughts were now completely wrapped up with the preparations for the coming climax in Japan.

Upon my arrival in Washington, at about noon on the day after the test, I talked to Harrison and we followed up his previous cable to Mr. Stimson with another one. The next day in the early afternoon he suggested that it would be a good idea if I wrote a full report and sent it over by courier to the Secretary at Potsdam. I agreed, for I realized that the cables, because of their guarded brevity, had not given Mr. Stimson as much information as he might need.

Because of various uncancelable appointments, I could not work on the report that afternoon. When I learned that the courier was scheduled to leave for Potsdam the following morning at about two o'clock, I made arrangements to hold the plane until the report was ready. After writing the first paragraph, which set the general tone, I outlined to Farrell the ground it should cover. This enabled him to start at once on the first draft of the report as well as to write the section that was to be attributed to him.

At about 6:30 that night I started on the balance of the draft and went over with Farrell the part he had written. After we had completed a rough draft, we revised it and then continued to polish it throughout the evening.

Because it was so highly secret, only Mrs. O'Leary and one other fully cleared secretary were allowed to do any of the typing. They had been at work since eight o'clock that morning with only two short breaks for lunch and for sandwiches in the early evening, and by midnight, when the final draft was ready, they were so tired that every page was torture. Shortly after two o'clock the report was finally ready and signed, and was then taken directly to the courier plane.[5]

In Potsdam it was delivered to Colonel Kyle, Mr. Stimson's aide,

[5] See Appendix VIII, page 433.

who placed it in his hands at 11:35 A.M. on July 21. He and Harvey Bundy read it immediately, and then made the earliest possible appointment with President Truman, for 3:30. At 3:00, finding that General Marshall was available, the Secretary had him read it and talked with him about it. He then went over to the President's house and read it to Mr. Truman and Secretary Byrnes. After that, accompanied by Bundy, he conferred with Churchill and Lord Cherwell. This conference was soon interrupted but was resumed the next morning, when Churchill read the complete report for the first time.

Mr. Stimson's diary for Sunday, July 22, 1945, is most enlightening:

Churchill read Groves' report in full. He told me that he had noticed at the meeting of the Three yesterday that Truman was much fortified by something that had happened, that he had stood up to the Russians in a most emphatic and decisive manner, telling them as to certain demands that they could not have and that the United States was entirely against them. He said, "Now I know what happened to Truman yesterday. I couldn't understand it. When he got to the meeting after having read this report, he was a changed man. He told the Russians just where they got on and off and generally bossed the whole meeting." Churchill said he now understood how this pepping up had taken place and he felt the same way.

The receipt of the news of the successful test at Alamogordo and the reassurance that we would be ready for the first bombing of Japan on July 31, weather permitting, froze the previously tentative decision that it was now time to issue the Potsdam ultimatum to Japan. Just how differently the ultimatum would have been worded without this knowledge, no one knows, but with the news of our success in hand President Truman and Mr. Churchill were able to see it dispatched with a great deal more confidence than otherwise might have been the case.

CHAPTER 22

OPERATIONAL PLANS

The Alamogordo test had not set aside all doubts about the bomb. It proved merely that one implosion-type, plutonium bomb had worked; it did not prove that another would or that a uranium bomb of the gun type would. We had made every possible component test we could think of. We were reasonably sure of each one. We knew we could bring the U-235 portions of the bomb together in such a way that, if the theories of atomic energy were correct and U-235 behaved as plutonium had, the bomb should go off. But still no test had been made of the complete bomb. Nevertheless, the indications for success were strong enough so that no one urged us to change our plans of dropping the first gun-type bomb in combat without prior test. In any case, we simply had to take the chance, because the production of U-235 was so slow, even compared to plutonium, that we could not afford to use it in a test. We could now establish firmly that the first atomic bomb drop on Japan would be sometime around the first of August.

The major portion of the U-235 component for this bomb began its journey overseas on July 14 when a convoy consisting of a closed black truck, accompanied by seven cars with security agents, left Santa Fe for Albuquerque. From Albuquerque, the bomb was flown in an Air Force plane to Hamilton Field just outside San Francisco, where it was picked up and carried to Hunter's Point. The parts of the bomb, which were packed in a large crate and a small metal cylinder, were in the custody of Major Furman from my office, and Captain Nolan, a radiologist at the Los Alamos Base hospital.

Early in the morning of July 16, the bomb—minus the last necessary bit of U-235—was put aboard the cruiser *Indianapolis,* which sailed almost immediately. The *Indianapolis* had a fast but uneventful trip through Hawaii to Tinian. The only untoward incident during the voyage grew out of the fact that Furman and Nolan were traveling in the guise of field artillerymen, which led to some very searching and embarrassing questions by the ship's gunnery officers, which they were wholly unprepared to answer.

The *Indianapolis* arrived at Tinian on July 26 and discharged her cargo that same day. She then put out to sea, headed for the Philippines, but her voyage came to a sudden and tragic end on July 30 when she was attacked by a Japanese submarine and went down with some nine hundred of her crew.

In looking back at this phase of the operation, I can only feel intense relief that our cargo was safely delivered to its destination, for, as we later learned, the *Indianapolis* was a very poor choice to carry the bomb. She had no underwater sound equipment, and was so designed that a single torpedo was able to sink her quickly.

I had arranged with Lieutenant General H. L. George, the head of Air Transport, for the final parts, including some U-235, of the Hiroshima bomb to be flown from Albuquerque to Tinian. Because I did not want to risk having the plane disappear in flight with an extremely valuable though small piece of U-235, I asked for two large cargo planes in perfect condition, and the best possible crews. I told George that the cargo would be almost infinitesimal and that the second plane was wanted just in case of need, to tell us where the first one crashed—if it did. George had had his orders from Arnold and he simply said, "Whatever you want, you get."

A few days later he told me, "I just had one of my senior staff officers in here telling me about that operation of yours using two C-54's to carry a few hundred pounds of cargo. The mildest words he used were 'unreasonable' and 'idiotic.' I finally told him that we would continue to furnish whatever the MED asked for and that he didn't have to say any more." I never asked George how much Arnold had told him but he always gave us the utmost in co-operation.

One of our security officers at Los Alamos, who rode one of the planes, Second Lieutenant R. A. Taylor, Jr., told me later of his trip to Tinian. He was responsible for a box about the size of two orange crates which contained certain assembly parts for the bomb. The other plane carried Lieutenant Colonel Peer de Silva, the head of security at Los Alamos, with the final small key component of U-235 for the bomb. En route to San Francisco (Hamilton Field), the first stop, the two planes were to keep in close contact and if de Silva's plane went down the other was to spot the point and circle until help came.

When the cargo arrived at Albuquerque for loading it was under heavy armed guard. The two plane crews were flabbergasted when they saw the small boxes that were to comprise the entire freight.

The flight to San Francisco was uneventful but when they were about forty-five minutes out over the Pacific, de Silva's plane had to return to the mainland because of engine trouble. Contrary to what I had counted on, Taylor's plane, instead of returning too, continued directly to Hawaii and arrived there several hours before de Silva's plane came in. Despite our desire to keep the flight inconspicuous, a radio tip—probably sent unofficially—alerted Hawaii to a very important flight, and Taylor was met at the airfield by a group of senior officers who expected at least one high-ranking general with entourage instead of a single second lieutenant. Taylor had no written orders with him, since they were in de Silva's hands, and soon found the local personnel determined, in accordance with custom, to load the empty plane to capacity. Fortunately, the plane crew had been convinced before departure from California that the trip was most unusual and with their support he was able to keep the ship unloaded until de Silva arrived. From then on the trip was uneventful.

De Silva had been instructed previously on what to do if any essential request were refused. He was to say that his was a highly secret mission under General Marshall's direct personal control and that, if there was any doubt, a telephone call should be made directly to General Marshall for verification. I was sure that no one would be likely to make such a check, and, if by chance someone did, I hoped that General Marshall's office would spot it as part of my operations and ask me about it.

This procedure was used only on rare occasions during the entire project. It invariably stopped all further argument.

Our basic operational plans were to launch the atomic attack using the gun-type bomb as soon as the final bit of U-235 reached Tinian by air and weather permitted. As soon as possible after this, a second attack would be initiated. This time an implosion-type bomb would be dropped. The controlling factor on the second bomb was the date by which a sufficient amount of plutonium could be processed and delivered to Tinian. After that, all that was needed was suitable weather.

Soon after my return from Alamogordo, I had learned that General Spaatz was in town making preparations to go to the Pacific where he was to take field command of the 20th Air Force. I got hold of him and together we went over the situation. He quickly agreed with and confirmed all the arrangements I had previously made with LeMay.

On July 23, I prepared the final written directive for the coming operations out of Tinian. It read as follows:

To GENERAL CARL SPAATZ, CG, USASTAF:

1. The 509 Composite Group, 20th Air Force, will deliver its first special bomb as soon as weather will permit visual bombing after about 3 August 1945, on one of the targets: Hiroshima, Kokura, Niigata and Nagasaki. To carry military and civilian scientific personnel from the War Department to observe and record the effects of the explosion of the bomb, additional aircraft will accompany the airplane carrying the bomb. The observing planes will stay several miles distant from the point of impact of the bomb.

2. Additional bombs will be delivered on the above targets as soon as made ready by the project staff. Further instructions will be issued concerning targets other than those listed above.

3. Dissemination of any and all information concerning the use of the weapon against Japan is reserved to the Secretary of War and the President of the United States. No communiqués on the subject or releases of information will be issued by Commanders in the field without specific prior authority. Any news stories will be sent to the War Department for special clearance.

4. The foregoing directive is issued to you by direction and with the approval of the Secretary of War and of the Chief of Staff, USA. It is desired that you personally deliver one copy of this directive to General MacArthur and one copy to Admiral Nimitz for their information.

Signed: T. T. Handy[1]

One of our big problems had been whether to use each bomb as it became available or to store up a sufficient number to permit their employment in quantity. Most professional soldiers will go to almost any length to avoid piecemealing away their resources. However, in this case, we felt that the considerations of time and expected power justified our using the weapons as they became available. If our assessment of these factors had been erroneous, our position might very well have been extremely embarrassing, for our projected production rates called for one Fat Man of the type we had tested at Alamogordo (but not by an air drop) to be delivered at Tinian on about August 6, with a second one to be ready about August 24, and additional ones arriving in increasing numbers from there on. As I have said, we had no assurance that a uranium bomb, the Thin Man, would work at all. Yet, that was the weapon that we employed against Hiroshima, since we had used up our entire immediate supply of plutonium in the test bomb.

On July 24, I sent a memorandum to General Marshall at Potsdam to obtain his final approval of our plan of operation. It was about two pages long. Attached to it were a small map of Japan and near-by Asia (cut from a large *National Geographic* map); a one-page description of each of the four targets (Hiroshima, Nagasaki, Kokura, and Niigata); a one-page draft of a necessary action by the Joint Chiefs releasing the previously reserved targets (all but Nagasaki, which had never been reserved) to the Commanding General, Army Air Forces, for attack only by the 509th Group, 20th Air Force; and a copy of a tentative draft of a directive to Spaatz. This apparently made no mention of Nagasaki.[2] That particular target had only re-

---

[1] General Handy was Acting Chief of Staff during General Marshall's absence in Potsdam.

[2] The original was destroyed in Potsdam (to avoid any chance of breach of security). A copy, possibly made afterward, in my files omitted Nagasaki in the directive.

cently been added in place of Kyoto and in some way I must have
failed to list it in the enclosure. The cabled draft that was sent later
for his definite approval did include it.

The memorandum gave the probable date of the first bombing as
between the first and tenth of August. It was to be as soon after the
first as the Little Boy could be assembled, and the weather permitted
visual bombing. We expected that this would be about August 3, but
it depended upon how soon we would have the final amount of
U-235.

The memorandum stated that there would be officers in each
bombing plane whose familiarity with the design, development and
technical features of the bomb would qualify them to render final
judgment if any emergency required a deviation from the plan. The
Senior Technical Officer on the first flight would be Captain Parsons
of the Navy.

General Marshall was also furnished with the probable readiness
dates of the implosion bomb, and I explained that there would have
to be a gap of at least three days between successive bombs, no
matter what type was used. I did not give him the reason for this:
we needed the time to assemble the bomb, and nothing should be
hurried. Judging by the Alamogordo test, I said, we now expected this
bomb to have an explosive force exceeding the equivalent of ten
thousand tons of TNT and possibly reaching as high as thirty thousand
tons.

I went on to forecast our delivery rates after the third bomb, and to
say that all bombs would be delivered by the 509th Composite Group
on targets having high priorities at the time of delivery. I added that
all instructions concerning the targets would be issued through the
Commanding General, U.S. Army Air Forces. This was in accord
with the policy agreed to by Arnold and me some months previously
that control over the weapons' use would remain in Washington
during the foreseeable future.

My statement that all instructions would be issued through the
Commanding General, USAAF, was also a clear indication to General
Marshall that the orders for our operations against Japan would

originate with me, be approved by him, and be issued to Spaatz over Arnold's signature.

In two brief paragraphs, I gave the organization for the coming operation:

Major General L. R. Groves has overall direction of the atomic fission bomb project. Brigadier General T. F. Farrell, General Groves' Deputy, and Rear Admiral W. R. Purnell, the Navy member of the Military Policy Committee, will be at the base prior to the first mission to co-ordinate the project with Army and Navy Commanders in the theater.

The Air Forces operations will be under the command of General Spaatz, Commanding U. S. Army Strategic Air Forces. The 20th Air Force is under the command of Major General LeMay,[8] The 509th Composite Group is under the command of Colonel Paul Tibbets.

General Marshall's approval of this plan put our operation fully into motion.

On July 31 (July 30 in Washington), Farrell sent me a rather lengthy cable. He had arrived at Guam the day before and had had a meeting with LeMay that afternoon. His cable included this sentence: "1 August is interpreted by LeMay and Farrell as coming within the intent of the directive." He said that the bomb could be ready 2300 E.W.T., 31 July, and added: "LeMay needs 11 hours more which would be 1 August, 1000 E.W.T." I have been unable to find any trace of a reply to this cable in which I either confirmed or objected to the LeMay-Farrell interpretation.

If I did not reply, it was an error, for I should have done so in order to avoid any possible misunderstanding that might cause a delay. The fact that I fully agreed with their interpretation makes no difference. If the cable had said July 30, I would have reacted sharply and at once. I knew from Secretary Stimson that President Truman wanted to be sure that Japan had enough time to answer the Potsdam ultimatum. I had told Mr. Stimson before his departure that we did not expect to be ready until July 31, and we both understood that unless I received orders to the contrary, we would wait for that date even if unexpectedly we should be ready a day or two sooner. The difference in dates between Japan and Washington was

[8] LeMay became Chief of Staff under Spaatz on August 2, 1945.

overlooked in all of our arrangements. It should not have been; I should have caught the omission.

It was, and always had been, my understanding that our target date was the first of August, and I was surprised in rereading the official order recently to find mention of August 3. I am sure I had no reason for using this date rather than August 1, except that at the time the order was actually signed by General Handy we did not think we could deliver the bomb before that date.

This surmise is fully supported by the wording of my August 6 report to General Marshall on the bombing of Hiroshima. It opened with the sentence: "The gun type bomb was ready at Tinian on 31 July awaiting first favorable weather." This report went on to say that originally we had expected that the first favorable weather would be on August 3, but that in fact it was not until the sixth that the bomb could be used.

Several writers have attempted to read hidden meanings into the date of August 3. In doing this, they have ignored entirely the actual wording of the order which was, "After about 3 August." They have also ignored the fact that the word "about" is thoroughly understood in the American Army. Official travel regulations of that period even defined "about" as normally including a period of four days before and four days after the specific date cited.

Everyone concerned understood this, and Farrell's cable was sent only because of his desire to be certain that there could be no misunderstanding. Even then he did not put it as a query, but as a statement of his and LeMay's understanding.

Soon after Farrell's arrival in Guam, and after the bomb components were ready, while we were waiting only for favorable weather, Spaatz sent a cable reminding us of the reported location of POW camps in and near the target areas. He asked whether the locations of these camps would upset his orders, particularly with respect to the selection of targets. There were no POW camps listed in the vicinity of Hiroshima, although there was one reported about a mile outside of Nagasaki, which was supposed to contain several hundred Allied POW's.

Handy received the message and turned it over to me. As I studied it, I was bothered by the fact that the intelligence estimate that both Spaatz and we were using was apparently incorrect in its details. If it was correct, the camp was on the west side of Nagasaki Bay; yet it seemed much more likely that it would be on the other side, which was much closer to the docks where it was believed that the prisoners were being worked. No matter which location was right, however, it did seem likely that at the probable time of an explosion POW's would be working in the dock area and would be fully exposed to the expected hazards.

Spaatz' query was not an easy one to answer. I am sure he would not have sent it if the control of the operation had not been so closely held in Washington, for it involved a decision that normally would have been made by the commander in the field. Handy felt that the decision should be made by the Secretary of War, but later agreed with me that we should tell Spaatz to disregard the reported camps; however, we decided that I would show the Secretary the outgoing cable before sending it. This would free him of the burden of making the decision, as I felt we should, since the burden was my own. At the same time he would have an opportunity, if he chose to take advantage of it, to overrule us before any harm was done. A cable was therefore prepared, telling Spaatz that there was no change in the targets because of the POW situation; that he could, however, adjust the aiming points, which were already his responsibility, in such a way as to decrease the possibility of hitting any POW camp.

I took the cable in to Mr. Stimson, as well as the one we had received from Spaatz, and said I was showing them to him for his information. I added that this was our responsibility and we were not passing it on to him. I did not emphasize that he could change it if he wished, though I told him that I was sending it as soon as I left his office. His only reaction was to thank me for showing him the cable before it was sent.

Many small problems arose as the date of the first bombing approached. Two years before, the War Department had issued orders prohibiting persons with knowledge of future military operations from flying over enemy-held territory. The reason for this order was sound—to prevent leakage of information through possible prisoners of war—but it created an impossible situation for us. We had to have Parsons and Ashworth and their two technical assistants aboard the bombing planes. We also needed Tibbets and his best pilots on these flights, and we wanted to have certain scientists in the observation planes. We simply could not operate under such a restriction.

When this problem was brought to my attention in Washington, I immediately secured War Department authority to disregard the prohibition for both civilian and military personnel concerned with Centerboard, as our overseas operation had been named, provided their flights were essential. It was provided further that they had to be specifically authorized by me or by General Farrell in the case of personnel not permanently assigned to the 509th Group, and by the Commanding General, U.S. Army Strategic Air Forces, or a senior officer designated by him, in the case of permanent group personnel. These authorizations had to be specifically contained in official written orders. Without such waivers we would have been seriously handicapped, not only in the bomb-carrying plane, but in the instrument plane and the observation plane as well.

CHAPTER 23

HIROSHIMA

August 1 came and passed; the weather was not favorable over Japan and LeMay did not think that it would be wise to undertake the mission under those conditions. The six crews that might be used were given special instructions on the procedures they were to follow, and at another briefing on August 4, Parsons explained the effects they could expect when the bomb exploded. Most of them knew by now that they were dealing with a special type of bomb, but Parsons' statement that the force of the explosion would be equivalent to that of twenty thousand tons of TNT came as a complete surprise.

During the period of waiting, the special air-sea rescue plans were settled. They emphasized that no other aircraft would be permitted within fifty miles of the target during a period of from four hours before until six hours after strike time. Not even for rescue operations would this restriction be lifted. Special air-sea rescue facilities were to be provided by both Army and Navy planes and by submarines, and in spite of the necessary restrictions we had placed upon it, rescue coverage in this operation would be far better than average.

In my discussions with Arnold and LeMay and in my instructions to Farrell, I had made it plain that on this first flight it was essential for us to be able to talk to Parsons afterward no matter whether the flight was a success or not, and particularly essential if it was unsuccessful. It was extremely important that we should know exactly what had happened. I added that aside from this consideration, while the lives of the men in these planes were precious, they were no more so than those of any other plane crew.

Other air attacks on Japan were to be carried out on the same day as our mission, to divert any Japanese defense actions that might endanger our operation. Hiroshima would be the primary target, with Kokura Arsenal and Kokura the secondary targets, and Nagasaki the tertiary target. The aiming point for Hiroshima was close to the Japanese Army Headquarters.

Hiroshima was a highly important military objective. The Army Headquarters was located in a castle. Some 25,000 troops were in its garrison. It was the port through which all supplies and communications passed from Honshu to Kyushu. It was the largest city, excepting Kyoto, that was still undamaged by American air raids. Its population was believed to be over 300,000, and it was a beehive of war industry, carried on in moderate-sized plants and in small shops as well as in almost every home.

We would use a total of seven planes. One would be sent to Iwo Jima to serve as a spare in case the bomb-carrying plane developed mechanical troubles on the flight from Tinian. Three planes would go ahead, one to each target area, to appraise the local weather and to relay the information back to the bomb-carrying plane, which would be accompanied by two observer planes to the general vicinity of the target. One of these carried special measuring and recording instruments, including some that would be dropped near the target to radio back their readings.

Radar was to be used as an aid but the actual bombing was to be accomplished visually. If this proved to be impossible, the bomb was to be brought back, probably to Iwo Jima, as the plane's gas supply might not permit the return to Tinian. We were anxious to avoid having it come down at other air bases, for in case of a landing accident we wanted personnel on the ground who would be aware of the special precautions that would have to be taken.

These arrangements were complicated, but not difficult for an organization that had been as carefully organized and trained as the 509th Group. Provisions were also made for the strike photographs to be taken by the 3rd Photo Reconnaissance Squadron and two photo crews were briefed on their assignment by the 509th's intelligence officers.

On the morning of the fifth, there were indications that the weather would be good on the following day. Since we needed twenty-four hours' notice before take-off so that the bomb could be assembled and checked out, LeMay confirmed that the mission would probably take place on August 6. Immediately after assembly, the Little Boy was placed on its trailer enshrouded in canvas for concealment and moved to the loading pit, from which it was raised into the B-29. The final tests were run and the bomb was ready for the take-off by the early evening of the fifth.

Until the moment of take-off the aircraft and the bomb within it remained under the continuous surveillance of security guards and representatives of the various key technical groups. A final briefing was conducted at midnight. Then came a preflight breakfast, followed by religious services, and the bomb was airborne.

The pilot of the *Enola Gay*—the B-29 that flew the mission—was Colonel Tibbets; Major Thomas Ferebee was the bombardier; Captain Parsons was the weaponeer; and Lieutenant Morris Jepson was the electronics test officer.

Because of the many technical details involved it was essential that there be no possible confusion as to responsibilities. For this reason Norstad, as Chief of Staff, had written to the Commanding General XXI Bomber Command on May 29, 1945:

"In actual delivery it is desired that the B-29 airplane which carries the bomb also carry two military officer specialists. The senior officer specialist will be qualified by familiarity with the design, development and tactical features of the bomb, to render final judgment in the event that an emergency requires deviation from the tactical plan."

Parsons had decided with Farrell's approval to complete the final assembly of the bomb after takeoff. His purpose was to minimize the hazards of a crash on Tinian. I had previously said that I was opposed to this as unwise, because it was unnecessary and because it would be very difficult to do it in cramped conditions in the plane. I was not informed of the plan until it was too late to interfere.

The progress of the mission against Hiroshima is well described in the log which Parsons kept during the flight.

6 August 1945   0245[1] take-off
0300   started final loading of gun
0315   finished loading
0605   headed for Empire from Iwo
0730   red plugs in[2]
0741   started climb. Weather report received that weather over primary and tertiary targets was good but not over secondary target
0838   leveled off at 32,700 feet
0847   electronic fuses were tested and found to be O.K.
0904   course west
0909   target Hiroshima in sight
0915½   drop bomb

The original scheduled time was 0915. Thus, in a flight of some seventeen hundred miles taking six hours and a half, Colonel Tibbets had arrived on target only one-half of a minute off schedule.

The 20th Air Force order covering the operation prescribed a turn of 150° after the bomb was released in order to gain a maximum distance from the point of explosion; such a turn, our studies indicated, could be made without undue risk to the plane and its crew.

Immediately after the bomb was dropped from 31,600 feet, the plane began its getaway maneuver. The flash was seen during this turn and fifty seconds after the drop, the shock waves hit the plane. There were two of these, the first the direct shock wave and the second the reflected wave from the ground. By that time the plane was fifteen miles away from the burst. Parsons' log continued:

Flash followed by two slaps on plane. Huge cloud
1000 still in sight of cloud which must be over 40,000 feet high
1003 fighter reported
1041 lost sight of cloud, 363 miles from Hiroshima, with the aircraft being 26,000 feet high

[1] Tinian time (August 5, 11:45 A.M. Washington time).
[2] These plugs armed the bomb so that it would detonate when released.

The crews of the strike and the two observation aircraft reported that five minutes after release a dark gray cloud of some three miles in diameter hung over the center of Hiroshima. Out of the center of this grew a column of white smoke which rose to a height of 35,000 feet, with the top of the cloud being considerably enlarged.

Four hours after the strike, the photo reconnaissance planes found that most of the city of Hiroshima was still obscured by the smoke cloud, although fires could be seen around its edges. Unfortunately, no report reached me from these planes, except merely that they were unable to get pictures. Pictures taken the following day showed that 60 per cent of the city was destroyed.

The area devastated at Hiroshima was 1.7 square miles, extending out a mile from ground zero. The Japanese authorities estimated the casualties at 71,000 dead and missing and 68,000 injured.

The most important result achieved by the Hiroshima bombing was not the physical damage, although over 50 per cent of the buildings were totally destroyed, nor was it the fifteen to twenty thousand Japanese soldiers who were killed or severely wounded, nor was it the thousands of other people killed and injured. The important result, and the one that we sought, was that it brought home to the Japanese leaders the utter hopelessness of their position. When this fact was re-emphasized by the Nagasaki bombing, they were convinced that they must surrender at once.

Farrell had informed me by cable on Saturday, August 4, that the weather prediction was favorable for our operations, and that if it remained so the plane would take off on Sunday, at approximately noon. (Washington times and dates are used throughout this description of Washington events.) General Marshall was notified, as was Secretary Stimson, through Harrison.

I went to my office rather early on Sunday morning and found a cable telling me that the take-off was still scheduled for that day. I remained there, awaiting a report of the take-off or of its postponement. I fully expected to receive some word on the situation at Tinian by not later than 1:30 or 2:00 P.M. By that time I had finished the work I wanted to do, and realized that there was nothing I could do but to sit back and wait. There were several officers on duty who were also just sitting and waiting. In retrospect, I cannot understand why I did not make an early teletype inquiry of Farrell, except that this might have been thought to indicate a lack of confidence, and this was something I never did if I could help it.

I finally gave up, recognizing that there was no useful purpose to be served by my sitting there fretting. I decided that all of us in my office would be better off if I went out to play tennis for an hour, so I told the Duty Officer, Major Derry, who was regularly assisting me in the operational phases of the project, how to reach me without delay, and departed. I took with me another officer to sit beside the telephone at the courts. This had been my regular practice on occasions when I might be hard to get hold of, ever since our first separation plant began operations. It not only ensured that I could be reached promptly in case of need, but it also kept me from constantly wondering whether someone might not be trying to reach me. This officer called my office every fifteen minutes to ask if there was any news, and we returned after about an hour or two to find that there was still no word from Tinian.

At about five o'clock, Derry informed me that he had just had a telephone call from General Marshall, who had returned from a weekend at his home in Leesburg, and wanted to know if there was any word on the operation. Derry answered that there was not, but that I was there and he would call me to the telephone. General Marshall's reply was typical of his thoughtfulness throughout my association with him: "I don't want you to bother General Groves. He has enough to think about without answering any unnecessary queries."

Not anticipating this dearth of information, I had arranged for my wife and daughter and George Harrison to have dinner with me at the Army-Navy Club. I had thought that we could do this between the take-off report and the strike report from the plane, which I had expected to receive about 7:30 P.M. On meeting them, I told Harrison that we had no news. Later, General Handy came by and asked me if I had any news and again I had to say no. When we were well along with dinner, I was called to the phone. It was then about 6:45. As I went, I noticed both Harrison and Handy had stopped eating and I could feel their eyes boring into my back. It was Derry with the first report that the plane had left on schedule. This report was about six hours late, but still it was a relief to have something. On my return, I told Handy and then Harrison.

A few minutes later I left for my office. As my family drove me downtown, I told them that I would have to stay there all night. Despite the fact that it was the first time I had done this during the war, it did not give rise to any question or comment; their lifetime in the Army had conditioned them well.

In accordance with my directions, Derry had promptly informed General Marshall, through the Secretary of the General Staff, Colonel McCarthy, about the plane's departure. I was greatly disturbed and embarrassed by the evident breakdown in our communications, so I telephoned Major General H. C. Ingles, the Chief Signal Officer and an old friend, to ask him to look into the matter. He replied that he had already heard from General Marshall and that he did not need any additional stirring up from me. He had no idea what was the matter, but he was taking every possible step to correct the unknown trouble. Still no strike message came in.

The people in my office who knew the general situation, and this did not include everyone, gathered during the evening, both to be of assistance if necessary, and because of a natural desire to keep abreast of events. During the evening I worked on several papers that I wanted to study quietly. In order to ease the growing tension in the office, I made a point of taking off my tie, opening up my collar and rolling up my sleeves. While this was completely out of character for me, I did it for the specific purpose of creating a more informal, relaxed atmosphere. The hours went by, more slowly than I ever imagined hours could go by, and still there was no news. By now the strike message was three or four hours overdue.

I have never been certain whether our queries into the delay ever really got to the bottom of the matter or not. But it appeared that

NOW IT CAN BE TOLD

the trouble resulted from one of the mix-ups that so often occur at important moments. Messages to me had always been sent over an Air Force top secret channel which went from Tinian to Guam to Washington. On this occasion the message left Tinian and then somehow was put on an Army channel. This went to Manila and from there to Washington. Why, I have never learned.

At about 11:15, I received a telephone call from McCarthy, who said that General Marshall had asked him to find out if we had any news about the actual strike. I replied, "No, we have not received any." He said that General Marshall had told him that if any news came in later, McCarthy was not to call him then, but to tell him in the morning.

About fifteen minutes later the strike message did come in. Parsons reported (in special one-time code, of course):

Results clearcut, successful in all respects. Visible effects greater than New Mexico tests. Conditions normal in airplane following delivery.

Received at the same time was this message relayed from the plane:

Target at Hiroshima attacked visually. One-tenth cloud at 052315Z.[3] No fighters and no flak.

As soon as we had decoded these messages, I called McCarthy. He asked me what I thought he should do in view of General Marshall's previous instructions. I replied that that was his problem, but if I were in his place, I certainly would call General Marshall right away. He followed my advice, as it was what I am sure he wanted to do anyway. The only comment he received was, "Thank you very much for calling me."

Naturally our people were greatly excited over the news, but I retired to my office and wrote the rough draft of a report for delivery to General Marshall in the morning. I planned to amplify it when I received the expected longer, more detailed report after the plane returned to Tinian. As soon as I completed the draft, I went to sleep on

[3] Greenwich time (6:15 P.M. Washington time).

the cot that had been brought into my office, after telling the Duty
Officer to call me when the next message came in. At about 4:30 A.M.
he wakened me to deliver the detailed hoped-for cable from Farrell,
which had been dispatched after the plane returned to Tinian. This
message read:

Following additional information furnished by Parsons, crews, and observers on return to Tinian at 060500Z.[4] Report delayed until information
could be assembled at interrogation of crews and observers. Present at
interrogation were Spaatz, Giles, Twining, and Davies.

Confirmed neither fighter or flak attack and one tenth cloud cover with
large open hole directly over target. High speed camera reports excellent
record obtained. Other observing aircraft also anticipates good records
although films not yet processed. Reconnaissance aircraft taking poststrike photographs have not yet returned.

Sound—None appreciable observed.

Flash—Not so blinding as New Mexico test because of bright sunlight.
First there was a ball of fire changing in a few seconds to purple clouds
and flames boiling and swirling upward. Flash observed just after airplane
rolled out of turn. All agreed light was intensely bright and white cloud
rose faster than New Mexico test, reaching thirty thousand feet in minutes
it was one-third greater diameter.

It mushroomed at the top, broke away from column and the column
mushroomed again. Cloud was most turbulent. It went at least to forty
thousand feet. Flattening across its top at this level. It was observed from
combat airplane three hundred sixty-three nautical miles away with
airplane at twenty-five thousand feet. Observation was then limited by haze
and not curvature of the earth.

Blast—There were two distinct shocks felt in combat airplane similar
in intensity to close flak bursts. Entire city except outermost ends of dock
areas was covered with a dark grey dust layer which joined the cloud
column. It was extremely turbulent with flashes of fire visible in the dust.
Estimated diameter of this dust layer is at least three miles. One observer
stated it looked as though whole town was being torn apart with columns
of dust rising out of valleys approaching the town. Due to dust visual
observation of structural damage could not be made.

Parsons and other observers felt this strike was tremendous and awesome even in comparison with New Mexico test. Its effects may be
attributed by the Japanese to a huge meteor.

[4] Greenwich time.

After the message was decoded, I amplified and revised my previously prepared draft, and while it was being typed, changed my uniform. Then about 6:15, I called George Harrison at his hotel and told him that I was on my way to General Marshall's office and would be there shortly before seven, which was the General's regular hour of arrival.

I was waiting when the General arrived, and immediately handed him the written report, which was about two pages long. Within a minute or two we were joined by General Arnold, and then by Harrison. After a brief discussion, General Marshall called the Secretary of War on the telephone. Mr. Stimson was then at Highhold, his house on Long Island, for he had come back from Europe quite worn-out, and had gone home for a day or two of rest. The telephone circuit over which we talked was specially designed to be secure. After receiving the facts, he extended his very warm congratulations for me to convey to all concerned.

General Marshall expressed his feeling that we should guard against too much gratification over our success, because it undoubtedly involved a large number of Japanese casualties. I replied that I was not thinking so much about those casualties as I was about the men who had made the Bataan death march. When we got into the hall, Arnold slapped me on the back and said, "I am glad you said that—it's just the way I feel." I have always thought that this was the real feeling of every experienced officer, particularly those who occupied positions of great responsibility, including General Marshall himself.

It was quite evident to all of us, as well as to the Secretary up on Long Island, that our hope of ending the war through the development of atomic energy was close to realization.

General Marshall said that he would prefer to have me remain in the Pentagon that morning instead of returning to my office on the Washington side of the river. He said there were a number of things that might come up and he wanted me to be available. He added, "The Secretary of War is away. Why don't you take over his office while you are over here?" This was most convenient, since it was immediately next to General Marshall's.

The major problem facing me then was the Presidential release, which we hoped to make at 11 A.M. It had been customary for the Japanese to announce the results of American bombing raids even before the planes had returned to their bases. In order to ensure the greatest possible impact on the Japanese Government and people, we wanted the announcement of our bomb to be made from Washington, and made promptly.

To help prepare this press release and the many others that would be necessary, back in the spring we had secured the services of William L. Laurence, a science reporter for the *New York Times*. While we had in the project a number of competent men with sound newspaper backgrounds, they already had more than enough to do, and it seemed to us, in any case, that it would be much better to bring in an outside newspaperman who would have a more objective touch. Our first thought had been to borrow Jack Lockhart from the Office of Censorship. Lockhart felt that he could not be spared from his present assignment, and suggested that we get someone who had a better background in scientific reporting, recommending Laurence as the best man he could think of for the job.

We already knew quite a bit about Laurence. His article on the possibilities of atomic power, which had been published by the *Saturday Evening Post* in 1940, had been of considerable interest to us. In fact, early in 1943, we had asked the *Post* to report to us at once any request received for this particular back number and to delay mailing it until they received our instructions. Actually, no request for that issue was ever received.

Laurence had an excellent reputation as a scientific writer, and after careful consideration and learning all we could of his background, I made an appointment to see Edwin L. James, the managing editor of the *New York Times*. I asked him to have Laurence available in case our conversation necessitated his presence.

On the appointed afternoon, I visited James' office, accompanied by Major W. A. Consodine, an experienced lawyer from my office who dealt with intelligence and counterintelligence matters and who had also been a professional reporter for a number of years. Without disclosing the purpose of the project, I said that it was of extreme importance, that we needed a newspaper writer for a number of months, and that we wanted Laurence. I said that he would be working for us and there would be no rights or privileges of any kind accruing to the *Times* while he was with us. Consodine added that this would not prevent them from using his name in a by-line, but that the same story would be distributed to all news media, which could not be expected to give credit to him or the *Times*. We asked James for his frank opinion of Laurence, which confirmed the opinion we had already formed after investigating his background. A few minutes with Laurence himself only made us more certain than ever that this was the man we wanted.

In discussing the arrangements, it seemed desirable for security reasons, as well as easier for the employer, to have Laurence continue on the payroll of the *New York Times,* but with his expenses to be covered by the MED. I asked James to keep Laurence's assignment as secret as possible, but I never asked him if any other persons on the *Times* knew where Laurence was or what he was doing. James did let word get out that Laurence was on a special assignment and he later arranged, about August 1, for a story in the *New York Times* with a London date line and carrying Laurence's by-line. This was helpful in throwing off suspicion and, as we hoped it would, calmed the anxiety of his wife, who was wondering where he was and why he would not tell her what he was doing.

Laurence's work in the project started with a thorough review of our work in general. After undergoing an indoctrination in Washington, he visited Oak Ridge, Hanford and Los Alamos. His appearance at Los Alamos created quite a stir because a number of scientists there recognized him as a newspaper reporter and, until they finally learned his mission, were quite disturbed by thoughts of how he could have managed to infiltrate the project. In general, his first task was to write drafts of the releases as quickly as possible.

He was an eyewitness to the explosion at Alamogordo, and not long after that was sent overseas to Tinian, arriving just too late to be included in the observation plane for the Hiroshima bombing, but not too late to witness the take-off preparations. When the Nagasaki mission took off, he was an observer in the instrument plane that accompanied the bomber. On this occasion, he served as a pool correspondent, with the only limitation on his report consisting of a check on his dispatch for security before its release. It was this dispatch that led to a Pulitzer Prize for the best news story of the year, a justly deserved award.

Laurence had assisted us in preparing the White House press release weeks before the Alamogordo test, under the guidance of the Interim Committee,[5] and it had been approved by both Secretary Stimson and the President. The statement was quite long and emphasized the tremendous destructive power of atomic explosions. Some minor revisions were made after the Alamogordo test.

In addition to the White House statement, a release had been prepared for the Secretary of War, to be issued shortly thereafter. Other releases pertaining to certain details of the project were also ready at various key points throughout the country. They, of course, had to be kept under the most rigid control. The purpose of this wide distribution was to permit release at the locations where our work had been carried on.

Unfortunately, we had not, by the morning of August 6, received any report of the damage inflicted on Hiroshima. All we had was Parsons' visual estimate of the power of the bomb, which presumably was not at variance with the information received by the observing plane's instruments, and Farrell's very condensed report on the interrogation of the crews. We were thus fairly certain of the force and the height of burst, but only of that. Looking back, it all seems simple today, but it was not then. I felt it most important to get the announcement out; yet the announcement was predicated for its effectiveness upon enormous damage to the target. There were suggestions that we should delay our announcement. There were suggestions that we

[5] The Interim Committee was composed of nine civilians appointed by President Truman on Secretary Stimson's recommendation in the spring of 1945. They were to draft essential postwar legislation, prepare for the White House release of news, and advise generally on the steps needed to prepare the future handling of atomic energy in the United States.

should soften it. There were suggestions that I should get further information from Farrell on Tinian. There were innumerable suggestions. These did not come from General Marshall, but from various people in the office of the Secretary of War.

Mr. Robert A. Lovett, for whose astuteness I had the highest regard, advised softening our claims, using a very appealing argument, to wit: that the Air Force (he was then Assistant Secretary of War for Air) had proclaimed the complete destruction of Berlin on a number of occasions. He added, "It becomes rather embarrassing after about the third time."

Hoping to avoid any repetition of Mr. Lovett's experience, I did try to discuss the situation with LeMay over the telecon from Washington to Guam, and through him, with Farrell over radio telephone to Tinian. I soon learned, however, that by the time LeMay could be reached it would be too late for us to revise the release.

I realized then that if any changes were to be made they had to be made quickly and before I could obtain any confirmation. The pages involved would have to be remimeographed and then the completed announcements would have to be delivered to the White House. This would all take at least an hour. I finally compromised by making a minor change in the first paragraph, which I felt would not lessen the impact of the news on the Japanese, and would still leave us a loophole in case the bomb had not had the anticipated destructive force. My principal concern about an overstatement was that it might weaken the effect of the announcement on the Japanese.

During this time I was bending every effort to get in touch with Farrell on the telecon:

Washington: Have the officer in charge come to teletype and give us his name. We have a most urgent message which takes top priority by order of Secretary of War. . . .

Is your officer there? Instruction 1. Set up a circuit with Tinian immediately for standby. Instruction 2. Get General LeMay to teletype immediately for urgent conference requested by the Secretary of War. If he is not on the Island, get General Giles, or General Twining, or General Spaatz, in the order listed. Utmost urgency, please acknowledge. Instruction 3. Inform us immediately which officer is coming and how long it will take him to get there.

GUAM: Minute, please. Have to check to see if we can accept.

WASHINGTON: This circuit at the present time takes all priority and no traffic will be accepted except by order of the Secretary of War.

GUAM: OK. Fine.

WASHINGTON: Instruction 4. Contact Brigadier General T. F. Farrell on Tinian and have him go to the teletype there and stand by at request of Secretary of War.

Farrell had been awake for many hours during the operations; first for some twenty-four hours during the assembly of the bomb and the preparation for the plane's departure, and then he had stayed awake throughout the time of its flight and for some hours afterward. When he finally went to bed, it was at the other end of the island from the communications center, where he could not be reached easily. Yet, in spite of his exhaustion, for which a medical officer had given him some sleeping pills, when we finally got hold of him, Farrell rose to the occasion in a manner which most younger men might envy.

Finally, at about ten o'clock, I received the welcome news:

"General LeMay is present now."

Present in Washington: Present in Guam:
General L.R. Groves   General Curtis LeMay

GROVES: Important that we make press announcement here immediately. Has there been any confirmation of the information reported by the plane in Farrell's message to me this morning? Did you get that?

LEMAY: I am not familiar with General Farrell's message.[6]

GROVES: Can we guess?

LEMAY: Only confirmation of crews reports is pictures taken by K-20 camera from the tail gunner's position in the strike airplane. Target area of approximately three miles in diameter is covered by a grey dust like smoke. This smoke slopes to the center forming a mushroom of white smoke that rises to about 27,000 feet. This picture taken approximately three minutes after strike report the target completely covered with smoke

---

[6] It was characteristic of LeMay that he made no caustic comment, as he might well have done, about being kept in the dark about the information that had been sent to Washington.

and a column of dense white smoke rising to about 30,000 feet with a lesser concentration rising to 40,000 feet. F-13 arriving 4 hours later reports smoke columns still there. Have oblique pictures but expect no detail from them; there will be report of these pictures in two hours. Is this the information you want?

GROVES: Generally, yes. Do you have any estimate as to the results of F-13 trip over the target? As to the area and extent of damage or has it been impossible for them to get any idea due to smoke? Are cloud conditions satisfactory for photographic work?

LEMAY: No but pictures may show something. They have obliques only as they were briefed not to fly in cloud and smoke over the target. The F-13's reported the column still there upon their arrival four hours after bombs away.

GROVES: Does Farrell have any idea as to when pictures may be taken from the standpoint of the smoke columns? I should have thought that this column would have disappeared from the immediate target area within an hour of two.

LEMAY: The F-13 crews 4 hours after bombs away saw several fires near the dock area in the fringes of the cloud. Their magnitude could not be determined because of density of the smoke cloud.

GROVES: Will you ask Farrell thru your relay setup if he sees any reason for not releasing information to the public in the United States without delay?

LEMAY: General Farrell sees no reason why information on Hiroshima strike should not be released to American Public at once and strongly recommends such release.

GROVES: Will you please convey my congratulations and appreciation to the people out there, both mine and yours, and to you personally, my very best.

This information, while it did not confirm, did not upset my guess that the city had been destroyed. I saw no reason to alter my original decision to make the release, and so I took no steps to stop it, although there was still time.

As I look back on it, I feel that I should have discussed the matter with General Marshall or at least asked Arnold or Hull for their advice. At the time, however, I saw no reason why I should not make the decision on my own responsibility. It was true that the release had been previously approved by President Truman, but I think it only reasonable to believe that this was done on the assumption that we would have definite knowledge of the damage.

At the White House, everything was rather humdrum that morning. The press had been informed that there would be an important announcement by the President at eleven o'clock. White House correspondents are accustomed to such notices, and this one did not particularly arouse their interest. Many of them did not bother to go themselves, but sent assistants. Everything changed, however, when the President's Press Secretary arose and read the first few sentences. As the words "more than twenty thousand tons of TNT" came out of his mouth, there was a tremendous rush of reporters for the releases, which were on a table at the exit from the room, and then to the telephones and their offices.

The newspaper world was as astonished as everyone else. Because this news broke during the academic vacation period, many of the scientific reporters were away. They were immediately recalled. Then as the impact of this event began to be more fully realized, most newspapers published our releases in their entirety. This is one of the few times since government releases have become so common that this has been done.

As the papers came out, an important committee of the General Motors Corporation was holding a meeting in New York City. The members, among them Walter Carpenter, the president of du Pont, had left instructions that they were not to be disturbed. Shortly after noon I reached Carpenter on the phone and told him to get a newspaper and read the news. I added my warmest thanks for what he and du Pont had done. He sent a clerk out for a paper, and when it was brought in someone suggested that Charles Kettering, the famous inventor, be pressed into service to read the story on the grounds that he was a scientist and could understand and explain it. Carpenter said nothing for he did not know how much he could say about du Pont's role in the project. When Kettering was well along in the story, he read that du Pont had been responsible for a considerable part of the work. He dropped the paper abruptly and made a few harsh comments to Carpenter about letting him make a fool of himself. Carpenter explained that he didn't know how much security permitted him to say, so all he could do was read the papers like the rest of them.

In that he was typical of most of the members of the project. Their work was recognized at last, but they still were not free to discuss it.

CHAPTER 24

THE GERMANS HEAR THE NEWS

Shortly before supper on the sixth of August, Major T. H. Rittner, the British officer-in-charge at Farm Hall, where the German scientists were quartered, informed Otto Hahn that an announcement had just been made by the BBC that an atomic bomb had been dropped on Japan. Hahn was completely shattered by this news and said that he felt personally responsible for the deaths of hundreds of thousands of people, since it was his original discovery that had made the bomb possible. He went on to tell Rittner that he had contemplated suicide when he first saw the full potentialities of his discovery, and now that these had been realized, he felt that he personally was to blame. After bracing himself with alcoholic stimulants he became calmer, and went down to supper, where the news was announced to the assembled guests.

It was greeted with incredulity. The discussion revolved excitedly about how the United States must have produced Element 93.

HAHN: An extremely complicated business; for "93" they must have a machine which will run for a long time. If the Americans have an uranium bomb then you're all second-raters. Poor old Heisenberg.
HEISENBERG: Did they use the word uranium in connection with this atomic bomb?
HAHN: No.
HEISENBERG: Then it's got nothing to do with atoms, but the equivalent of 20,000 tons of high explosive is terrific. . . . All I can suggest, is that some dilettante in America knows it has the equivalent of 20,000 tons of high explosive and in reality, it doesn't work at all.
HAHN: At any rate Heisenberg, you're just second-raters, and you may as well pack up.

HEISENBERG: I quite agree. . . . I am willing to believe that it is a high pressure bomb and I don't believe that it has anything to do with uranium, but that it is a chemical thing where they have enormously increased the whole explosion.[1]

I took pleasure in Dr. Hahn's statement: "If they have really got it, they have been very clever in keeping it secret." Soon there crept into the discussion the question of morality, which has been so violently debated in all scientific circles ever since. The words of the German scientists speak for themselves on this subject:

WIRTZ: I'm glad we didn't have it.
WEIZSÄCKER: I think it's dreadful of the Americans to have done it. I think it is madness on their part.
HEISENBERG: One can't say that. One could equally well say, "That's the quickest way of ending the war."
HAHN: That's what consoles me.

There followed long and intensive discussions on how we could have built the bomb. In the light of later claims to the contrary, Heisenberg's remarks at this point were most significant: "There are so many possibilities, but there are none that we know. That's certain."

All of them agreed that they could have succeeded had they been able to make the necessary effort. They finally broke up on a note of wishful thinking and solace:

HAHN: Well, I think we'll bet on Heisenberg's suggestion that it is a bluff.
HEISENBERG: There is a great difference between discoveries and inventions. With discoveries one can always be skeptical and many surprises can take place. In the case of inventions, surprise can really only occur for people who had not had anything to do with it. It's a bit odd after we have been working on it for five years.

At nine o'clock the guests were assembled to hear the official radio announcement. They were completely stunned to learn that the news was, in fact, true. Immediately, intensive discussions began on the magnitude of the American effort. My first impressions of Goudsmit's

---

[1] All reported conversations were translated from the spoken German.

shrewdness as an interrogator were confirmed by Bagge's remark: "Goudsmit led us up the garden path."

The Germans seemed most impressed that we were able to accomplish the vast amount of work that they realized we must have done, and that they had been unable even to begin under the Third Reich.

KORSHING: That shows at any rate that the Americans are capable of real co-operation on a tremendous scale. That would have been impossible in Germany. Each one said that the other was unimportant. . . .
HEISENBERG: One can say that the first time large funds were available in Germany was in the spring of 1942, after that meeting with Rust, when we convinced him that we had absolutely definite proof that it could be done.

Heisenberg lamented his inability to devote an effort to the German nuclear project that was commensurate with the effort made on the V-1 and V-2 missiles. However, in the final analysis he seemed to realize that this was in large part the fault of his own group.

HEISENBERG: We wouldn't have had the moral courage to recommend to the government in the spring of 1942, that they should employ 120,000 just for building the thing up.

Then up cropped the apology that the reason the Germans had not succeeded was because they had not really wanted to succeed.

WEIZSÄCKER: I believe the reason we didn't do it was because all the physicists didn't want to do it, on principles. If we had all wanted Germany to win the war we could have succeeded.
HAHN: I don't believe that, but I am thankful we didn't succeed.

In the course of the evening Gerlach said that the Nazi party seemed to think that they were working on a bomb. The party people in Munich were going around from house to house on April 27 or 28, 1945, telling everyone that the atomic bomb would be used the following day. But the most surprising statement came from Heisenberg. He wondered how we were able to separate the two tons of

U-235 needed for a bomb. This confirmed Goudsmit's belief, founded on his interrogations, that the Germans had not thought of using the bomb designs we used. Ours took advantage of fast neutrons; the Germans thought that they had to moderate them as in a pile. In effect, they thought that they would have to drop a whole reactor, and to achieve a reasonable weight they would need this enormous amount of U-235.

More discussion followed and finally the group broke up for the night. However, conversation by pairs continued far into the early hours of the morning. In the course of these exchanges it became apparent that Gerlach was the only one who was completely grief-stricken about their lack of success. The others seemed glad to claim that their goal had been atomic power rather than a bomb. A prophetic remark on the international implications of the bomb came from Weizsäcker.

WEIZSÄCKER (to Wirtz): Stalin certainly has not got it. If the Americans and the British were good imperialists they would attack Stalin with the thing tomorrow, but they won't do that. They will use it as a political weapon. Of course, that is good, but the result will be a peace which will last until the Russians have it, and then there is bound to be war.

A feeling swept over the group that perhaps now they would be allowed to go home; then a reaction to the apology advanced by Weizsäcker began to grow:

BAGGE: I think it is absurd for Weizsäcker to say he did not want the thing to succeed. That may be so in his case, but not for all of us.

The next morning our guests read their newspapers avidly, and devoted themselves throughout the remainder of the day to drawing up a memorandum to clarify the press reports on their progress in atomic development. In its final form, this memorandum read:

As the press reports during the last few days contain partly incorrect statements regarding the alleged work carried out in Germany on the atomic bomb, we would like to set out briefly the development of the work on the uranium problem.

1. The fission of the atomic nucleus in uranium was discovered by Hahn and Strassmann in the Kaiser Wilhelm Institute for Chemistry in Berlin in December, 1938. It was the result of pure scientific research which had nothing to do with practical uses. It was only after publication that it was discovered almost simultaneously in various countries that it made possible a chain reaction of the atomic nuclei and therefore for the time a technical exploitation of nuclear engines.

2. At the beginning of the war a group of research workers was formed with instructions to investigate the practical application of these energies. Towards the end of 1941, the preliminary scientific work had shown that it would be possible to use the nuclear energies for the production of heat and thereby to drive machinery. On the other hand it did not appear feasible at the time to produce a bomb with the technical possibilities available in Germany. Therefore, the subsequent work was concentrated on the problem of the engine for which, apart from uranium, heavy water is necessary.

3. For this purpose the plant at the Norsk Hydro at Rjukan was enlarged for the production of larger quantities of heavy water. The attacks on this plant, first by the Commando raid, and later by aircraft, stopped this production towards the end of 1943.

4. At the same time, at Freiburg, and later at Celle, experiments were made to try and obviate the use of heavy water by the concentration of the rare isotope U-235.

5. With the existing supplies of heavy water the experiments for the production of energy were continued first in Berlin, and later at Haigerloch (Württemberg). Towards the end of the war this work had progressed so far that the building of a power producing apparatus would presumably only have taken a short time.

Throughout the remainder of the period during which they were in our custody, the professors continued to speculate about how we had succeeded in producing the bomb. Heisenberg and others of his colleagues periodically gave scientific lectures. V-J Day led to expressions of considerable relief and intensified their feeling that they would soon be going home.

We were now in a dilemma about what to do with these men. We did not want them to come to America or to remain in England for they would inevitably learn a great deal about our work and would not for sometime make any contribution in return. We did not want them to come under Soviet control, as with their background they would be of great value to the Russians. The only satisfactory solution was to return them to West Germany and make certain that conditions for them there would be such that they could not be tempted by Russian offers. This would take time to arrange.

At about this time they began to have a few visitors. Sir Charles Darwin was the first and he was soon followed by Professor P. M. S. Blackett. Blackett came ostensibly to discuss a revival of science in postwar Germany, but, in the course of their conversation, he proposed announcing all the details of our work. Heisenberg replied promptly and emphatically that this should not be done, because the Russians would never co-operate in any attempt at international control.

I had made arrangements for one of my lieutenants to deliver letters from the scientists to their families in Germany and pick up the replies. This did a great deal to improve their morale. It was interesting to note throughout this period that the professors considered themselves still to be *Herrenvolk* and continued to feel that uranium was a German monopoly. They were most indignant that the wife of one of them, who lived then in the French zone, was cooking for French soldiers. However, gnawing away at them all the time, consciously and subconsciously, was the distress of knowing that others had succeeded where they had failed.

HEISENBERG: Well, how have they actually done it? I find it is a disgrace if we, the professors who have worked on it, cannot at least work out how they did it.

After Blackett's visit the professors, particularly von Weizsäcker, became interested in the impact of atomic energy upon diplomacy.

WEIZSÄCKER: There is a certain trend in the world which is now beginning to appear; let us call it "internationalism." There are quite a number of people, especially in England and America, who think that way and I don't know at all whether they're doing their countries any good. But, they are the people to whom it is best for us to attach ourselves, and we'll have to support that. Those people who don't want to keep any secrets about the atomic bomb are the people who are useful to us.

On October 2, Heisenberg, Hahn and von Laue met with a number of British scientists at the Royal Institution in London, to discuss their return to Germany. It was made clear to them at that time that we wanted the German atomic scientists to live in either the American or British zones of occupation. The professors replied that this might present problems, since they would have to have laboratory facilities available if they were to perform any useful work and that the institutes existing in the United States and British zones were limited and inadequate.

In the meantime, we had been having trouble with Marshal Montgomery, who flatly refused to accept any of the scientists in the British zone under the condition that we specified—protective custody without confinement. Montgomery stated most emphatically and quite understandably, I might add, that anyone who was to be placed under his custody for safekeeping must be confined and kept under close surveillance. Negotiations on this point dragged on and the professors became apathetic. They discussed the possibility of withdrawing their parole, but eventually they became overwhelmed by their frustration.

HEISENBERG: I just think that talks with the Captain are somehow futile. He listens to us, and then passes it on to the Commander, already with certain reservations. Then they have a talk about it, air their feelings a bit as to how unpleasant the whole thing is, and with that things have really come to an end as already the Commander does not pass this on to a higher authority any more. Perhaps if we are very lucky the Commander tells the competent Colonel or General who is sitting here in London. It is impossible that it will ever reach America where a decision will be made.

It may not be too late yet for Dr. Heisenberg to derive a little solace from knowing that his words to the Captain always reached me in Washington promptly, and that I took action to secure his immediate release as soon as we had accomplished our purpose with him.

A brief furor was stirred up on November 16, when the scientists read in the newspaper of the award of the Nobel Prize to Dr. Hahn. A considerable amount of tension arose when this announcement was not immediately confirmed. When the invitation finally did arrive from Stockholm, Hahn was required to reply that he accepted the prize, but would be unable, for the present, to receive the award in person.

On the twenty-second of December, the scientists were finally notified that they were about to return to Germany. By this time we had been able to adjust matters so that reasonably suitable laboratory facilities were available for their use in the American and British zones. It is noteworthy that not a single one of these men left for the East despite the quite attractive offers they must have received from the Soviet Union. Just before they left, I was most amused to read a report from Captain Brodie, who was Major Rittner's assistant at Farm Hall, entitled "Wirtz Hauls Down His Colours, or He Who Fights and Runs Away, Lives to Fight Another Day." The Captain reported that on the day of their departure, the professors decided that, all things considered, their treatment had not been so bad after all, and that perhaps now would be a good time to mend their political fences.

WIRTZ: There is a lot to be said for the Commander after all, no matter how much we may have cursed him. In any event, it may be wise to be in his good books. We never know when we may have another use for him.

CHAPTER 25

NAGASAKI

Throughout July and into August, until the first atom bomb was dropped on Japan, the center of dramatic interest, naturally, was Tinian. At the same time, however, our test program at Los Alamos for the second implosion bomb had been going on at an accelerating rate. The main problem was that the company manufacturing certain essential parts for a nonatomic assembly in the Fat Man had been unable to meet delivery schedules. This reduced the number of tests possible on that particular assembly. It also prevented efficient over-all testing, since many tests had to be made twice, once with all components except the missing one and then at a critically late date with a complete assembly minus the nuclear components. In fact, the first live tests of the missing part were conducted only a few days before the actual dropping of the bomb on Nagasaki.

After it had been finally processed and formed into the exact shapes at Los Alamos, the plutonium for the Fat Man was flown to Tinian in a special C-54. Some of the other vital parts for two additional Fat Men were flown out in two B-29's belonging to the 509th, which had been held at Albuquerque especially for this purpose. In all cases, the plutonium shipments were accompanied by special personnel to guard against accident and special precautions were taken to ensure that if a plane carrying any of the plutonium did crash, we would have a fairly good idea of where it went down.

Like the Little Boy, the first Fat Man was to be used in combat as soon as we had enough fissionable material. Toward the end of July, the bomb was rescheduled to be dropped on August 11, which was an improvement over our previous target date of August 20. By August

7, it had become apparent that we could probably slice off another day from our schedule.

Admiral Purnell and I had often discussed the importance of having the second blow follow the first one quickly, so that the Japanese would not have time to recover their balance. It was Purnell who had first advanced the belief that two bombs would end the war, so I knew that with him and Farrell on the ground at Tinian there would be no unnecessary delay in exploiting our first success.

Good weather was predicted for the ninth, with bad weather to follow for the next five days. This increased the urgency of having the first Fat Man ready still another day earlier. When the decision to do so was reached, the scientific staff made it clear that in their opinion the advancement of the date by two full days, from the eleventh to the ninth, would introduce a considerable measure of uncertainty. I decided, however, that we should take the chance; fortunately all went well with the assembly, and the bomb was loaded and fully checked by the evening of August 8.

Six Pumpkin-carrying planes were assigned various targets in Japan for the eighth, but because of weather only two of them reached their primary targets; three of them reached secondary targets, and one aborted and returned to Tinian. In the field order for the second atomic mission there was nothing to indicate the extraordinary nature of the bomb, although anyone reading it would realize that this was by no means a routine assignment.

There were only two targets designated this time: Kokura, primary; and Nagasaki, secondary. Niigata was not made a third target because of its great distance from the other two cities. To increase the chance of using the primary target, the strike plane, no matter what the weather report, was ordered to pass close enough to it to make certain that visual bombing was not possible before it went on to the secondary target. To avoid any chance that the photo planes would arrive too early, they were required to check in with both Iwo Jima and Tinian before proceeding past Iwo Jima. In case of doubt, due to inadequate information, they were to photograph both targets.

The Kokura arsenal was one of the largest war plants in Japan. It produced many different weapons and pieces of war equipment. It extended over almost two hundred acres and was supported by numerous machine shops, parts factories, electric power plants and the usual utilities.

Nagasaki was one of Japan's largest shipbuilding and repair centers. It was important also for its production of naval ordnance. It was a major military port. The aiming point was in the city, east of the harbor.

The strike plane and the two observing planes took off shortly before dawn on the ninth. Major Charles W. Sweeney was pilot of the strike ship, Captain Kermit K. Beahan was the bombardier, Commander Ashworth was the weaponeer, and Lieutenant Philip Barnes was the electronics test officer.

It was not possible to "safe" the Fat Man by leaving the assembly incomplete prior to take-off, as had been done in the case of the Little Boy. There was considerable discussion among the technical staff about what would happen if the plane crashed, and possibly caught fire, while it was taking off. They realized that there would be a serious chance that a wide area of Tinian would be contaminated if the plutonium were scattered by a minor explosion; some thought that there was even a risk of a high-order nuclear explosion which could do heavy damage throughout the island's installations. Of course, we had gone into all this at length during our preliminary planning, and on the basis of my own opinion, as well as that of Oppenheimer and my other senior advisers, that the risk was negligible I had decided that the risk would be taken.

As happens so often, however, there was constant interference by various people in matters that lay outside their spheres of responsibility. Throughout the life of the project, vital decisions were reached only after the most careful consideration and discussion with the men I thought were able to offer the soundest advice. Generally, for this operation, they were Oppenheimer, von Neumann, Penney, Parsons and Ramsey. I had also gone over the problems at considerable length with the various groups of senior men at Los Alamos, and had discussed them thoroughly with Conant and Tolman and with Purnell and Farrell and to a lesser degree with Bush. Yet in spite of this, some of the people on Tinian again raised the question of safety at take-off at the last moment. Their fears reached a senior air officer, who asked for a written statement to the effect that it would be entirely safe for the plane to take off with a fully armed bomb. Parsons and Ramsey signed such a statement promptly though with some trepidation, possibly with the thought that if they were proven wrong they would not be there to answer. Ramsey then advised Oppenheimer at once of the various design changes that must be made to ensure that future bombs would in fact be surely safe.

One very serious problem came up just before take-off, which placed Farrell in the difficult position of having to make a decision of vital importance without the benefit of time for thought or consultation. Despite all the care that had been taken with the planes, the carrying plane was found at the last moment to have a defective fuel pump, so that some eight hundred gallons of gasoline could not be pumped to the engines from a bomb bay tank. This meant that not only would the plane have to take off with a short supply of fuel, but it would have to carry the extra weight of those eight hundred gallons all the way from Tinian to Japan and back. The weather was not good, in fact it was far from satisfactory; but it was good enough in LeMay's opinion, and in view of the importance of dropping the second bomb as quickly as possible, and the prediction that the weather would worsen, Farrell decided that the flight should not be held up. Just before take-off Purnell said to Sweeney, "Young man, do you know how much that bomb cost?" Sweeney replied, "About $25 million." Purnell then cautioned, "See that we get our money's worth."

Because of the weather, instead of flying in formation, the planes flew separately. To save fuel, they did not fly over Iwo Jima but went directly to the coast of Japan. Their plan was to rendezvous over the island of Yokushima, but this did not work out. The planes were not in sight of each other during their overwater flight and only one of the observation planes arrived at the rendezvous point. The missing plane apparently circled the entire island instead of one end of it, as it was supposed to do according to Sweeney's plans. Although Sweeney had identified the one plane that did arrive he did not tell Ashworth. Unfortunately, because it did not come close enough, Ashworth was unable to determine whether it was the instrument-carrying plane, which was essential to the full completion of the mission, or the other, which was not. Sweeney's orders were to proceed after a short delay of fifteen minutes but he kept waiting hopefully beyond the deadline. The result was a delay of over half an hour before they decided to go on to Kokura, anyway.

At Kokura, they found that visual bombing was not possible, although the weather plane had reported that it should be. Whether this unexpected condition was due to the time lag, or to the difference between an observer looking straight down and a bombardier looking at the target on a slant, was never determined.

After making at least three runs over the city and using up about forty-five minutes, they finally headed for the secondary target, Nagasaki. On the way they computed the gasoline supply very carefully. Ashworth confirmed Sweeney's determination that it would be possible to make only one bombing run over Nagasaki if they were to reach Okinawa, their alternate landing field. If more than one run had to be made they would have to ditch the plane—they hoped near a rescue submarine.

At Nagasaki, there was a thick overcast and conditions at first seemed no better for visual bombing than at Kokura. Considering the poor visibility and the shortage of gasoline, Ashworth and Sweeney decided that despite their positive orders to the contrary, they had no choice but to attempt radar bombing. Almost the entire bombing run was made by radar; then, at the last moment, a hole in the clouds appeared, permitting visual bombing. Beahan, the bombardier, synchronized on a race track in the valley and released the bomb. Instead of being directed at the original aiming point, however, the bomb was aimed at a point a mile and a half away to the north, up the valley of the Urakami River, where it fell between two large Mitsubishi armament plants and effectively destroyed them both as producers of war materials.

On the way to Okinawa warning ditching orders were announced;

but the plane made it with almost no gas left. Sweeney reported there wasn't enough left to taxi in off the runway.

The Nagasaki bomb was dropped from an altitude of 29,000 feet. Because of the configuration of the terrain around ground zero, the crew felt five distinct shock waves.

The missing observation plane, which fortunately was the one without the instruments, saw the smoke column from a point about a hundred miles away and flew over within observing distance after the explosion. Because of the bad weather conditions at the target, we could not get good photo reconnaissance pictures until almost a week later. They showed 44 per cent of the city destroyed. The difference between the results obtained there and at Hiroshima was due to the unfavorable terrain at Nagasaki, where the ridges and valleys limited the area of greatest destruction to 2.3 miles (north-south axis) by 1.9 miles (east-west axis). The United States Strategic Bombing Survey later estimated the casualties at 35,000 killed and 60,000 injured.

While the blast and the resulting fire inflicted heavy destruction on Nagasaki and its population, the damage was not nearly so heavy as it would have been if the correct aiming point had been used. I was considerably relieved when I got the bombing report, which indicated a smaller number of casualties than we had expected, for by that time I was certain that Japan was through and that the war could not continue for more than a few days.

To exploit the psychological effect of the bombs on the Japanese, we had belatedly arranged for leaflets to be dropped on Japan proclaiming the power of our new weapon and warning that further resistance was useless. The first delivery was made on the ninth, the day of the Nagasaki bombing. The following day General Farrell canceled the drops, when the surrender efforts of the Japanese made any further such missions seem ill-advised.

Throughout the week following the Hiroshima strike, we were wrestling with a royal foul-up in the handling of press stories from the Pacific Theater. It will be recalled that Spaatz had been directed not to issue any communiqués on our mission. The purpose of this ban and on news stories as well was to avoid lessening the psychological effect of the bomb. Uncertain of how much the Japanese people would learn from their own government, we did not want to weaken the impact of the White House statement. Then, too, there was the chance that an announcement made without full knowledge of the background facts could arouse hope in Japan that other bombs would not follow. Our commanders in the Pacific were not in possession of the facts. Moreover, as far as we knew, there were no newsmen in the theater, besides Laurence at Tinian, who had any background in the atomic field.

We intended to remove the ban as soon as the situation cleared. Before the bombing, we had sent our people definite instructions that would guide them in briefing the overseas commanders on the security problems involved in clearing news stories, when this responsibility was released to the theater.

Frankly, we did not fully anticipate the difficulties that these initial restrictions would cause. The public relations section that had been set up in my office could have handled the expected number of messages. The difficulties arose from three sources: the large number of war correspondents on Guam, the extraordinary number of words each wanted to send, and a completely unforeseen bottleneck in communications, which we should have learned of in advance, but did not. Because the dispatches were still subject to clearance upon arrival in Washington, they could not be sent over the normal press communication channels, but instead had to be sent over a single security channel. This meant that only one message could be sent at a time.

Within twenty-four hours of the White House announcement, the War Department Information Office was under great pressure from all newspapers for relief. The hopelessness of the situation was apparent to everyone involved. It was also clear to me that with the reports already received there was no longer any need to maintain such close control. The restrictions we had imposed were lifted that evening and clearance power passed from Washington to Guam.

Major John F. Moynihan, Farrell's press assistant, was invaluable, particularly during those hectic early days when our plans for handling press dispatches proved to be so hopelessly inadequate. Had he not been there, we would not have been ready to release the control from Washington to the theater when we did.

After August 11, all information was handled by the theater forces.

Although we had certainly miscalculated in planning our press operations, we were at least well prepared in a different but related respect; and for a long time I gave thanks daily for the Smyth Report. As far back as early 1944, Conant and I had discussed the necessity of having ready some account of the operations of the MED, and of the preparatory work accomplished by the OSRD. It would be a formal record of our activities, particularly scientific and administrative, giving credit where credit was due, and would, we hoped, meet the heavy demands for information that were certain to deluge us, once the bomb proved successful. We also wanted something that would serve as a guide for persons connected with the project in their discussions with outsiders, to help them avoid unwitting breaches of security.

The Military Policy Committee approved, and in April, 1944, we asked Dr. Henry D. Smyth, of Princeton University, to prepare the report. Smyth had been long and closely associated with the project, both with the MED and the OSRD. He had served on the Uranium Committee beginning in the summer of 1941, and had been one of the first to support Ernest Lawrence's suggestion of the possibility of a large-scale electromagnetic process. In the fall and winter of 1943-44, he acted first as an associate director, and later as a consultant at the Chicago laboratories. In our opinion, he was eminently qualified to carry out this difficult task.

Smyth, who was exempted from all normal compartmental rules that might interfere with his work, visited each of the various elements of the MED, conferring with the key people and collecting data. As the rough drafts of his chapters emerged, he went over them with Conant and me. Each part of his manuscript was cleared with the persons directly responsible for the activities it dealt with. Within a year, twelve of the proposed thirteen chapters had been completed in preliminary draft form. After a thorough review of Smyth's work, Conant and I agreed that it should be prepared for public release, and set June 30 as the target date for its completion. We asked Tolman to help Smyth put the manuscript into final form for publication.

At the same time Smyth and Tolman, at my request, prepared the rules to be used in determining whether material could or could not be included in the published report. These, after discussion with Conant, were approved, with minor modifications. Under the rules, Tolman was appointed as the principal reviewer of the manuscript, to be assisted by W. S. Schurcliff, who served principally as an editor, and Dr. Paul C. Fine.

Many changes in the original draft became necessary as our security criteria were applied to it. Copies of pertinent sections were given a final review by scientists in the various parts of the project, both for factual content and for security considerations. In order to speed up this process, officer couriers delivered the copies, and generally waited until the review was completed. Each person was asked to give us a written memorandum stating whether the material he had read was acceptable and met with his approval, and to suggest any changes that he thought should be made. We were particularly anxious that everyone be accorded the recognition he deserved. This, we felt, would lessen the chances of future security breaks. The only hitch in our plans occurred at Los Alamos, where Oppenheimer was unable to spare the time required for a thorough review of the section dealing with his laboratory. However, we decided to go ahead with the report as written, which had taken into account comments from Los Alamos on an earlier draft, rather than to hold up publication.

The report was completed on July 28, but not before we had had to fly some fully cleared MED stenographers up to Washington from Oak Ridge. As a part of the final editorial review, Tolman reread every word of the report with extreme care and annotated it by marking all possible material involving security and citing the rule under which each of these particular passages had been allowed.

Finally he and Smyth went over these notes together and prepared a memorandum in which they stated that nothing in the report violated the rules that had been established.

On August 2, when the Smyth Report was ready for submission to the printer, a meeting was held in Mr. Stimson's office. Present were the Secretary, Harvey Bundy, Conant, Tolman, George Harrison, James Chadwick, Roger Makins (a British Minister in Washington, then handling atomic matters for his government), Colonel Kyle, the Secretary's aide, and myself. Mr. Stimson opened by saying that he felt the decision on publishing the report could well await the return of the President from overseas; he thought that in the meantime the two statements to be issued by the President and himself would satisfy all demands for information until the President could reach a decision. Conant advised Mr. Stimson to release the report, saying, among other things, that "its publication will help us defend against the inevitable cry for more information about the project." Faced with this clamor, we would almost certainly have serious breaches in our security. Conant and I both emphasized that the assistance provided by the report to any nation capable of duplicating the bomb would be negligible at best.

After the conference broke up, Chadwick read the report carefully and was quite disturbed by its contents. However, in the course of a meeting with Tolman and me, his fears were laid to rest. Two days later he summed up the situation quite precisely by saying: "I am now convinced that the very special circumstances arising from the nature of the project, and of its organization, demand special treatment, and that a report of this kind may well be necessary in order to maintain security of the really essential facts of the project. To judge how far one must go in meeting the thirst of the general public for information, and the itch of those with knowledge to give it away, so as to preserve security secretly on vital matters is indeed difficult, but so far as I am competent to express an opinion, I find myself broadly in agreement with my United States colleagues."

Dr. Chadwick went on to recognize that the information divulged in the report would be useful to foreign governments and others interested in atomic energy development, "not because any one item of information is particularly important, but from the illuminating effect of a well-arranged, coherent and well-written presentation on the development of the many aspects of this project." He continued: "At the same time, I would agree that such assistance to possible competitors is not as much as one might think at first sight; it is indeed more apparent than real.

"I have tried to form an estimate of this assistance as a saving in time to a competitor making a serious attempt to develop the . . . project. I believe that the saving might amount to a few months, say three. It could hardly be more."

Even before the Hiroshima bombing, a thousand copies of the Smyth Report had been prepared, using the Pentagon's top secret reproduction facilities. These copies were kept securely locked up in my office.

Soon after the President gave his approval, the report was released for the Sunday morning newspapers of August 12, 1945, and for use by radio broadcasters after 9:00 P.M., EDT, on August 11. The release was made through the Press Branch of the War Department Bureau of Public Relations, and was accompanied by the following statement:

Nothing in this report discloses necessary military secrets as to the manufacture or production of the weapon. It does provide a summary of generally known scientific facts and gives an account of the history of the work and of the role played in the development of different scientific and industrial organizations.

The best interests of the United States require the utmost co-operation by all concerned in keeping secret now and for all time in the future, all scientific and technical information not given in this report or other official releases of information by the War Department.

The following addition should be made to paragraph 12.18 of the Smyth Report: "The War Department now authorizes the further statement that the bomb is detonated in combat at such a height above the ground as to give the maximum blast effect against structures, and to disseminate the radioactive products as a cloud. On account of the height of the explosion, practically all of the radioactive products are carried upward in the ascending column of hot air and dispersed harmlessly over a wide area. Even in the New Mexico test, where the height of explosion was necessarily low, only a very small fraction of the radioactivity was deposited immediately below the bomb."

As expected, with the publication of the report, objections began to come in from a number of scientists in the project. Smyth examined each one carefully, and where they were justified prepared the appropriate corrections and additions for inclusion in later printings. Any objections that were not met represented nothing more than differences of opinion between individuals. On the whole, and considering the rather difficult conditions under which it was prepared, the Smyth Report was extraordinarily successful in its efforts to distribute credit fairly and accurately. It would have been impossible to have prepared any document for publication covering the work of the Manhattan District that every reader would have found to his liking. But the fact is that all those who had the greatest knowledge of the subject were nearly unanimous in approving its publication as it was finally written. And there can be no question that it excellently served its purpose as an essential source of accurate information, particularly for news-hungry America in the early days after Nagasaki.

Immediately after the Nagasaki drop, I went to see General Marshall about our future operations against Japan. By that time, as I have said, I had become convinced that the war would end just as soon as the Japanese could surrender. In view of the policy that Mr. Stimson had laid down when he deleted Kyoto from the target list, I did not want to provide any basis for later claims that we had wantonly dropped a third bomb when it was obvious that the war was over. Yet our production facilities were operating at such an accelerating rate that the materials for the next bomb would be ready for delivery to the field momentarily. General Marshall agreed completely with my appraisal of the situation and we decided that we should hold up all shipments of fissionable material until the thirteenth. Then if there was no surrender, shipments would be resumed.

When that deadline came, unfortunately, neither the General nor

Mr. Stimson was available, being deeply involved in the negotiations for the armistice. However, it seemed to me that under the circumstances it would be a terrible mistake for us to send overseas the ingredients of another atomic bomb.

I discussed the situation with Handy. He said that it was absolutely impossible for me to see the Secretary or the Chief of Staff that day. In that case, I said, I would continue to hold all fissionable materials in the United States and would appreciate it if he would tell Mr. Stimson and General Marshall this at his earliest opportunity. Some days later, General Marshall commented that he was glad I had taken that action.

Throughout the time that the surrender negotiations were going on, our entire organization both at Los Alamos and on Tinian was maintained in a state of complete readiness to prepare additional bombs in case the peace talks should break down. During this period, seven Pumpkin missions were flown against Japan in preparation for further atomic attacks, if they should become necessary.

When peace came, we closed out our activities on Tinian as quickly as possible. The civilian scientists left the island on September 17. Kirkpatrick and Ashworth remained behind to dispose of the project's property. Anything that might in any way disclose information about the bomb was either to be returned to Los Alamos under tight security guard or taken out to sea and scattered in deep water. This policy was generally observed, although there was the usual slackening off of security immediately after the first bombing, and certain exterior portions of the bombs were seen by outsiders. However, most, if not all, of the recognizable items were parts of the Pumpkin bombs, so this did not constitute a serious breach of any essential security measure.

The aim of the Alberta group had been to assure the successful combat use of an atomic bomb at the earliest possible date after the necessary fissionable material was available. That it was eminently successful was shown by the fact that the first combat bomb was ready for use against the enemy within seventeen days after the

Trinity test. These seventeen days were spent largely in accumulating, preparing and shipping the additional active material needed to complete the Little Boy bomb. Actually, Alberta's procedures proved to be so efficient that it could have had the first atomic bomb ready for the drop on July 31, if the weather had been favorable for its use—within two days of the time the last of the active materials was delivered. The second bomb was used in combat only three days after the first, though it was a completely different model and much more difficult to assemble.

Faced as we were with innumerable uncertainties in our operations against Japan, it had always been comforting to know that the 509th Group was willing and able to perform any task that was humanly capable of achievement. Yet the group had its problems, too. When the war was over, and Tibbets was asked what was the greatest difficulty that the 509th had encountered, he said it was the uncertainty; he never knew from one day to the next whether the plans of the previous day were still in effect. In spite of this, his men went about their work with quiet competence and accomplished their mission in the face of greater unknowns than had ever confronted a military organization.

Only four days after the first bomb was dropped, the Japanese began surrender negotiations. The letdown that had followed the success of the Trinity test became evident on a larger scale throughout the project. With the war ended, or about to end, many of our people began to discuss the future consequences of our work. The thoughts that they expressed were not particularly new, but until then, there had been little time to spend on nonessential conversation. Since 1939 they had been busy. Now they all realized for the first time that atomic energy was a fact and not a theory and they realized, too, that a nuclear war could never be fought on this earth without bringing disaster to all mankind. This had been immediately evident to everyone who witnessed the Trinity test.

When it finally came, V-J Day was a sober and thoughtful occasion for most of us who had labored so hard and so long to help bring it about. We had solved the immediate problem of ending the war, but in so doing we had raised up many unknowns. Our feelings at this time were eloquently summed up by Oppenheimer when, on October 16, 1945, I presented the laboratory with a Certificate of Appreciation from the Secretary of War:

It is with appreciation and gratitude that I accept from you this scroll for the Los Alamos Laboratory, for the men and women whose work and whose hearts have made it. It is our hope that in years to come we may look at this scroll, and all that it signifies, with pride.

Today that pride must be tempered with profound concern. If atomic bombs are to be added as new weapons to the arsenals of the warring world, or to the arsenals of nations preparing for war, then the time will come when mankind will curse the names of Los Alamos and Hiroshima.

The peoples of this world must unite or they will perish. This war that has ravaged so much of the earth has written these words. The atomic bomb has spelled them out for all men to understand. Other men have spoken them, in other times, of other wars, of other weapons. They have not prevailed. They are misled by a false sense of human history who hold that they will not prevail today. It is not for us to believe that. By our works we are committed to a world united, before this common peril, in law, and in humanity.

PART III

CHAPTER 26

THE MED AND CONGRESS

Shortly after V-J Day, Secretary Stimson resigned and Secretary Robert P. Patterson was appointed in his stead. Patterson had known what the MED was trying to do and in a very general way was familiar with our operations. A major part of our money had come from an appropriation under his control for expediting the procurement of war materials. He told me once, early in the project, that while he had no responsibilities in connection with it, he stood ready at any time to give me whatever assistance he could. Actually, I did not have to call on him for help very frequently. Whenever I did, he responded immediately, and his services were invaluable.

On one occasion, he told me he was worried about the amount of money that he had authorized me to use, and felt that after the war he might well be asked, and quite properly, to explain to the Congress whether this money had been wisely spent. He went on to ask if I would have any objection to having a representative of his go over the project, particularly the vast construction operations at Hanford and Oak Ridge, so that he would have assurance from his own man that all was well. I said I would be very glad to have an inspection made, but that it would be impossible to obtain any one man outside the project who could adequately cover it all. There were no scientists who understood such major construction and almost no engineers who would understand the kind of science with which we were involved.

He said that he was not interested in checking the scientific phases of the work, for obviously he had to take the advice of people who did understand it. He did feel, however, that the construction operation, about which many rumors had reached him, could be gone over adequately. I agreed and told him I would be glad to have him select a competent man to do this. He asked whether I would object to Jack Madigan, one of his special assistants, and a distinguished engineer of many years' experience. I assured him that Madigan would be most satisfactory to me, and that I would arrange for him to see whatever he wanted to throughout the project.

After a trip of several weeks, Madigan returned and told Patterson that he was ready to report. As Patterson told me afterward, he replied, "I'll have to see you this afternoon because I have to leave for a Cabinet meeting where I am sitting in for Mr. Stimson in five minutes." Madigan replied, "It won't take me five minutes—I can give you my report in thirty seconds." Intrigued by this, Mr. Patterson told him to go ahead, and Madigan reported, "If the project succeeds, there won't be any investigation. If it doesn't, they won't investigate anything else." Later that afternoon, he made a more detailed report, which satisfied the Secretary that there was no waste, that extravagance was not a keynote of the project, and that, in fact, it was being run on as economical a basis as anything of its size could be.

For reasons of military security, we had always made a determined effort to withhold all information on the atomic bomb project from everyone, including members of the Executive Department, military personnel and members of Congress, except those who definitely needed it and who were authorized to receive it. As a result our methods of obtaining funds had always been rather unorthodox.

During the early days, because of our rapidly changing plans, it had not been possible to establish any regular budgeting procedures. We were allocated funds that were already available to the War Department on an "as required" basis. For fiscal years 1945 and 1946, however, we had to ask for new funds. These requests were concealed in other requests for appropriations. During the entire period, we were allocated approximately $2,300,000,000 of which $2,191,-000,000 were expended through December 31, 1946.

In view of the unusual procedures by which we obtained funds and that we had to follow in spending these large sums of public money, I derived particular satisfaction from the testimony that the Comptroller General of the United States gave before the Senate's Special Committee on Atomic Energy in April of 1946. He said:

> We have audited, or are auditing, every single penny expended on this project. We audited on the spot, and kept it current, and I might say it has been a remarkably clean expenditure. . . .
> . . . the very fact . . . that our men were there where the agents of the Government could consult with them time after time assured, in my opinion, a proper accountability. . . .
> . . . from the very beginning, he [General Groves] has insisted upon a full audit and a full accountability to the General Accounting Office . . .

Before July 1, 1945, the majority of our money came from two sources—Engineer Service, Army; and Expediting Production; the latter being under the direct supervision of the Under Secretary of War. In justifying our requests for these funds, we were handicapped not only by the very size of the project and its many uncertainties, which made it impossible to budget in advance, but by the overriding need for secrecy, in the spending as well as in the getting.

Judge William P. Lipkin, then a finance officer with the rank of captain, recently told me that he remembers vividly what happened when he once questioned a rather sizable MED voucher that passed over his desk for payment. His superior told him firmly, "You will forget that you know anything about it. Just forget that you spoke to me about it. Just pay all MED bills and discuss the matter with no one."

We had a bad moment in late 1943, when Congressman Albert J. Engel informed the Under Secretary of War that information had reached him concerning our construction at Oak Ridge. He requested further details and stated his intention of visiting the site in the near future. In reply, he was told that this work was highly secret, and that the information he wanted could not be given to him; eventually, he was persuaded to forget his contemplated visit. Although we received numerous Congressional inquiries from time to time, in every instance the member concerned accepted our explanations, with some reservation, no doubt, and observed the War Department's request for secrecy.

We realized from the start that this could not go on forever, for our expenditures were too vast and the project was too big to remain concealed indefinitely. And, as always happens in the case of any large construction job, rumors and distortions of the facts abounded, and could understandably become a source of concern to any Congressman who heard them. We decided the only thing to do was to brief the leaders of both the House of Representatives and the Senate, so that they would understand and support the needed appropriations.

Consequently, on February 18, 1944, Secretary Stimson, General Marshall and Dr. Bush (we all felt it would be better if I were not present) met in the Speaker's office in the Capitol, with Speaker Sam Rayburn, Majority Leader John W. McCormack and Minority Leader Joseph W. Martin, Jr. Mr. Stimson reviewed the general state of the project and discussed the financial situation, including expenditures, available monies and estimated future requirements. He gave them our general program of construction, talked of the various possible procurement efforts and indicated an approximate schedule for the completion of our work. General Marshall talked of the project's relation to America's over-all strategic war plans, and Bush outlined the scientific background and explained the potentialities of the weapon.

The Congressmen indicated their approval without reservation. They said that, while the amount of money needed was large, they were in full agreement that the expenditures were justified, and that they would do everything possible to have the necessary funds included in the coming Appropriations Bill. It would not be necessary, they said, to make any further explanations to the Appropriations Committee.

It was agreed at this time that Rayburn would be given advance notice of how our requests for appropriations would be inserted in the bill. He would pass this information on to McCormack and Martin, and the three of them would then tell a few members of the Appropriations Committee that they had gone into the subject with

Secretary Stimson and General Marshall and that these items should not be questioned. The other members of Congress would be given only the most general reasons for the need to accord special handling to our requests for funds.

Later, at a similar meeting, Senators Alben W. Barkley, Wallace H. White, Elmer Thomas and Styles Bridges (respectively the Majority and Minority Leaders and the Chairman and Senior Minority Member of the Military Sub-Committee of the Appropriations Committee) were given essentially the same information. In his presentation, Bush quite optimistically stated that there were no scientists in either Britain or the United States, associated with our project, who did not believe that we would be successful. He went on to say that, while we had devoted our entire attention to the military use of atomic energy, he thought that atomic fission would ultimately prove to be of great benefit to the human race. The Senators showed a quick appreciation of the importance of the project and said they would have no trouble in handling our appropriations in much the same way as the House planned to.

Nevertheless, most of the members of Congress remained completely in the dark about our work.

Finally, in early February, 1945, when it was necessary to transfer other War Department funds to "Expediting Production" as a first step in making them available to us, Mr. Engel objected vigorously and demanded a detailed justification. Mr. Stimson handled this by inviting Engel to a meeting with me in his office, where we showed him some general information on unit costs of real estate, roads and housing. We were careful, however, not to give him any idea of the costs of our industrial facilities. This measure succeeded, for the moment, in warding off any charges based on rumors of great waste and unreasonably high costs.

Yet it was obvious that we would have more trouble in the future unless we gave fuller information to a few members of Congress. I urged, therefore, that we approach the Congressional leaders with the suggestion that, if they agreed, we invite a very small number of carefully selected legislators to visit Oak Ridge and, if they desired,

NOW IT CAN BE TOLD

Hanford as well. With President Roosevelt's approval, Mr. Stimson thereupon made an appointment for the two of us to discuss the matter with the House leadership.

This meeting never took place, for President Roosevelt died the day before it was to be held. For some reason, the War Department was not notified of the President's death immediately. Eventually, Secretary Stimson and General Marshall were asked to come to the White House, and were informed of it there. I learned of it from an officer whose wife heard it over the radio. When I telephoned the Secretary's office to verify my surmise that the meeting would not be held as scheduled, I learned to my amazement that this was the first intimation they had received of Mr. Roosevelt's death.

Later, President Truman and the House leadership concurring, we invited Congressmen Cannon, Snyder, Mahon, Taber and Engel to visit the Clinton Engineer Works at Oak Ridge in May. They appeared to be entirely satisfied with what they saw, and like everyone else were impressed with the hugeness and complexity of the installation. They did not raise a single question about possible waste; indeed they seemed thoroughly convinced that we had spared no pains to economize wherever it would not interfere with getting the job done. We told them about Hanford, but they indicated that they saw no necessity to go out there.

During their inspection of Oak Ridge, they were shown everything that I thought would be of interest to them. In order to permit complete freedom of discussion, I took the two senior members with me, Nichols took two more, and the fifth was taken by one of Nichols' principal assistants. Our cars were driven by senior officers who knew all about the Oak Ridge Project, so that there could be no security difficulty in carrying on an intelligent conversation at any time.

The only time I thought that there might be some question raised about the work came after our return to Washington. As I was saying good-by to them at the airport, Congressman Taber, who had long been renowned for his interest in keeping down government expenditures, said: "General, will you come over here a minute—I want to ask you a question."

My first thought was: "Well, here it comes," and so I was utterly astounded when Mr. Taber said: "There is only one thing that worries me, General. Are you sure that you are spending enough money at Oak Ridge?"

I assured him that I thought I was, but I hastened to add that in a project such as this no one could be absolutely sure; if we played it safe, we could never hope to win; chances had to be taken. Perhaps it would have been wiser to build a larger plant, but I did not think so. This rather vague answer seemed to put Mr. Taber's fears at rest. When I reported our little talk to Mr. Stimson, he was as amazed as I had been.

The Senate's special committee to investigate the national defense program became interested in various phases of the MED at an early date and we had some difficulty in keeping it from looking into the project. Any investigation of this sort would have endangered security and seriously impeded our progress, for it would have taken a great deal of effort and time on my part, as well as on the part of other key people in the project, to prepare the necessary answers for the committee.

Senator Truman, the chairman of the committee, agreed at Secretary Stimson's request to delay any investigation of the project as a whole until either security permitted it or the war was over. Later there was a proposal that two staff members of the committee, then both in uniform (General Frank Lowe, and Lieutenant Colonel Harry Vaughan), be sent to Pasco, Washington, near Hanford, to look into the question of alleged waste in the construction of housing, roads and other matters not connected directly with any process secret. Secretary Stimson replied that we would have to accept the responsibility for any waste or improper action that might otherwise be avoided by the work of the committee. Mr. Truman accepted the Secretary's decision in good grace. Throughout the chairmanship of Senator Meade, his successor, the same co-operative spirit prevailed.

After V-J Day, we arranged for Senator Mitchell of Washington, Senator Ferguson of Michigan and Senator Kilgore of West Virginia to visit the Hanford Engineer Works. Colonel Donald E. Antes, who represented me on this occasion, told them that the only figure he could give them was the over-all construction cost—$350 million.

They were also informed that I would be glad to appear before the committee and answer any questions that I could without violating security. The Senators and their party were the first civilians not directly connected with the project who had ever been permitted to enter a process building. Senator Ferguson summed up the reaction of the committee as a whole with the words: "Good job, well constructed, and there is no evidence of embellishments or extravagance of funds anywhere."

CHAPTER 27

THE DESTRUCTION OF THE JAPANESE CYCLOTRONS

A brief but violent flurry broke out at the end of November, 1945, when the news services carried accounts of the destruction of five Japanese cyclotrons. Two were in the Institute for Physical and Chemical Research at Tokyo, two were a part of Osaka Imperial University, and one was at Kyoto Imperial University.

There followed a period of very heavy correspondence for my office and that of the Secretary of War, and a number of papers were generated by the General Staff for internal consumption. Many scientific and intellectual societies joined individual citizens in protesting the destruction. They compared us to the Germans who burned the Louvain Library in 1914 and again in 1940, and denounced the Army's act as wanton and a crime against mankind—which it was not—and went on to label it stupid—which it most certainly was.

When the storm broke, I immediately tried to find out just how this situation had come about. Even today, with the benefit of considerable hindsight and after much thought, it is still impossible for me to reconstruct completely the chain of events that combined in this instance to produce a very serious error. However, although the complete story is not known, and probably never will be, it is possible to piece together enough of it to provide a valuable insight into some of the problems that grow out of any operation as large as that conducted by the War Department at the end of World War II.

The General Staff, on September 5, 1945, had issued instructions that all enemy war equipment be destroyed, except new or unique items, which were to be saved for examination. This directive was most explicit in requiring that "Enemy equipment not essentially or exclusively for war which is suitable for peacetime civilian uses should be retained."

This was amplified by a Joint Chiefs of Staff cable on October 30 to the commanders in the Pacific area and China, instructing them to seize all facilities for research in atomic energy and related matters, and to take into custody all persons engaged in such research. Although I do not believe I was consulted in the preparation of these instructions, I did eventually see the copy that was furnished to the Army for its information, and I most certainly would have concurred in the message as sent if I had been consulted. When the JCS cable arrived in my office, I called in one of my officers and went over its contents with him. We had long known of the existence of the five cyclotrons, and I was particularly anxious that special measures be taken to ensure that they were properly secured, in accordance with the JCS cable, but not destroyed. Yet, in the light of what followed, it is evident that I did not make my intentions clear.

On November 7, a message to General MacArthur was prepared by my office to go out in the name of the Secretary of War. The next day, this message was cleared through Mr. Patterson's office as a routine matter, without being brought to his personal attention. It was dispatched over his own special channels, which I sometimes used. As sent, the message ordered the destruction of the five cyclotrons in which we were interested, after all available technical and experimental data had been obtained from them.

On November 24, General MacArthur's headquarters in Tokyo reported by cable to the JCS that the cyclotrons had been seized on the twentieth, and that their destruction had commenced on the twenty-fourth. This report is especially interesting for two reasons: first, it referred only to the JCS cable, which ordered seizure—not destruction—of research facilities (apparently Tokyo was becoming confused at this point); and second, it was given unusually wide distribution, one copy being designated for me, among others.

The destruction of the cyclotrons should have come to an abrupt halt at this precise moment. However, of the nine persons in authority who were sent copies of General MacArthur's cable of November

THE DESTRUCTION OF THE JAPANESE CYCLOTRONS

24, not one—myself included—actually saw it. Every one of the nine copies seems to have been noted, initialed and filed by subordinate staff officers, in spite of all the detailed instructions that existed in every headquarters to prevent just such an occurrence.

In my own office, the confusion stemmed at least in part from the fact that the officer handling this matter was fairly new to the project, and was unaccustomed to our way of operating. Any officer with more experience in the project would have questioned my apparent desire to have the cyclotrons destroyed, and if I had persisted would have urged me to reconsider my decision. I am sure that the other responsible officials who were similarly unaware of what was going on were faced with similar problems.

And so the destruction of the cyclotrons continued to completion. Not until the story broke in the newspapers did this policy matter finally come to the attention of persons at the policy-making level.

The first realization that something was wrong came in General MacArthur's headquarters when, on November 28, a cable arrived in Tokyo from the General Staff, requesting that one of the cyclotrons be returned to the United States for re-erection and study. As these five units had just been reported destroyed, the discrepancy was immediately brought to General MacArthur's attention and he promptly dispatched a cable to General Eisenhower, personally, to acquaint him with the possibility that conflicting instructions had been emanating from the War Department. Apparently, Eisenhower (who by now had succeeded General Marshall as Chief of Staff) never saw this message, for no action was taken at that time.

The story broke on November 30, when the United Press reported that the cyclotrons had been destroyed by our occupation forces in Japan. Almost immediately, under a Tokyo date line, the New York *Herald Tribune* quoted sources within General MacArthur's headquarters as saying that orders from General Eisenhower directing that the cyclotrons be destroyed had been carried out.

These same sources went on to state that the destruction of cyclotrons had been initiated on orders from Washington and implied, I think correctly, that the authorities in Tokyo had complied with them most reluctantly. This report led to inquiries from the United Press, asking the War Department to clarify the differences that seemed to have sprung up between Washington and Tokyo.

The Tokyo dispatch had an immediate effect on the General Staff, and on December 2, General Hull cabled Lieutenant General R. K. Sutherland, MacArthur's Chief of Staff, that no such instructions as had been reported in the press from Tokyo had ever been issued by the War Department. He did indicate, however, that some part of the misunderstanding might attach to the War Department through its total failure to comment on MacArthur's cable of November 24 announcing that the destruction had begun.

On December 3, MacArthur replied to Hull's cable of the second with one directed to Eisenhower. In it he pointed out that he had received special instructions from the Secretary of War to destroy the cyclotrons, and went on to mention his attempt of November 28 to bring the matter of conflicting instructions to the personal attention of the Chief of Staff. He expressed his opposition to the destruction of the cyclotrons, and closed by stating that it had been necessary to issue the Tokyo news release in order to contradict false allegations made in the American press, since no action had been taken in Washington to correct the impression that the responsibility for the decision to destroy the cyclotrons lay with the occupation forces.

On the same day, General Hull, acting for the Chief of Staff, assured General MacArthur that he had acted correctly and indicated that the situation resulted entirely from a failure to co-ordinate outgoing messages in Washington.

We could not correct the destruction. There remained for us a very bad public relations situation which had been greatly aggravated by the wide variations between the statements from Tokyo and those from Washington. At that time, Secretary Patterson was considering a proposal that the War Department's response to criticism of the cyclotrons' destruction should be along the line that:

> . . . the action taken was in implementation of the established policy of the United States that the Japanese should be prevented from engaging in any activity related in any way to war making.

In order to ensure peace for generations to come we desire to eliminate to the maximum extent possible, the Japanese war-making potential. While it is recognized that a cyclotron may be used for scientific research in other fields, it is essential to the carrying out of atomic bomb research which our government believes should be prohibited to naturally belligerent and dishonest nations.

We cannot afford to pay the price twice that we have had to pay once. . . .

This did not appeal to me at all, and after we had discussed the matter at some length, Patterson finally approved the following explanation for release:

General MacArthur was directed to destroy the Japanese cyclotrons in a radio message sent to him in my name. That message was dispatched without my having seen it and without its having been given the thorough consideration which the subject matter deserved. Among other things, the opinion of our scientific advisers should have been obtained before a decision was arrived at.

While the officer who originated it felt that the action was in accord with our established policy of destroying Japan's war potential, the dispatch of such a message without first investigating the matter fully was a mistake. I regret this hasty action on the part of the War Department.

Secretary Patterson signed this statement himself. Only the general policy of the War Department precluded the mention of my name as being the responsible officer. I have always felt that Mr. Patterson took great pains in this case, as in every other, to avoid any appearance of passing the buck to me.

The press as a whole seemed quite surprised by this frank and open admission of error. As a result, this incident quickly lost its news value and the clamor soon subsided.

The lessons to be learned from this affair are many. They were brought home most forcibly to me; and others both within and without the military service can benefit from them just as well. The commander must always make his intentions unmistakably clear to his subordinates; I did not do so in this case. Even the most successful and competent organization will not continue automatically to produce perfect work when large numbers of new people are brought into it without previous experience or training; my confidence in the MED as an organization probably exceeded its capability at this time. There are many dangers inherent in turning over high-level matters to staff officers who have not been thoroughly prepared for their responsibilities; no matter what it cost us in effort, we should have provided formal training for these early replacement officers. Yet, over all these other lessons looms the basic truth that was demonstrated here again, that honest errors, openly admitted, are sooner forgiven.

CHAPTER 28

TRANSITION PERIOD

It had long been in my mind that with the end of the war we would lose the great majority of officers assigned to duty in the MED, including many on whom we had depended the most heavily. Like every other military command when V-J Day arrived, we were faced with the question of how fast we could reduce our work to peacetime needs and release our wartime personnel, yet still maintain a going concern.

As I explained earlier, the military side of the Manhattan Project was run almost exclusively by nonregular officers. Within a year after the project was set up, we had reduced the regular contingent to two—the District Engineer, Colonel Nichols, and myself. Later, as the war progressed, we picked up a few more as they returned from foreign service, but we never asked to have a man returned for this purpose.

At the war's end, most of our temporary officers had been in the service long enough to be entitled to prompt discharge. They were anxious to return to civil life, and the majority had excellent positions awaiting them. But replacing them was not going to be simple.

I determined that I would need about fifty regular officers to run the various elements of the MED. I wanted men who were young enough to break into the atomic field, but who were senior enough in rank to have demonstrated their ability to accept heavy responsibilities, and whose age would be an asset in their dealings with our scientific personnel, almost all of whom were extremely young.

It was important that the officers whom we selected command the respect of persons already in the project, not only of the officers they would be thrown with, but particularly of the scientists. Experience had shown that the scientists were most critical of anyone whose mental alertness did not equal or excel theirs. Slowness of comprehension or inability to keep all the pertinent facts in mind, once they were explained, was fatal, as not only military personnel, but also civilians of considerable capacity and reputation in their own fields had repeatedly discovered. For example, it was not always easy for a nonscientific mind to grasp the idea of a critical mass, and the fact that it was not possible to determine whether a bomb was feasible until a full-scale bomb was exploded. Failure to understand this was never forgiven.

Because we were in a hurry, we concentrated at first mainly on graduates of West Point. We wanted officers who as cadets had been highly regarded by the Academic Board not only for their scholastic achievements but for their other qualities. We preferred men who were among the first five or ten of their class, and we did not want anyone who stood below the first 10 per cent. A successful athletic career, demonstrating a more than average determination and will to win, was a particular asset.

Later, when it became necessary to secure a large number of young officers for duty at the new Sandia Base at Albuquerque for our bomb assembly teams, these requirements were relaxed slightly. Also, as time went on, it became possible to make a searching investigation of the background, educational and military, of other officers, not graduates of West Point, and to bring them in.

My request for highly qualified officers was not greeted with too much enthusiasm by the General Staff, whose position was very well put by General Handy, when he said there was no reason why I should have a solid group of the best officers in the Army; that there were other important things besides the Manhattan Project. Many officers I wanted were in rather important spots overseas. All of them were officers that no commander wanted to lose. Many were removed from the list when we learned of their current duties. We did not want to interfere with an officer's best interests if we could help it, and we did not want dissatisfied officers who were thinking about what might have been if we had not insisted on their assignment to us.

The situation came to a head when I' asked for Colonel K. E. Fields. This choice was vigorously opposed by General McNarney, who then commanded the American forces in Europe. Fields had been relieved from his command of an engineer group and placed in charge of the European Theater's athletic program. This was a very important job at that time because of the absolute necessity of maintaining morale in the Army of Occupation. Yet, I did not feel that it would be particularly disadvantageous either to the Army or to Fields if he were brought back; so I asked Handy to have the necessary orders issued.

As was customary in such cases, McNarney was queried and he objected strongly. Handy then told me that he could not go along with my request, that I was asking for too many good men, and that by my system of hand-picking I was getting more than my share. I replied that there was no place for anyone in the atomic field who was not a super-superior officer, that we simply could not use anyone else and, moreover, that the assignment to the MED of anyone who was less than super-superior would lead to adverse reactions among our scientific personnel, and through them, among the rest of the academic world and the press.

My next step was to talk to the Chief of Staff, General Eisenhower. He more or less went along with Handy on the theory that, after all, I should not expect to be able to hand-pick my entire organization.

A few days later, in the course of a conversation with Secretary Patterson the matter of personnel came up. When, in response to a query, I told him that I was having trouble getting an adequate number of competent people, he asked General Eisenhower to come in and to bring Handy with him. Handy defended his position vigorously, stating that I had been demanding the best and was not willing to accept anything but the best, that this was unfair to the other commanders, and that he felt that I should be turned down. Eisenhower supported him.

Although the discussion was quite forceful, and occasionally even heated, not once was any reference made either directly or indirectly, then or afterward, to the fact that this was a matter that lay within the province of the Chief of Staff. After about five minutes, Mr. Patterson said, "I agree with Groves," and went on to say: "I want him to have as many officers as he decides he needs and of the quality he thinks he needs, and I want him to have complete freedom of choice."

I should point out in this connection that throughout my entire association with the MED, and later with the Armed Forces Special Weapons Project, there was never a time when the Chief of Staff, either Marshall or Eisenhower, raised any question about, or indicated any displeasure over, the fact that the Secretary of War took up so many matters directly with me, and generally did so without their prior knowledge. Although this led to occasional embarrassment for me, and I am sure that more than once it caught them unawares, they were both big enough men to overlook such minor deviations from ordinary procedures.

So it was that the superb reserve officers of the project's wartime years were replaced by equally fine professionals. Only the high quality of our regular officers enabled us to weather the difficult period of demobilization between V-J Day and the activation of the Atomic Energy Commission.

Unfortunately, the policy of super-selection has not endured. Its demise is one of the principal reasons why the Army is no longer supreme within the military establishment in the area of nuclear warfare. Indeed, by insisting upon putting the best people in command of units in the field, and giving staff vacancies in the nuclear weapons field equal rather than preferred treatment, the field Army commanders, though they may have solved their immediate problems, have brought upon themselves a rather bleaker future.

Even greater than the officer problem was that of the demobilization of the scientific staff. Many of these men had come from academic jobs and were anxious to return to them. The senior men wanted to get back to their universities in order to re-establish themselves in their departments. The more junior men wanted either to take up the unusually fine positions that had been offered to them or to complete their work for advanced degrees. There was also the unrest that always occurs whenever a large group that has been held under a tight rein is suddenly released from control. Many men had stayed on long after they would have preferred to leave for other fields. With the great interest in atomic physics, many comparatively junior men were receiving offers from academic institutions far beyond anything they could previously have expected to get after even twenty years of experience.

There was also the usual feeling that the great goal had been achieved and that there was nothing to look forward to. Primarily for this reason, I was certain that most of the senior men would want to leave promptly. There were also a few whom I was anxious to see leave for security reasons. In the main, these men had been engaged in the atomic project before the Army entered the field, and had been retained in the belief that from a security standpoint it was safer to keep them than it was to let them go. I also felt it likely that the American people would demand very rigid security clearances in the future, and that this was a time when anyone not possessing a perfect pre-MED record could leave the project without stigma of any kind, and with full recognition, well deserved, of having done a top-notch job for our country.

We did not think that the breaking up of the big laboratory at Columbia would be a serious handicap to us; its work was done. The Berkeley laboratory, I felt, would continue as long as Ernest Lawrence lived, provided it received proper financial support from the government. The Chicago laboratory had already been considerably reduced in size, as many of the scientists had moved on to Los Alamos; and with the need to develop the peaceful uses of the atom it could be reasonably maintained.

It was particularly important to continue the Los Alamos laboratory so that the nucleus of a staff for future weapon improvement would always be available; this presented a much more difficult problem. Oppenheimer had told me that he wanted to leave as soon as he could, and we discussed a possible successor for him. It would have to be someone with sufficient prestige to secure the co-operation of his colleagues at Los Alamos, and the assistance of distinguished scientists throughout the country—particularly of those who were now leaving the project. Also, he should not be one of the top four or five men in the project, for I wanted him to feel that this was a great opportunity for him.

After much thought and considerable discussion with Oppenheimer and others I asked Dr. Norris Bradbury to take the position. Bradbury had spent several years at Los Alamos and had played an important part in the development of the gun-type bomb. Also, he was a Navy reserve officer, a circumstance I thought would help him in maintaining smooth relations between the civilian scientific staff and the military administrative officers.

Bradbury accepted with the understanding that he would not want to remain at Los Alamos indefinitely, and would, after not more than five years, return to academic life. Fortunately, he is still there, and the results of his work are a source of great pride to those of us who had a part in his selection. His performance has exceeded our expectations, and those were extremely high.

Bradbury's appointment was well received, and greatly facilitated our efforts to build up a thoroughly competent staff in the laboratory. The new divisional heads were selected by Bradbury, with the advice of Oppenheimer and others at Los Alamos. Because of the excellence of their choices, my part was purely confirmatory. These men, like Bradbury, had all demonstrated qualities of leadership as well as technical competence, and held the respect of their fellows and of their superiors.

It was not long before the question of whether to relocate the laboratory or to build up Los Alamos as a permanent installation came to the fore. A number of people strongly urged that we abandon the Los Alamos site and re-establish the laboratory in southern California. They proposed that the work be concentrated in laboratories to be constructed in the general Los Angeles area, and that all experimental work requiring room be conducted in the vicinity of Inyokern, across the mountains in the desert area. No other site was seriously considered.

I decided against moving the laboratories for several reasons. Chief among them was a strong feeling that to separate the organization into a comfortable laboratory setup in an urban area, while all major experimental work was done at a point where living was not easy, would not only result in considerable operational difficulty, but also in a great deal of lost time and expense. We also had a major investment in Los Alamos—a site that had proved to be entirely satisfactory in every way, except for its distance from any large center of population, and this was a disadvantage I thought would largely disappear now that the war was over.

In order to make all concerned realize that Los Alamos would be an enduring affair, the decision had to be made without delay. It would not have had to be made by me if the Congress had passed necessary legislation for the peacetime setup with reasonable promptness, but as the hearings dragged on, it became impossible to wait any longer. One of the first steps was to initiate the construction of new permanent family housing at Los Alamos, which convinced all doubters that the site would not be abandoned. Nevertheless, this was the kind of decision I did not feel that I should be making, for it definitely committed my successors, whoever they might be, for years to come.

Another pressing problem during the demobilization period involved the thermal diffusion plant. This plant had been built in an abnormal hurry. While its construction had been justified at the time because it enabled us to speed up the production of the Hiroshima bomb, it did not seem wise to keep such an expensive plant going when time was no longer the all-controlling consideration. I therefore decided to close it down, though I knew that once it was shut down it would be extremely difficult to put back into operation.

This, of course, meant a sizable reduction in employment, but as people left our other plants for civilian jobs, there were plenty of vacancies for those who wanted to continue to work at Oak Ridge.

Another decision that had to be made without any guidance from Congress was in connection with the gaseous diffusion program. Here, we had started in early 1945 to extend the plant facilities in order to make certain that we would have an ample supply of uranium in case the bombs were not so effective as we expected them to be, and we would therefore need a much greater number of them. I decided that this work should be continued and the plant finished.

At the same time I was receiving little if any guidance from the Executive Branch, since both the Secretary of War and the Chief of Staff, having only recently come into the atomic picture, felt that my background enabled me to make the necessary decisions better than they could. Nevertheless, throughout this period and until the date of my retirement, I had no difficulty in seeing both Mr. Patterson and his successor, as well as General Eisenhower, whenever I felt there was a need. As a matter of fact, I saw even more of them and discussed our problems in much greater detail with them than I ever had with Secretary Stimson and General Marshall.

When I first started to explain the atomic program to Mr. Patterson and General Eisenhower, they said that they did not want me to give them any secret information if I could help it, particularly about the production rates and the number of bombs on hand. As General Eisenhower put it, "I have so many things to deal with that it puts an undue burden on me to be given any secret information, as I am then forced to think constantly about what is secret and what is not. In a project such as this, where knowledge is held to such a few people, it makes it particularly difficult." Not until there was an actual threat of severe criticism from one Senator that the Secretary and the Chief of Staff were placing too much reliance on me was there any change in this policy. Even then, both expressed their regret that they had to have the information. Personally, I was relieved, since I had always thought they should have it.

During this trying period I was no longer able to turn to the Military Policy Committee for advice, for it had gone out of existence at the end of the war. However, with the exception of General Styer, who had left Washington for foreign duty, its members were always available for unofficial advice and consultation whenever I needed them.

I continued to be concerned about Los Alamos, where there was a natural but definite relaxation of activity. While plans for the future were being made, I arranged for members of the scientific panel of the President's Interim Committee to go out to the laboratory to discuss the technical potentialities of atomic energy. A series of lecture courses was organized—called the Los Alamos University —to give the younger staff members a chance to learn some of the things that they had missed during the war years, about both the actual development of the bomb at Los Alamos and recent scientific progress. It was during this period, too, that under the editorship of Hans Bethe, the Los Alamos technical series was prepared. These papers set down a systematic record of the laboratory's work during the war.

At the same time, a number of the scientists were doing some theoretical work on future bomb designs. Weapons production was being continued and plans made to carry on the work on the Fat Man, including certain changes that we felt would improve its explosive power.

I asked Oppenheimer to visit Los Alamos and tell me what I could do to help Bradbury during this difficult transition period, because I thought that since he had previously been so close to the situation and to the people he could get to the heart of matters better than I could. There was no question but that Bradbury needed all the help that I could give him if he was to have an adequate operating laboratory by June, 1946, which was the goal we had set. There was also the Bikini test to get ready for.

After his visit in November, Oppenheimer wrote to me, recommending two projects for my immediate approval, and asked for my guidance on two others. He urged me to expedite the declassification of information insofar as it was possible (plans for this were already under way); he also recommended that we arrange more efficient channels for classified communication between Los Alamos and the other MED activities, and that qualified scientists at other sites be authorized to discuss their work with their counterparts at Los Alamos.

Bradbury had told Oppenheimer that the senior regular officers who had recently been assigned to Los Alamos were not really equipped to supervise the more specialized technical aspects of the work, regardless of their excellent engineering and military training. I knew this and had not put them there for that purpose, but rather to familiarize them with the broader phases of the development. Bradbury did think they would be most useful at Sandia, our location in Albuquerque where the nonexplosive parts of the bomb were being assembled. He said that military personnel should not be assigned to the laboratory except at his specific request and that no attempt should be made to provide military personnel for responsible positions in the various physics or gadget (bomb components) divisions. This had never been our intention, since none of our regular officers had the necessary scientific knowledge for those jobs.

In his letter, Oppenheimer also made the point that there simply was not enough suitable housing to provide for both the laboratory staff and the Army contingent required for the nontechnical phases of the operation. The laboratory staff alone, at a minimum, came to something like fourteen hundred, with the word minimum emphasized.

Many things that had been merely troublesome in the past were intolerable in a permanent peacetime community; yet nothing had been done about any of them in the three months after V-J Day. Bradbury recommended that the censorship of mail be discontinued, and that no insurmountable obstacles be put in the way of having friends and relatives for visits on the post, provided that this would not require increased housing, and that, in certain specific cases, the visits be approved by the Intelligence Office.

Bradbury also said that proselyting of his staff by men who were at Los Alamos but who had plans for going elsewhere put a severe strain on his ability to keep the laboratory together. The University of Chicago was probably the worst offender, but others were also engaged in wholesale recruiting campaigns, disregarding the hurt to Los Alamos. Oppenheimer suggested, first, that we encourage the departure of the offenders, and second that Bradbury be enabled to offer to those men he wanted to keep definite reassurances about the technical program, about future living conditions and about their salaries.

Oppenheimer told me that Bradbury found it hard to explain "a statement attributed to you that you had lost your first and second quality scientists and were in danger of losing your third, fourth and fifth." This was like many another statement attributed to me in that I had never said it or anything resembling it. It was true that we had lost some scientists of great reputation, and that we had lost many of the heads of the various groups, but Los Alamos had been so strongly staffed that even the men whose reputations were not so widespread were nevertheless possessed of great ability and, in my opinion, were fully capable of doing an excellent job and of carrying forward new ideas just as vigorously as would the slightly older men of greater reputation. To tell the truth, with the bomb now an accomplished fact and the war over, I had a feeling that the new leaders would approach the work ahead with greater enthusiasm than their predecessors might.

After looking into the matter for me, General Farrell recommended that I: (1) telephone Oppenheimer and Bradbury to say that I would formally approve Bradbury's plans for the technical programs as soon as I received a letter from him on the subject; (2) informally approve his plans by telephone; (3) tell Bradbury that Parsons would keep him current on plans for the Bikini test and would bring the Los Alamos laboratory into the test as the principal technical operating agency; (4) tell Bradbury that I would discuss the super-bomb with him as soon as I had his report on it; (5) ask Bradbury for a letter outlining what he would like done to improve liaison between Los Alamos and the rest of the project; (6) inform Bradbury that officers would not be assigned to Los Alamos for technical duty without consulting him; (7) tell Bradbury that Farrell would go out to Los Alamos immediately to discuss matters with him; and (8) tell Bradbury I would like to have him visit Oak Ridge, Hanford, Chicago and Washington at his convenience in order to get an over-all view of the project.

I immediately put these recommendations into effect. No effort was spared in making certain that Bradbury knew that we were backing him to the hilt and were doing everything within our power to make his lot an easier one.

In the months immediately following and, in fact, until the middle of 1946, a major portion of the effort at Los Alamos went into Operation Crossroads—the Bikini tests. Although the Navy Department (Vice Admiral W. H. P. Blandy) had over-all responsibility for the operation, technical responsibility rested with the MED. Our people produced, assembled and tested the weapon components. They controlled the timing of Test Able and detonated the bomb for Test Baker.[1]

Out of these tests came a great deal of valuable information. We were able to plan carefully for the collection of much technical data, and for the first time were in a position to confirm much that we could only infer before. The tests clearly established that atomic weapons could easily rout any major beach attack and that a capital ship could not operate in an atomic war. They also demonstrated that the radioactivity generated by an underwater atomic explosion would result in major casualties wherever the contaminated spray reached.

The job of safeguarding the forty thousand men of the Joint Task Forces at Bikini was given to Captain George Lyons, of the Navy Medical Department, who had previously been associated with the MED. The nature of this operation required the division of medical duties into highly specialized fields. One of these was radiological safety. This was our job, planned and carried out under the direction of Colonel Warren. It included the protection of the entire task force as well as of the natives of the atoll and the surrounding area. Measures also had to be taken to assure that there would be no radiological damage to the people of Japan or the United States.

There were two specific categories of danger: first, from the deto-

---

[1] There were two tests at Bikini—Able, where the bomb was exploded well up in the air, and Baker, where a bomb was set off under water.

nation itself, and second, from radiation persisting after the detonation. The latter would come chiefly from fission products: fissionable material itself and the matter—salt water, dust, debris and such—in which radioactivity had been induced.

In January, we organized a group at Oak Ridge to train officers and men in the proper use of instruments to detect radiation; they were taught also how to interpret radiation in terms of actual damage. Later that month, they underwent additional training at the Universities of Chicago and Rochester and at the Philadelphia Navy Yard. In March, they witnessed demonstrations at Los Alamos and made several field trips, including one to the Trinity site. These men served as monitors in the photographic planes, the reconnaissance planes, the rescue planes and in all the various types of watercraft that participated in the Bikini test.

During these transitional months between the war's end and the takeover by the Atomic Energy Commission, we launched a wholly new project—supplying radioactive isotopes to qualified researchers. Congressional delay in passing legislation made things particularly awkward for us in this venture, because all the work of the MED had been paid for by funds specifically appropriated for military use. Since the production of radioactive isotopes was for peacetime use, we actually had no authority to work in this area.

We had always been mindful that there would be these by-products of atomic power, but we had thought their major value would be in research—medical, biological and agricultural. We expected that they would have some industrial use, but not very much. Not long after the end of the war, however, we realized that they would probably prove to be of much greater value than we had anticipated.

I felt that they should be made available for research as promptly as could be arranged. Their production would require months, since the material had first to be radiated in a pile and then reduced to the form in which it would be used. Out of my discussions with people in whom I had the utmost confidence came a final plan for radioactive isotope distribution within the United States. The first shipment was made to St. Louis, Missouri, where it was used in research on bone cancer. Soon afterward, isotopes were made available to scientists in other countries under suitable restrictions that prevented their being used for research in military areas.

I did not approve of giving isotopes to researchers free of charge. In the first place, I did not think that it was in my province to give away material belonging to the United States; and, secondly, I thought the material would be much more carefully used if it were paid for than if it cost nothing. In setting up the price schedules— which later proved to be excessive in some instances, too low in others —our principle was to recoup for the United States the actual operating costs of producing the isotopes, with nothing to be charged for capital investment or depreciation. While everything about this undertaking was new, including the shipping of highly radioactive material to many different locations, it did not present too many problems.

Since the start of this program, the number of shipments has climbed enormously, so that in the last year the value of the radioactive isotopes to industry in this country has risen to somewhere between one-half billion and one billion dollars. It is not possible to calculate their value to research, and the surface in this area has scarcely been scratched.

The development of commercial power from atomic energy has not progressed so rapidly as I had hoped it would, although in the fall and winter of 1945 I estimated that it would be a matter of decades before we had economical atomic power for commercial uses under normal conditions. Even before the war ended we had started a group of scientists working on the problem. This was possible because, although it was a peacetime project, it was essential to keep the Chicago group together in case scientific support should suddenly become vital for the Hanford plant. We could not expect to hold onto such competent men if we merely let them sit idly by waiting for a call. For that reason alone, I thought it was wise to initiate studies of how best to proceed with the development of atomic power. This work was primarily under the control of Dr. Farrington Daniels, of the University of Wisconsin.

Before the end of 1945, the group had charted its course and was prepared to begin. Here I had to call a halt, for it seemed to me that the permanent peacetime organization that would succeed the MED should handle the practical development from the start. Unfortunately, it was over a year before the Atomic Energy Commission assumed responsibility, and then they did not proceed on the basis of Daniels' studies. All his work had been carefully reviewed, and both Nichols and I, as well as everyone else whom we consulted, considered it a sound basis for the initial approach. The result was a serious delay before any effective work was undertaken on this problem. Today, nuclear power is still uneconomical under normal conditions, but I believe that it will not be too long before it will start to take its place in the world's economy.

Nuclear power for special uses, such as the propulsion of submarines and other naval vessels, and the production of small power plants for the delivery of heat and power in isolated locations, has reached a quite satisfactory point, although undoubtedly a great deal of progress remains to be made. And, as no one needs to be told, since the end of the war the power of atomic weapons has been greatly increased and, in addition, tactical bombs of much lesser explosive power have been produced. The latter will enable a field commander to attack a much wider range of targets without destroying large areas and without danger of excessive fallout and ground and air contamination.

The advent of the hydrogen bomb has again entirely changed the military picture. This we always thought was inevitable; it was one of the reasons we worked on the hydrogen bomb, particularly on its theory, during the war years, though it was clear that the A-bomb could be completed much sooner and therefore should receive priority of effort.

As soon as the war was over, we took steps to provide for the indoctrination of a number of engineers in the atomic area. Among other moves we established a course at Oak Ridge where the engineers of some of the bigger companies, as well as some military officers, could be trained in what might be termed the practical end of atomic engineering.

One of the men who trained there was Captain H. G. Rickover, at the special request of Vice Admiral Earle Mills, the head of the Bureau of Ships in the Navy. Mills explained to me that he would probably soon retire from active duty and assumed that it would not be too long before I followed him. He said he was afraid that there would then be considerable danger that a development in which both he and I had long been interested—namely the atomic-powered submarine—would probably not be vigorously pushed unless we had suitable naval personnel convinced of its possibilities, trained in the fundamentals, and enthusiastic about its future. He was anxious to have a very determined man involved, and wanted my approval of the assignment of Rickover. First, though, he wanted me to know that Rickover was not too easy to get along with and not too popular; however, in his judgment, he was a man whom we could depend on, no matter what opposition he might encounter, once he was convinced of the potentialities of the atomic submarine.

Mills was a good judge of men.

CHAPTER 29

THE AEC

One of the original tasks of the Interim Committee (entirely civilian in membership[1]), which had been appointed in April to advise the President on the problems that would be raised by the development of the atomic bomb, was to prepare the draft of a legislative bill for the domestic control of atomic energy. Along with the Secretary of War, General Marshall and everyone else who held a responsible position in the project, I was most anxious to have legislation passed promptly, for now that the MED had done the job for which it had been brought into being it was important to have a reasonably clear-cut national policy laid down for our future guidance. At first I had even hoped that the President might call a special session of Congress to consider the matter. While I always regarded this possibility as most doubtful, I felt that the administration could and should prepare appropriate legislation so that it could be considered by the proper committees prior to the opening of Congress; then with the backing of the administration, a sound law could be enacted soon after Congress convened.

However, there were a number of persons who were pushing their own pet schemes. Typical of these was the proposal that we disclose all the details of our atomic development to the world, and particularly to Russia. This idea found one strong supporter in Washington

[1] Henry L. Stimson, Secretary of War, Chairman; George L. Harrison, James F. Byrnes, personal representative of the President; Ralph A. Bard, Under Secretary of the Navy; William L. Clayton, Assistant Secretary of State; Dr. Vannevar Bush, Director of the Office of Scientific Research and Development, and president of The Carnegie Institution; Dr. Karl T. Compton, Office of Scientific Research and Development, and president of The Massachusetts Institute of Technology; Dr. James B. Conant, Chairman of the National Defense Research Committee, and president of Harvard University.

in Secretary Ickes. At the same time a few politically ambitious people decided that they could advance their careers by displaying an interest in atomic energy matters.

It was recognized that any bill that might be passed promptly would be an interim measure, since it would take some time for the administration and the nation to understand fully the implications of the new force. For this reason the original draft prepared by the Interim Committee contained a provision that the law should be revised at the end of two years. The draft legislation was sent to the White House soon after V-J Day. In accordance with usual custom, it was then circulated for comment among the various members of the Cabinet. Normally, such matters are handled rather quickly, since each Secretary merely indicates that his particular department will not be affected (if that is the case), and, therefore, he has no comments. In this instance, all replies were received quite promptly, except that from the State Department. After several months had passed, Secretary Patterson, at my urging, requested that the matter be expedited, but it seemed extremely difficult to get any action out of the State Department. There appeared to be a concerted effort there to delay consideration of the draft. From what I know of the workings of that department at that time, I am sure that this procrastination was not due in any way to Secretary Byrnes. He had been a member of the Interim Committee and had been very much in favor of rapid action. With his long experience in government, he appreciated the extreme difficulties under which we would have to operate until the necessary legislation was passed.

We had at that time an operating force in excess of fifty thousand people, and our expenditures were running to about $100 million a month. As I have mentioned earlier, many important decisions remained to be made, and had to be made in accord with an established national policy. Yet with the ending of the war there was no national policy. The only guidance that I could obtain was that I should continue to operate the project as I thought best. Such broad powers were justified during the war, but they were not, in my opinion, justified once the war was over, and particularly after months and months passed by.

Finally, Secretary Patterson secured the President's consent to having the draft of the legislative proposals submitted to Congress. This draft was the so-called May-Johnson Bill. Hearings on it were held before the House Military Affairs Committee. Secretary Patterson and I appeared as witnesses, along with a number of other persons. The official views of the War Department, which were also the personal views of Secretary Patterson, General Marshall, General Eisenhower and myself, were set forth very clearly at that time. The record of what was said was clear and available for all to read. It left no question concerning our official position, our personal opinions or the fact that Secretary Patterson and I, individually, both felt that the War Department's position was sound.[2]

Essentially, our position could be stated as follows: First, the responsibility for the development of atomic energy should not remain in the War Department; second, it was not sound to put anything like as much power as I had had and still possessed into the hands of one man; and third, I did not subscribe to the philosophy that it was all right to have it in my hands, but not in the hands of a successor. Moreover, we both strongly felt that partisan politics must be kept out of atomic energy. Our statements had to be worded diplomatically because of our country's recent international political ventures, and because Mr. Patterson was a Republican Cabinet officer in a Democratic administration.

In spite of the record, since the very start of the postwar period there has been a continuous stream of propaganda calculated to lead the American people and the people of the world to believe that the War Department—and General Groves in particular—was determined to retain close control of atomic energy. The effect of this propaganda has been truly remarkable, despite the fact that it was entirely false, and was known to be false by those who first originated it and by many of those who constantly repeated it. It made no difference how many times they were corrected to their faces, often quite bluntly, or in writing; they continued to put the idea forward. Even today many of them keep up the sham and appear

[2] See Appendix IX, page 440.

likely to do so as long as they live. In retrospect, this seems to have been one of the most perfect brain-washing operations in modern times, and it has been particularly effective among the better-educated Americans. Those who continued to spread it even after the facts were brought to their attention were obviously doing so deliberately with the intent of gaining political advantage from propaganda that they could not help but know was absolutely without foundation.

At the initial hearing, one point was definitely cleared up for the public. During the questioning, I was asked by Congresswoman Clare Boothe Luce whether there was any truth to the rumor that the bomb could have been used much sooner, but that, for some obscure reason, President Roosevelt had not wished to have it used. Her questioning was very sharp and, as one would expect, very much to the point. I answered that the bomb was dropped as soon as it was available; that the first bomb which was tested at Alamogordo on July 16 was not ready until a few days before the actual test; and that up until that time we did not have a sufficient amount of plutonium or uranium with which to hold the test.

I also emphasized that we did not have enough fissionable material for the first bomb dropped on Japan until a few days before the actual operation; that, shipping the final portion by air, the bomb could not have been assembled and dropped prior to July 30; that I had informed the President through the Secretary of War that we would not be ready until the thirty-first of July; and that the Potsdam ultimatum had been issued on that basis. Therefore, the first time that the bomb could have been used without further reference to the President was July 31. I also testified that the bomb was dropped on the first day after the thirty-first of July that the weather permitted. To answer categorically, there had been no avoidable delay whatever in using the bomb once the fissionable material was available.

The May-Johnson Bill provided for a board of nine commissioners who would oversee the nation's atomic energy program in a manner similar to that followed by most boards of directors in large corporations. The duties of the commissioners would not have been so

arduous as to require them to devote their time exclusively to atomic
affairs. The Interim Committee felt, and I am sure they were right,
that this would enable the President to obtain the services of the
best available people—men such as Bush and Conant and the key
executives of large chemical and utilities companies who would
necessarily relinquish any conflicting private interests. Under this
plan, the day-to-day business of the atomic program would be conducted by the general manager, who would be selected and supervised by the commission.

The only features of the bill that in any way permitted military
participation in the proposed organization lay in the proviso, which
I still feel was absolutely sound, that military officers, active or
retired, should not be barred from service either as members of the
Commission or as its general manager. There was, however, no
requirement that there be any military representation on the Commission.

Although the bill was acted upon and passed promptly by the
House, the Senate did not move so quickly. This was because some
political opportunists took advantage of the fact that atomic energy
cut across a number of committee responsibilities and hence did not
definitely belong to any one committee. It was finally agreed that
there should be a special committee appointed to consider the problem, and to recommend legislation. A dispute immediately arose
over who should be the chairman of this committee. If it was an
investigating committee without any power to act on legislation,
then, in accordance with normal custom, it would be chaired by the
member of the majority party who had introduced the resolution
establishing the committee. This was Senator Brien McMahon, of
Connecticut. If, on the other hand, it was to pass on the advisability
of actual legislation, as was the case here, it should be chaired by
the senior member of the majority party on the committee. This
would not have been McMahon.

I have never been able to learn who it was on the White House
staff that put the idea across that the chairman in this instance should
be Senator McMahon, despite his lack of seniority. I have always

doubted whether President Truman was particularly interested in which course was taken, but as a former member of the Senate and a wholehearted advocate of its dignity and protector of its customs, he would have naturally been inclined to follow custom.

When the Senate committee finally met, Secretary Patterson and I again explained the views of the War Department, which were unchanged.[3] I urged prompt action on the grounds that it was not fair to me to have to make decisions that would determine the future of atomic energy for many years to come, without some guidance from the Congress regarding the course I should pursue. I also emphasized the difficulties I was having in holding our organization together in the face of its very uncertain future. Yet, despite our obvious need, the Senators were in no hurry to arrive at any decision.

The only reward for my efforts here came when one of the Senators, Mr. Byrd, I believe, stated that he would prefer to have me make as many of the controlling decisions as possible, since he thought that I could probably do it better than any newly appointed group unfamiliar with the background. He went on to add that as far as he was concerned it might be better for the United States just to leave the atomic energy program in my hands for all time to come. His idea that I should run it as I thought best was not objected to by any member of the committee, so I at least had a mandate, even though by default.

The May-Johnson Bill never got out of committee in the Senate. In its place, the McMahon Bill was offered allegedly as a means of ensuring that the development of atomic energy would not rest in military hands. After many vital changes in committee the Mc-Mahon Bill eventually passed both the Senate and the House and was enacted into law on July 31, 1946, almost a year after the war ended.

To make matters worse, the Commission provided for in the Act was not appointed promptly because President Truman was unable to secure the acceptance of his preferred choices for the chairman-

[3] See Appendix X, page 441.

ship. In fact, it was not until immediately before the November Congressional elections that the commissioners were named, and it was not until January 1, 1947, that they actually took over their responsibilities for atomic affairs. This period, from July 31 to the end of 1946, was a most difficult one, for everyone knew that I was in a caretaker's position, and they had no assurance that my views would be those of the Commission. After the commissioners were finally appointed, it was quite evident that my views would not be accepted without a long-drawn-out delay.

The Atomic Energy Act, as it was finally passed, contained only a few provisions that I would have preferred to see written differently. One of its shortcomings was its failure to provide that active military officers could serve either as commissioners or as the general manager. (Of the five general managers to date, three, Major General K. D. Nichols and Brigadier General K. E. Fields of the Army and Major General A. R. Luedecke of the Air Force, have been military officers, who have had to retire from active service to accept the post for which the commissioners, of their own volition, selected them. This has removed from the active list three extremely capable, vigorous officers. Particularly with the experience gained as general managers they would have been invaluable to the military establishment.) The most vital of its defects, however, was its basic concept that the Commission act as an executive body with all commissioners holding equal powers. Although the chairman acts as the spokesman of the Commission, what he says must reflect the views of the Commission, and for some years the other commissioners were normally present at any meeting with the President. During the chairmanships of Lewis Strauss and John McCone, this was not the practice. Their terms coincided with the presidency of General Eisenhower, who always looked to the chairman rather than to the group. Even so, I have always understood that while these chairmen have expressed the views of the majority they have at the same time informed the President of any strong minority views.

As I pointed out to the Senate committee, ever since the failure of the tribunes of Rome no executive group has ever functioned

well. I very much preferred—in fact, I thought it was essential—
that the Commission should function as a board of directors with
the general manager as the chief executive. I would not have objected if the chairman of the Commission had also been its general
manager, although I thought it better to keep the two jobs separate.
I said that in my opinion the only executive actions to be taken
directly by the Commission, without going through the general manager, should be those which involved grants of government funds
except where the grants were in payment for services rendered, and
the establishment of regulations that would affect the rights of American citizens. The former would include such matters as scholarships
and fellowships, and special grants to institutions for basic research
where specific results were not required. The latter would include
such matters as rules limiting the possession of radioactive material.
These, I felt, should not be under the control of a single individual.

Of the original members of the Commission, two, Sumner Pike
and William W. Waymack, had no background whatever in atomic
matters. Robert F. Bacher had been connected with Los Alamos
since early 1943, and was completely familiar with our work there.
He was a nuclear physicist of ability and understood the scientific
problems involved in the preparation processes. Lewis Strauss had
come into contact with atomic theories in the early days before the
war and was deeply interested in scientific matters generally. Although
not a physicist, he was familiar with a great many of our problems.
He had also had a wide experience in international affairs and in
banking, and had served as a special assistant to the Secretary of
the Navy during the war, and as such had become slightly involved
in atomic matters in 1945. David E. Lilienthal had been exposed
briefly to the problems of atomic energy while a member of the panel
of consultants to the Acheson Committee. He had had the advantage
of seeing many of our installations and of talking to many of the
scientific personnel. He had also been chairman of the TVA for a
number of years, and was thoroughly accustomed in the management
of government operations.

When the Commission was appointed, the members asked for a

meeting with the Secretary of War, the Chief of Staff and me, to discuss the turnover of the Manhattan Project. They made a point of asking that they not be required to take over their responsibilities until the first of January. This was contrary to the law, and to all normal procedures. It made my own position particularly difficult, for I was no longer simply a caretaker awaiting a final decision— I was a caretaker who could make no major decisions during a period when major decisions were vital.

In view of the commissioners' request, however, I agreed to remain in charge of the project. During the interim period, I placed at their disposal every assistance that the Manhattan District could render, and particularly urged them to make full use of Colonel Nichols. Furthermore, I told them that I would always be available to consult with or to advise them on any matters that might arise, whenever it was mutually convenient, and it would always be convenient for me. To my surprise, they took almost no advantage of this offer.

It was quite evident, and Secretary Patterson commented on it after the meeting, that some of the commissioners were definitely hostile to me, though why I did not know. This was unfortunate, but it did not deter me from giving them all the co-operation I could insofar as they would permit it. Throughout this period, I encouraged the people working in the Manhattan Project to remain with the Commission, at least for a reasonable length of time. I thought a continuity of personnel was essential to the best interests of the United States, since only these people could give the commissioners an adequate picture of the past. Moreover, they could provide invaluable assistance in the initial operations of the Commission, particularly until its members had time to gain an adequate comprehension of the problems they faced. (Many of these people are still with the Commission fifteen years and five chairmen later.)

One point on which I took a particularly strong position was the matter of security clearances. I considered this so important that I put my opinion into writing and when my original letter to the Commission appeared to be misunderstood, I reinforced it with another.

In this, I pointed out that the fact that certain people in the Manhattan Project had been cleared for employment should in no way be considered the sole basis for granting them clearance under the provisions of the Atomic Energy Act.

Prior to this Act, the clearance of people on the project was subject only to my judgment of whether their employment would be in the best interests of the United States. No other criteria existed. After the passage of the Atomic Energy Act, however, this criterion went out the window, for the Act was quite specific concerning clearance limitations. I wanted to emphasize the differences in procedure both to protect the commissioners and to put the War Department's position on record.

One very troublesome matter arose during the turnover when the Commission asked to have a formal inventory and transfer of property. I pointed out that the very magnitude of the project made this impossible, and that while I had no property responsibilities, the officers under me who did could not be expected to carry this responsibility for the months, and possibly even the years, it would take to make a complete check. As far as I was concerned, I was turning over to the Commission one Manhattan Engineer District, as is. This view was finally accepted simply because there was nothing else to do.

Shortly after the turnover to the Atomic Energy Commission, I was designated to organize the Armed Forces Special Weapons Project (AFSWP). This job had its genesis in a conversation I had had with Secretary Patterson following the passage of the Atomic Energy Act, when I had told him I was most anxious to be relieved of all connection with the project as soon as the Commission could take over. He said then that he was concerned about how the War Department would be able to fulfill its responsibilities resulting from the development of atomic weapons, and urged me to remain on active duty to handle the problems this would present. He added that if he had anything to say about it, he would insist on my staying on the job until I was retired for age, and then would ask to have me recalled to active duty for as long as I was physically and mentally able to be of assistance.

Since I had already spent about half of my military service in Washington, this was not an intriguing prospect, and I asked to be assigned elsewhere. It seemed to me that this was advisable both from the Commission's point of view and from mine. It is never sound to have a former boss sitting on the sidelines observing the operations of his successor; it is an impossible situation for both. Although Patterson refused to approve my request for another assignment, I agreed to stay on long enough to help set up an organization within the department to handle the military operations that would be involved in atomic matters, but I said that once it was firmly established I would quite probably apply for retirement. (As it turned out, I remained in command of this organization until my retirement in 1948.)

I considered that my major responsibility with the AFSWP was to carry out Mr. Patterson's directive that it be soundly organized and capably manned. It was a combined service organization in the truest sense. No artificial requirements were established for rank or branch of service. To be sure, I had two deputies, one from the Navy and one from the Air Force, but they were both men experienced in the Manhattan Project's operations. Throughout the rest of the organization, there was no attempt made to balance the different services. Each man was placed where he was best qualified. This sometimes resulted in entire sections being made up of people from one service.

The organization was deliberately overstaffed for its responsibilities, because it was primarily a training unit and I considered it essential to indoctrinate as many officers as possible in the various problems involved in military uses of atomic energy. We even went so far as to have a special course of lectures for all officers, given by a distinguished atomic physicist, Dr. George Gamow, from George Washington University. This was given during duty hours and attendance was compulsory.

We also organized a unit at Sandia Base near Albuquerque. Here groups of carefully selected young officers and senior noncommissioned officers were trained in the details of atomic bomb assembly and organized into teams thoroughly capable of doing the job under the kind of field conditions that had existed on Tinian. The unit was also responsible for procuring and developing the equipment the assembly teams would need in the field. Throughout, the aim was to give each man as much technical information as he could absorb. The whole purpose of the operation was to make absolutely certain that in case of war, or even the threat of war, the Defense Department would have at its instant disposal teams ready and trained to assemble atomic weapons.

CHAPTER 30

POSTWAR DEVELOPMENTS

British representatives in Washington had begun urging the preparation of an agreement to provide for postwar nuclear collaboration between the United States and the United Kingdom as far back as the spring of 1945. In later discussions on the subject after President Roosevelt's death, the British representatives on the Combined Policy Committee suddenly brought to our attention a document they referred to as the Hyde Park *aide-mémoire*. They told us that this paper summarized a conversation that they believed took place between President Roosevelt and Prime Minister Churchill at Hyde Park on September 18, 1944. This was the first intimation we had had of the existence of any such document, and I got the impression that the regular British representatives in Washington were every bit as surprised at its sudden emergence on the scene as we were.

Despite numerous thorough searches, we were unable to discover any trace of the *aide-mémoire* in any American files. Neither were we able to find any American who remembered seeing it or hearing of it. We were told that Churchill thought the agreement had been handed to Admiral Leahy, but Leahy said he recalled nothing in the form of an agreement and further that he recalled no discussion of any kind on atomic energy during the Hyde Park visit.

(Apparently he failed to consult his notes, for his book on the war years, *I Was There,* gave an account of a long discussion on atomic energy on the evening of September 19, between President Roosevelt and Churchill, although it did not mention the agreement.)

While the mutual confidence which had prevailed throughout the war continued undiminished, we were completely mystified by the

British references to this document. I am sure that, on their part, the British must have been annoyed by our insistence that we could find no copy of what they considered to be a valid and binding agreement. Where was it? Why had President Roosevelt never told any of us about this highly important document? This still remains a mystery.

At our request, Secretary Stimson was furnished with a copy of this paper by Field Marshal Wilson. Later Mr. Churchill sent us a photostatic copy of the original for our permanent records.

Not until many years later, when the Roosevelt papers were being catalogued at Hyde Park, was the mystery of the whereabouts of the missing Hyde Park *aide-mémoire* finally cleared up. At that time, a copy of it was found in a file of papers pertaining to naval matters. The misfiling was due, I suppose, to the fact that the paper referred to Tube Alloys, the British code name for the atomic project, and the file clerk must have thought it had something to do with ship boiler tubes. Just how such an important and highly secret paper could have been handed over for routine filing no one has been able to explain. Such a mistake, under less propitious conditions, could have had disastrous results. As it was, it did not seem to make too much difference.

The *aide-mémoire* was of particular significance because it contained the statement: "Full collaboration between the United States and the British Government in developing Tube Alloys for military and commercial purposes should continue after the defeat of Japan unless and until terminated by joint agreement."

The British naturally attached great importance to this particular phrase when they sought to have our wartime collaboration extended into the postwar era. They were particularly anxious to be relieved of the restrictions on British postwar commercial applications that had been imposed by Article IV of the Quebec Agreement. On our side, also, we felt a need for reappraising the secret wartime Quebec Agreement in the light of the postwar situation. To this end, Patterson wrote to Byrnes on November 1, urging that the State Department undertake a study of the problem.

Shortly afterward, Prime Minister Attlee arrived in Washington, accompanied by Sir John Anderson. There they engaged in conversations first with President Truman and Secretary Byrnes, for the United States, and Prime Minister Mackenzie King and Mr. Lester Pearson, representing Canada. Later, as the conversations developed, the most active participants were Dr. Bush, Sir John Anderson and Mr. Pearson. They eventually arrived at the Truman-Attlee-King Declaration of November 15, 1945. This declaration said, in effect, that while the free exchange of nuclear information would be desirable, it must be controlled and limited to peaceful purposes; to that end the three powers would restrict such exchanges until adequate controls were established by the United Nations.

At the same time it was decided that the Combined Policy Committee should be directed to prepare a draft of an agreement to supersede the Quebec and the Combined Development Agreements. To initiate this, the President directed Secretary Patterson to arrange for a conference with Sir John Anderson, and to have Sir John and me draw up a written agreement covering the decisions already reached at the White House. When Patterson told me this, I asked him what the decisions were. He was startled and asked me if I hadn't been at the meetings. I told him that I had not been asked, presumably because the State Department had not wanted me to be there.

After some discussion, Patterson telephoned Byrnes, who was also surprised to learn that I had been excluded from the discussions. He told us in a general way what he knew, but disclaimed any knowledge of specific decisions, since he had withdrawn from the meeting when a discussion of the details started.

I was thus faced with the difficult problem of writing a memorandum based upon decisions of which I had no knowledge whatever and which had been made by persons whose identity I did not at the time know. It was possible only because I was familiar enough with the subject to know what those decisions should have been, regardless of who arrived at them, or what they actually were.

Sir John and I agreed that the Combined Policy Committee and the Combined Development Trust should be continued and should

supervise whatever arrangements were later agreed upon; but that
the Quebec Agreement and possibly the Combined Development
Trust Agreement should be terminated and replaced by a new agreement more suitable to the postwar situation. To help the Combined
Policy Committee arrive at its recommendations, we prepared a
memorandum of intention listing the various points that the new
document should cover.

We emphasized that this memorandum was to serve merely as a
general guide and not as a commitment to basic policies. The main
points of its departure from the Quebec Agreement were the inclusion
of Canada as a partner and a provision that all atomic materials,
regardless of source (including those in the British Empire), would
be subject to allocation by the Combined Policy Committee.

The provisions regarding the exchange of information were essentially unchanged. One of the most important paragraphs in our
memorandum read:

There shall be full and effective co-operation in the field of basic
scientific research among the three countries. In the field of development,
design, construction and operation of plants such co-operation, recognized
as desirable in principle, shall be regulated by such *ad hoc* arrangements
as may be approved from time to time by the Combined Policy Committee as mutually advantageous.

This constituted our compliance with our quite hazy instructions.

We were also asked to prepare a short joint statement to be issued
by the President and the British and Canadian Prime Ministers. It
included the words "that there should be full and effective co-operation in the field of atomic energy" between the three countries. This
statement, known as the Truman-Attlee-King Statement, was signed
on November 16, 1945.[1] Secretary Byrnes did not see it before it
was issued, and he was later a severe critic of its necessity and of
its wording.

The words "full and" which appear in the first provision of this

[1] Not to be confused with the Truman-Attlee-King Declaration issued the
previous day.

statement were inserted at the insistence of Sir John Anderson, even though he stated that, in his opinion, they added nothing to the meaning. I strongly opposed the insertion as I felt that it considerably modified the meaning, and that the result was not in the interests of the United States. Secretary Patterson finally permitted the words to go in, stating that in his opinion as a lawyer it made little difference.

At the Combined Policy Committee meeting on December 4, Patterson offered the Groves-Anderson Memorandum as a heads of agreement for discussion and consideration. The committee thereupon appointed a subcommittee consisting of Roger Makins, Lester Pearson and myself, representing the United Kingdom, Canada and the United States, respectively, to prepare the new document that was to supersede the Quebec Agreement.

Before the next meeting of the Combined Policy Committee, however, I learned from one of the officers I had had study the situation that any such secret agreement would violate Article 102 of the United Nations Charter. I immediately brought this to the attention of Patterson and Byrnes, and as a result the American members of the committee concluded that no action should be taken that "could in any way compromise the success of discussions within the United Nations." It was thereupon agreed to refer the matter back to the signatories of the November 16 statement—the President and the two Prime Ministers.

In the meantime, the British had been pressing us for information that they considered essential to the success of their proposed atomic energy program. What they wanted amounted to practically all our development, design and production data, with the exception of those pertaining to the gaseous diffusion project, with which they were already familiar.

The scope of these data went far beyond any interchange contemplated by the Quebec Agreement, by the Truman-Attlee-King Declaration of November 15, by their Statement of November 16, by the Groves-Anderson Memorandum, or by the draft of the proposed new agreement. On April 16, 1946, Prime Minister Attlee stated this new British concept by telling President Truman that, in his view, "full and effective co-operation" could mean nothing less than full inter-change of information and the sharing of raw material. He went on to point out that any new agreement reached could be laid down by parallel instructions issued by the heads of the three governments.

President Truman replied on April 20, explaining that the Groves-Anderson Memorandum placed a different interpretation upon "full and effective co-operation" from that indicated by Mr. Attlee. He went on to say that he considered it inadvisable for the United States to assist the United Kingdom in the construction of atomic energy plants, in view of our stated intentions to press for international con-trol of atomic energy through the United Nations.

On June 7, Mr. Attlee again raised the question, insisting that continued collaboration would not be inconsistent with our advocacy of international control. He sought at length to justify all-out post-war collaboration as a natural outgrowth of our limited wartime collaboration. The President did not reply.

Again, on December 15, Mr. Attlee communicated with the Presi-dent in an effort to break what had become, since the passage of the McMahon Act, a complete stalemate insofar as interchange of infor-mation was concerned. As of January 1, 1947, when I turned over my responsibilities to the Atomic Energy Commission, the question had not been resolved.

In summing up the British contributions to the Manhattan Project, I would list:

1. *Encouragement and support at the highest official level.* This was most important and valuable, particularly in the beginning.

Late in 1941, the Uranium Section of the National Defense Re-search Committee, which was in direct charge of the American work at that time, had decided that our effort either should be discontinued or should be pushed forward at a more vigorous pace, with several millions of dollars to be put into research and many more millions to go for production facilities if our research yielded promising re-sults. This report was reviewed by a special committee of the Na-tional Academy of Sciences.

At the same time the British were conducting a similar review of the situation as they saw it. Their conclusions were far more optimistic than those of the American group. This British optimism, forcibly expressed by Mr. Churchill to Mr. Roosevelt, played an important part in the decision to expand the American laboratory effort. Without this early expansion before the fall of 1942, the later rapid development of scientific research and other work under the MED would not have been possible.

2. *Scientific aid.*

In the later stages of our work, the British sent over a number of atomic scientists, headed by Sir James Chadwick, a distinguished physicist and a highly valuable advisor. In addition there were about a dozen Canadian scientists in the Project. In view of our shortage of scientists, even this relatively small number was of help, particularly as they were all first-rate men. Unfortunately, among them was Fuchs, whose treachery impaired the total value of their otherwise splendid contributions.

3. *Preliminary study and some laboratory work on, and considerable enthusiasm about, the gaseous thermal diffusion method for separating U-235.*

Besides the methods for obtaining fissionable material that we finally carried through to the production stages, a dozen or so other methods were investigated in our laboratories. The British investigation of the gaseous thermal diffusion process was most helpful.

4. *Certain preliminary studies on the nuclear properties of heavy water.* However, as it turned out we did not find it necessary to employ heavy water in our project.

5. *Miscellaneous scientific and technical information.*

6. *Manufacture by the International Nickel Company at a plant in Wales of material needed to produce the gaseous diffusion equipment.* The company could not have produced this in one of their American plants without moving certain special machinery from Wales; the only alternative would have been to wait until new machinery could be made.

The most important discovery of technical value to come out

of the United Kingdom laboratories was that a certain kind of rubber was satisfactory for seals in the gaseous diffusion plant.

On the whole, the contribution of the British was helpful but not vital. Their work at Los Alamos was of high quality but their numbers were too small to enable them to play a major role.

On the other hand, I cannot escape the feeling that without active and continuing British interest there probably would have been no atomic bomb to drop on Hiroshima. The British realized from the start what the implications of the work would be. They realized that they must be in a position to capitalize upon it if they were to survive as a postwar power, and they must also have realized that by themselves they were unable to do the job. They saw in the United States a means of accomplishing their purpose. Their policy and their government's position were always consistent. Although we were able to keep our operations on a consistent basis during the war, American postwar policy often took off on tangents as new policymakers were brought into the picture.

Looking back on those war days, I can see that Prime Minister Churchill was probably the best friend that the Manhattan Project ever had. He seemed to be able to sense any lag in our work, caused usually by conflicts with other important war work. In such cases, a word from Number 10 Downing Street to the White House would give us new impetus. In addition to his other wartime achievements, Mr. Churchill emerged as our project's most effective and enthusiastic supporter; for that we shall always be in his debt.

An infinitely more difficult problem than our postwar arrangements with the British was the larger question of what international controls could be worked out for atomic energy. From the first, everyone connected with the project in a senior capacity had been conscious that this was something that would have to be dealt with eventually, but until the end of the war most of us were too busy to give it the consideration that it deserved. Mr. Stimson, though, did think all the way through the problem and its ramifications and, on September 11, 1945, addressed a remarkable proposal to the President, in which I wholeheartedly concurred.

The Secretary pointed out that while our atomic bombs would offset Russian strength on the continent of Europe, the Soviets would surely spare no effort to develop their own atomic arsenal. Unless the United States and the United Kingdom were to bring the Soviets into our position of control voluntarily, the Russians would only be stimulated in their efforts to catch up with us. Mr. Stimson estimated that it would take from four to twenty years for them to do so. He mentioned that their espionage activities during the war indicated that they had already started.

In any approach to the Russians, he said, we should exercise great care to avoid the appearance of negotiating with an atomic bomb at our hip, which could easily so embitter them that our efforts would come to nothing.

He said that the atomic bomb was not merely another more devastating weapon to be assimilated into our pattern of international relations; if it were, it could be treated like gas. Instead, it was a first step by man to release forces that were too dangerous to fit old concepts.

Any action to control atomic energy must be directed to Russia. He felt that the Russians would be more apt to respond honestly to a direct approach from the United States than to one made as part of a general international scheme. He said that we should approach the Russians directly and alone, with the consent of the British, to institute controls and limitations on the use of atomic bombs in war and to encourage the development of peaceful, humanitarian uses for atomic energy. We should not attempt to force any change in the internal government of Russia as a condition of sharing our knowledge with them.

Mr. Stimson's repeated admonition that any group of nations that had not demonstrated their power would not be taken seriously by the Russians apparently was not subscribed to by our State Department or the British. At any rate, nothing came of his proposal.

In a message to Congress in October, 1945, President Truman emphasized the problem of international controls in a single sentence: "In international arrangements as in domestic affairs, the release of atomic energy constitutes a new force too revolutionary to consider in the framework of old ideas—the hope of civilization lies in international agreements."

This subject was one of the principal topics of discussion in the meetings held in Washington between Mr. Truman, Prime Minister King of Canada, and Prime Minister Attlee of Great Britain, from November 10 to 15. They finally agreed that the initiative rested with "the three countries which possess the essential knowledge of atomic energy." They made it clear, however, that the responsibility for devising means to control the new discovery rested on all nations alike, or, as they put it, "upon the whole civilized world." They decided, therefore, to recommend that a commission be set up under the United Nations to draw up proposals for submission to that organization. This decision was embodied in the Truman-Attlee-King Declaration of November 15, 1945.

Meanwhile there had been, and continued to be, widespread public discussion about how the U.S. should conduct its atomic affairs. Much of the written material and the most vocal of the various protagonists took what would be called the liberal position. In general, they wished the United States to proceed with full confidence in the Russians and with good will toward all mankind. They also claimed, most erroneously, that the success of our project had been due entirely to international science, and implied that the United States had no particular rights in the matter.

Unfortunately, the scientific leaders in the project who normally would have been the spokesmen for their colleagues were preoccupied then with getting back to their peacetime occupations at their own universities and I have always felt they simply did not realize what was developing. The result was that a new and vociferous group of spokesmen arose from among the younger scientific people, few of whom had had any experience outside the academic world, and who even there had served in only very subordinate capacities. There were a few others, of course, some of whom sought personal prestige and some of whom wished to forward extreme social points of view. The propaganda emanating from these sources was eagerly seized upon by various ambitious political figures, and by a few people in the State Department who seemed to me more concerned about the momentary good will of other nations than about the welfare of the United States.

Finally, Secretary Byrnes formed a special committee to consider the problem and to advise him on how far the United States should go in offering to share its knowledge with other countries. For this committee he selected the Under Secretary of State, Dean Acheson; the Assistant Secretary of War, John J. McCloy; Vannevar Bush; James B. Conant; and me.

At the first meeting, Mr. Acheson immediately proposed that we appoint a panel to investigate the problem and to report back. He pointed out that we were all too busy to do this work ourselves. I objected on the grounds that at least three of us—Conant, Bush and I— knew more about the broad aspects of the problem that the Secretary wanted us to study than any panel that could be assembled. Besides, I had access to all the scientific assistance that might be needed on any particular point.

I was outvoted; and a panel was appointed with David E. Lilienthal as chairman; Chester Barnard, president of the New Jersey Telephone Company; Harry Winne, a vice president of General Electric; Oppenheimer; and Charles Thomas, of the Monsanto Chemical Company. Of these, only Oppenheimer had a complete awareness and understanding of the problems involved. Thomas was fully familiar with the chemical side of the work at Los Alamos, and to a lesser extent with certain of the other phases of the work. Winne had a good understanding of the problems involved in the electromagnetic process, for General Electric had provided valuable assistance in that operation, and he had been the responsible executive in charge. As far as I know Mr. Lilienthal and Mr. Barnard had little or no knowledge of the subject whatever.

The panel submitted a preliminary draft of its report to the committee early in March. This was thoroughly discussed and a number of vital changes were incorporated in the final report, which was presented "as a foundation on which to build." The plan, as recommended, called for step-by-step co-operation with the other powers rather than for the first steps to be taken by the United States. This is not our usual diplomatic approach.

The committee declined to recommend the publication of the report for two reasons. First, it was a report to Secretary Byrnes, who was best able to decide whether he wanted it published. Second, we did not feel it wise to disclose to the Russians just how far the United States was willing to go in sharing its knowledge before negotiations had even been arranged for. To the Russians, our final position would be their starting point.

Unfortunately, the report was published by the State Department on the grounds that certain portions of it had been leaked to the press and that it was therefore wise to make the entire text public. As in the case of many other leaks in Washington, everyone blamed someone else. The State Department blamed a Senate committee and the Senators blamed the State Department. While the newspapermen, in accordance with their code, refused to disclose their source, many of them did say privately that the Senate was not responsible.

The report was immediately interpreted as an official statement of United States policy.

During his service as head of the United States delegation to the United Nations, Bernard Baruch often commented on the impossibility of negotiating with the Russians when there was nothing left to negotiate. Naturally, he was bitter that the release of the report had ruined the chance of success. As everyone knows, the negotiations for international control dragged on endlessly, while the Russians were busily engaged in developing their own atomic bombs.

As Baruch has often said, "The Russians would not countenance any effective system of international control of nuclear energy. They were adamantly against any system of international inspection, control and punishment and giving up the veto on any such matters."

CHAPTER 31

A FINAL WORD

As memories of World War II have faded into history and as the considerations of national survival that determined its course have lost their urgency, many have come forward to question the desirability of our having developed atomic energy. Some people have even gone so far as to assert that the United States was morally corrupt in making the effort. When I consider the sources of these later claims, my first inclination is to ignore them and I would do so if it were not for the fact that if free men are to arrive at a fair appraisal and full understanding of the Manhattan Project, they must be acquainted with the points in its favor, as well as with the views of those who criticized it and of those whose responsibilities were small at the most.

The atomic bombings of Hiroshima and Nagasaki ended World War II. There can be no doubt of that. While they brought death and destruction on a horrifying scale, they averted even greater losses—American, English and Japanese. No man can say what would have been the result if we had not taken the steps that were necessary to achieve this end. Yet, one thing seems certain—atomic energy would have been developed somewhere in the world in the mid-twentieth century. Because of the great costs and difficulties involved and the apparently very small chance of success I do not believe the United States ever would have undertaken it in time of peace. Most probably the first developer would have been a power-hungry nation, which would then have dominated the world completely and immediately. If a peacefully inclined nation had been first, it might have tried initially to use atomic energy for peaceful purposes, but as others exploited its military uses, every country would soon have been forced, as has happened in recent years, to concentrate on the development of weapons. Not until each of the great powers had produced a full atomic arsenal would the threat of one-sided atomic war pass.

Once this state was finally achieved, and I feel that it has been, with sane national leadership, major war is impossible. All that stands in the way of effective international control is the acceptance by all the world's political leaders of this fact. But there is no reason why work on the peaceful uses of atomic energy cannot be pushed ahead vigorously.

The fact that during the years following the end of World War II an unquestioned superiority in atomic weapons rested with the United States has not only averted a catastrophic war without a complete surrender of our principles and our ideals, but it has enabled us to take the first steps toward the realization of a few of the many possible constructive uses of atomic energy. While it is tragic that the forces for destruction that we unleashed are stronger than man's present ability to control them, it is fortunate indeed for humanity that the initiative in this field was gained and kept by the United States. That we were successful was due entirely to the hard work and dedication of the more than 600,000 Americans who comprised and directly supported the Manhattan Project.

Looking back, I think I can see five main factors that made the Manhattan Project a successful operation:

First, we had a clearly defined, unmistakable, specific objective. Although at first there was considerable doubt whether we could attain this objective, there was never any doubt about what it was. Consequently the people in responsible positions were able to tailor their every action to its accomplishment.

Second, each part of the project had a specific task. These tasks were carefully allocated and supervised so that the sum of their parts would result in the accomplishment of our over-all mission. This system of compartmentalization had two principal advantages. The most obvious of these was that it simplified the maintenance of security. But over and above that, it required every member of the project to attend strictly to his own business. The result was an operation whose efficiency was without precedent.

Third, there was positive, clear-cut, unquestioned direction of the project at all levels. Authority was invariably delegated with responsibility, and this delegation was absolute and without reservation. Only in this way could the many apparently autonomous organizations working on the many apparently independent tasks be pulled together to achieve our final objective.

Fourth, the project made a maximum use of already existing agencies, facilities and services—governmental, industrial and academic. Since our objective was finite, we did not design our organization to operate in perpetuity. Consequently, our people were able to devote themselves exclusively to the task at hand, and had no reason to engage in independent empire-building.

Fifth, and finally, we had the full backing of our government, combined with the nearly infinite potential of American science, engineering and industry, and an almost unlimited supply of people endowed with ingenuity and determination.

When I was a boy, I lived with my father at a number of the Army posts that had sprung up during the Indian wars throughout the western United States. There I came to know many of the old soldiers and scouts who had devoted their active lives to winning the West. And as I listened to the stories of their deeds, I grew somewhat dismayed, wondering what was left for me to do now that the West was won. I am sure that many others of my generation shared this feeling.

Yet those of us who saw the dawn of the Atomic Age that early morning at Alamogordo will never hold such doubts again. We know now that when man is willing to make the effort, he is capable of accomplishing virtually anything.

In answer to the question, "Was the development of the atomic bomb by the United States necessary?" I reply unequivocally, "Yes." To the question, "Is atomic energy a force for good or for evil?" I can only say, "As mankind wills it."

APPENDIX I

MEMORANDUM FOR THE CHIEF OF ENGINEERS
SUBJECT: Release of Colonel L. R. Groves, C.E., for Special Assignment

1. It is directed that Colonel L. R. Groves be relieved from his present assignment in the Office of Engineers for special duty in connection with the DSM* project. You should, therefore, make the necessary arrangements in the Construction Division of your office so that Colonel Groves may be released for full time duty on this special work. He will report to the Commanding General, Services of Supply, for necessary instructions, but will operate in close conjunction with the Construction Division of your office and other facilities of the Corps of Engineers.

2. Colonel Groves' duty will be to take complete charge of the entire DSM project as outlined to Colonel Groves this morning by General Styer.

   a. He will take steps immediately to arrange for the necessary priorities.

   b. Arrange for a working committee on the application of the product.

   c. Arrange for the immediate procurement of the site of the TVA and the transfer of activities to that area.

   d. Initiate the preparation of bills of materials needed for construction and their earmarking for use when required.

   e. Draw up the plans for the organization, construction, operation and security of the project, and after approval, take the necessary steps to put it into effect.

BREHON SOMERVELL
Lieutenant General
Commanding

Some of these instructions were never carried out because, as the work progressed, they no longer seemed appropriate. No working committee was ever established; and it proved impracticable to transfer all activities to the Tennessee site. Also, the information then available did not permit

* The then code name for the atomic energy project.

us to draw up bills of materials any more than we could arrive at realistic before-the-fact plans for the organization, construction, operation or security of the project. These were developed and put into effect as the work progressed, and while their more important phases were often discussed with the Military Policy Committee, they were never submitted to anyone for formal approval.

APPENDIX II

The Metallurgical Project also had a number of activities at other sites. It was responsible in a very general way for overseeing certain metallurgical support we received from the Battelle Institute and the Massachusetts Institute of Technology; for chemistry support from the University of California; for the very important metal work done by Dr. F. H. Spedding at the State University at Ames, Iowa, and later, but above all, for the operation of the Clinton laboratory at Oak Ridge and for the work of a special group later established at Hanford to provide scientific advice and counsel on the ground to du Pont.

Among the principal contributors to our efforts from the Metallurgical Project were Drs. Enrico Fermi, Eugene Wigner, James Franck, Glenn Seaborg, Samuel Allison and Joyce Stearns in Chicago; Martin D. Whitaker and R. L. Doan at Clinton; Frank H. Spedding at Ames; John Chipman at MIT; and Norman Hilberry at Hanford, who was the chief representative at that point.

APPENDIX III

By order of the President all government patent rights arising from any invention involving war developments were to be put in the custody of the Director of the OSRD. When I became responsible for the MED, Dr. Bush already had a patent section in his office under the direction of Captain R. A. Lavender, USN, Retired.*

It seemed best to me and it was agreeable to Bush to continue to have

* As a Lieutenant Commander, Lavender was the radio operator on the NC-3, one of the three Navy seaplanes which attempted the first Atlantic crossing.

Lavender handle all of the MED patent work, under Bush's general administration. At the same time, I was vitally concerned with the general patent policies, with security, and with the numerous patent field offices in the MED.

This resulted in Lavender's having to report to and to satisfy both Bush and myself. As far as I ever knew, there was never the slightest conflict in this operation, because of Lavender's capacity as an administrator and as a patent lawyer and because there was always complete agreement between Bush and me on patent policy, which remained essentially the same as that inherited from the previous OSRD operation.

The mission of the Patent Division was to make certain that the government's interests were properly protected in inventions made as a result of research and development paid for by government funds. It also sought to secure the maximum control of patents vital to the use of atomic energy in the years to come, whether for peace or for war purposes.

The problem was complicated by the extreme security requirements and by the brand-new scientific theories involved. Lavender was faced with an almost complete lack of precedents for our type of invention in patent law. There was also the necessity for dealing with large companies, all of whom had developed quite rigid patent policies over the years.

In this brand-new field, there were no trained, qualified patent attorneys and there were also no precedents for any patent organization handling such a large number of inventions from so many different, highly technical sources.

The compartmentalization of activities within the MED added to the difficulties of the Patent Department, as did the attitude thoroughly ingrained in all personnel in the Manhattan Project that our primary purpose was to complete the development as quickly as possible. Compromises from the patent standpoint had to be made constantly.

One of the basic problems was the necessity for secrecy in the United States Patent Office. This was arranged by the Commissioner of Patents and worked satisfactorily throughout.

In all contracts the patent provisions were written with Lavender's advice and approval. We also looked to him for interpretation after the contracts were in effect. Every contract, whether it was made with a university or an industrial organization, contained one of several standard patent clauses. Each required that any inventions made under the work called for under the contract be reported to the government. The government would then elect whether or not it would file applications for patents thereon.

Our patent problems fell into three general categories, and a different contract paragraph was used for each. These comprised: (A) Where the nature of the contract was such that any invention might be related directly to, or concern some phase of, the basic research and development of the project. Here the government retained exclusive rights. (B) Where the work related only indirectly to the project, but was related to a field of activity within the normal business of the contractor. Here, while the government retained full rights, the contractor retained a non-exclusive, royalty-free license. (C) Where the work was a mere refinement of work already completed by the contractor. Here the government retained the right for its own uses on all of the patents, but the contractor had a license and the sole right to grant sublicenses.

Usually a great deal of information was furnished by the government to the contractor and normally all rights in any inventions became the property of the government.

Because of the fact that a number of British and Canadian scientists and engineers visited various project sites and exchanged certain technical information with the American scientists and engineers, procedures were established to avoid any conflict and to protect the interest of the individual government concerned.

Some time after the Quebec Conference of 1943 it was determined that the exchange of invention information on the project would not come under the provisions of the patent interchange rules which were already in effect. The filing for patent protection in foreign countries was a serious problem. This was handled very carefully in order that we would not at some later date find ourselves severely handicapped by reason of patents filed by persons not under our control.

The magnitude of the patent problem is indicated by the fact that some 8,500 project technical reports had to be reviewed, over 6,000 notebooks examined and something like 5,600 inventions looked into.

Probably no man had a greater insight into the technical details of the entire project than Lavender and his chief deputy, Roland A. Anderson, for they examined all the detailed inventions made.

APPENDIX IV

To define and evaluate the physical dangers that many of the people working in the project might be risking, and to find methods of prevention and treatment, we vastly expanded the program of medical research that had been going on ever since the United States became interested in atomic energy.*

The MED's Medical Section was headed and organized by Colonel Stafford Warren, with Captain H. L. Friedel as his executive officer. Its duties were spelled out in a memorandum from Nichols to Warren, dated August 10, 1943. At that time Nichols made the Medical Section responsible for performing, contracting for, supervising and maintaining liaison with all medical research necessary to the project. In carrying out these responsibilities, the Medical Section was to utilize to the full the facilities of existing research agencies; care was to be taken to avoid unnecessary duplication of work.

The most urgent problem was to determine the toxicity of the materials we were using: primarily, uranium and plutonium compounds; the related heavy elements, such as radium, polonium and thorium; and certain accessory process materials, such as fluorine and beryllium. This required the study of the manner in which the materials might be introduced into the body, whether by ingestion, inhalation, skin absorption or in other ways.

There had to be a careful analysis of the nature of the biological changes that produced physiological, histopathological or biochemical evidences of damage. The nature of radiation injuries and the mechanisms by which they were produced had to be studied, since this provided the information upon which to base protective measures before and therapeutic treatment after exposure. The effectiveness of various ways to remove hazardous dust and reduce skin contact and to prevent ingestion and inhalation had to be measured. The effectiveness of ointments, chemicals and other protective measures had to be assessed. Treatments for acute and chronic poisoning had to be tested.

As we got more and more into the operating phases of the project, we came upon many additional and wholly unexpected dangers. In the electromagnetic and diffusion processes, for example, hazards were created by the concentration of the uranium and by the accessory materials. The reactors held even greater dangers from alpha, beta and gamma rays; from neutrons; from plutonium metal and its compounds; and from the various radioactive fissionable products resulting from the operation of the pile.

To provide adequate safeguards required intensive study of plant processes. Such measures included the clinical survey of all exposed personnel, the monitoring of hazards by special instruments and the analysis

* This research was conducted at the Universities of Chicago, Rochester, Columbia, and also at the University of Washington, where the effects upon fish of acute and chronic exposure to radiation were investigated. Special projects were carried out at the Clinton laboratories in conjunction with the Monsanto Chemical Company's work there, at Los Alamos and at Western Reserve.

and investigation of graphite piles and other production equipment. Protective measures had to be taken to minimize the effects of any catastrophe that might occur; investigative and decontaminating equipment and techniques had to be developed; provisions had to be made for the treatment of possible casualties, and we could not overlook the possible contamination of supply systems, populated areas, or even agricultural land, both within our reservations and without.

The problems with which the Medical Section was confronted were considerably aggravated by the need for their resolution before work on many of our research programs could be undertaken. At its inception, the Industrial Medical Division had to deal not only with laboratories and plants which were wholly without precedent, but also with materials whose degrees of toxicity were virtually unknown.

Ionizing radiation in almost all instances produces no immediately observable biological effects. It was necessary, therefore, to take every precaution to avoid overexposure. These precautions had to be determined largely by careful assessment of the known effects of X-rays, radium and other radioactive elements employed in clinical treatment; and by some meager animal experiments. Later these were augmented by extensive animal experimentation performed under very careful controls. Once the tolerance was determined for each species tested, it was extrapolated to arrive at estimated permissible values for man. A large factor of safety was applied to the estimates before they were established for the employees on the project.

In the same way, permissible limits were established for each of the other possible hazards, always minimizing the risks by setting them on the side of safety. Some idea of the scope of the Medical Section's work in this area can be gained from a realization that there were fifteen different contractors performing hazardous work in connection with the experimentation, development of processes, and actual operations of one phase of metal-shaping alone.

At Hanford the fumes and gases emitted by the pile and by the separation process exhausts were examined carefully to prevent any danger to the plant and its surrounding areas. A number of specially designed instruments placed in strategic locations in the vicinity of the installation were used to make continuous checks on the amount of radioactive gases in the air. Similar measures were taken at our other facilities. No concentrations above the permissible levels were ever detected during the period of MED's operations.

The pile process contained a number of potential hazards, among which radiation was by far the greatest. A number of protective measures were taken, such as the use of special table coverings to guard against the contamination of work benches, the use of protective clothing and equipment, and the rotation of employees through places where high radiation levels might be encountered. The methods which we employed provided highly effective protection to those who worked on the pile processes.

Similar precautions were taken at the Clinton laboratory, as well as at various other operating plants throughout the District. In the gaseous diffusion process, all operating employees who could possibly become exposed were required to have special medical examinations in addition to the regular routine ones. These were to determine the physical condition of the worker before and after his employment, to make certain that he had not been adversely affected. And most important, they were to enable us to take prompt corrective measures at the first indication of trouble.

At the same time, we had to meet the rapidly growing demands for the more conventional medical and public health services. At Oak Ridge, for instance, we had to build a hospital because those in the surrounding communities could not possibly handle the load that we would impose on them. Starting out with a hospital of fifty beds, we wound up in the spring of 1945 with 310, plus a complete out-patient dispensary. Even at that, toward the end, we were overloaded and frequently had to put patients in the corridors. In addition, we had to provide dormitories and other facilities for the nurses and hospital workers.

For security reasons, the hospital was run on a closed staff basis under the leadership of a very able surgeon, Major Charles E. Rea. Only doctors regularly assigned to the hospital were permitted to care for patients. Industrial physicians employed by the various contractors at Oak Ridge were accorded courtesy privileges and occasionally acted as consultants.

We used civilian doctors at first to provide medical service at Oak Ridge, but in November and December of 1943, the entire medical staff, with the exception of one man who was not eligible, was commissioned in the Army Medical Corps. The advantage of this was that it made available to us all of the services of the Surgeon General of the Army, including the procurement of personnel and material. It also enabled us to use Medical Corps funds, as was proper, in providing treatment for military personnel who were assigned to the project, made it easier to handle security problems and made it possible for us to retain those doctors who displayed exceptional professional ability. On the other hand, we never submitted any of the usually required reports that might in any way reveal the nature, scope or military importance of our work. All such reports were retained in the files of the MED Medical Section.

The number of doctors assigned to Oak Ridge rose to a high of fifty-two in July, 1946, when the number of nurses reached almost 150. The nurses were all civilian employees.

A dental program was instituted along with the medical program and was continually adjusted to meet the needs of the community. Here, civilians were used, since dentists were not too hard to find and since they would have no contact with classified material. The number of dentists at Oak Ridge increased from two at the start to twenty-nine by March, 1945, decreasing subsequently to twenty-five.

The local medical societies adjacent to Oak Ridge and to Hanford were most co-operative in every respect. They were always consulted whenever we instituted medical and dental services. By this means, we avoided many potential difficulties regarding the services offered and the fee schedules.

Our policies concerning medical fees, personnel and services had to be revised from time to time in the light of our experience and the ever-growing demand for them. At first only first-aid and emergency treatment was provided, but as the facilities grew, our medical services expanded with them. At Oak Ridge, a prepayment plan for medical care was instituted. The charge for a family membership was four dollars per month, which covered hospital, diagnostic and physician services.

Medical service at Hanford was tailored to meet the needs of the two distinct phases of the work at that site: construction and operations. During the construction phase, medical service was provided to meet the needs for a temporary community. This was in accordance with the normal du Pont policies, which were quite adequate for our purposes. Initially, only first-aid and emergency medical service was furnished, but it soon grew into a hospital with clinical service and an industrial medical section. At Richland, the requirements were somewhat different, involving as they did the needs of a permanent community.

To complete the medical services furnished at Oak Ridge, several civilian psychiatrists were employed. These gave such consultation and advice as were needed.

APPENDIX V

Oak Ridge, or the Clinton Engineer Works, was located in Anderson and Roane counties in Tennessee. It consisted of industrial plants, their supporting facilities and the town of Oak Ridge. The installation occupied a rectangular-shaped Government Reservation, ninety-three square miles in area. Its workers were concentrated chiefly in Oak Ridge, which attained a peak population of about 75,000 during the summer of 1945. A maximum employment of 82,000 was reached in May, 1945. Thereafter both population and employment declined steadily until, by the end of 1946, they were down to 42,000 and 28,000 respectively.

In the course of our work, Oak Ridge grew to be Tennessee's fifth city by population, and the Clinton Engineer Works became the second largest consumer of TVA power. The area's motor transportation system was the largest in the Southeastern United States. Because it was a closed military area, we had to provide all the normal community, municipal and other government services.

The town's architectural planning was handled entirely by Skidmore, Owings and Merrill, except for its utility systems, which were designed by Stone and Webster. Construction at the site began in February, 1943. Throughout the entire period, we had to limit our activities to meeting only essential requirements. The first phase of our planning program called for the construction of more than three thousand family quarters, several apartment buildings and dormitories and numerous trailer parks, along with all the community facilities. The second and third phases were carried out along similar lines, but on a far greater scale. Aside from keeping all design as simple as possible, the factors bearing most heavily on our planning were minimizing construction costs, and using noncritical materials and noncritical labor. We were always guided by the realization that the construction of the town was secondary to the construction of the plants. For this reason, the construction of the town was stretched out until April, 1944.

The first housing became ready in the summer of 1943. Initially the town was operated by government personnel. This was not satisfactory for a number of reasons, but particularly because it was certain to result in Nichols' having to devote a great deal of personal attention to this operation, which, after all, was a side issue. It would also require the attention of many of his most competent people.

After considerable discussion and study, Nichols and I decided to bring in an outside managing operator and asked the Turner Construction Company to take over the job.

The scope of its responsibilities included the operation of the laundry and cafeterias; the transportation, water, sewage, electrical and heating systems; the nonmedical features of the hospitals; the schools; the public safety forces; and the housing, both of individual family types and dormitories. As I fully expected, the entire operation was well handled.

Naturally, there was never enough family housing, so assignments were based on quotas assigned to the major operating groups. House rentals and service rates were based on fair charges for utilities, service, maintenance and amortization. For the trailers, dormitories and hutments, the charges were based on fair rentals, giving due consideration to the interests of the project in obtaining desirable personnel. In general, commercial facilities at Oak Ridge were limited to those required to provide the daily necessities of the residents, without very much in the way of luxury items. They were grouped in neighborhood shopping centers and commercial areas.

The school system was based on a policy that the community should be provided with educational standards of high quality. The schools became a part of the Anderson County school system, but in order to maintain the standards we wanted, we supported the county in obtaining federal funds. Additional physical school facilities were provided from MED supplementary funds. By the end of 1946, when the population had decreased considerably, the school system included over 7,000 pupils, 285 teachers and the usual administrative assistants.

We also had to provide for the recreation and welfare of the inhabitants of Oak Ridge. These activities were generally handled through the Oak Ridge welfare organization, which was set up to be self-supporting. We provided motion picture theaters, soda fountains, snack bars, athletic fields and even a weekly newspaper. Religious activities were conducted in two Army-type chapels. Later, other buildings were made available for that purpose. Eventually, Oak Ridge possessed every service that might be found in any typical American community.*

To avoid too much monotony in the external appearance of the housing and to make it as acceptable as possible to its occupants, we originally employed nine different designs, all of which provided more or less equivalent units. Each of these provided for single-family units, except one which put four in the same building. They ranged in size from one to three bedrooms. Later, multifamily houses were built, and finally we went to prefabricated houses. These were all of different capacities. The prefabricated models ranged from one-bedroom through two- and three-bedroom units.

The hutment camp units housed five men each. They had been used elsewhere on other construction projects both in the United States and abroad and were generally considered quite successful. Two such camps

---

* Those responsible for planning Oak Ridge's construction and operation were Lieutenant Colonel Robert C. Blair, director of planning from August, 1942, to July, 1943; Lieutenant Colonel Thomas T. Crenshaw from July, 1943, to May, 1944; and Colonel John S. Hodgson, May, 1944, to January, 1946. Lieutenant Colonel Warren George made the initial preparations for constructing the central facilities in November, 1942, as Area Engineer, and later then served as chief of the Construction Division until the fall of 1943.

were built, one for white and one for colored workers. A utility building for every group of twelve huts and each camp had its own cafeteria, recreation building, barbershop and other facilities.

There was also a trailer camp that accommodated nearly four thousand government-owned trailers, which had been transferred from other agencies, and 269 privately owned trailers. In addition to these, there were camps which were operated as part of the general facilities, as well as a number of camps established in the diffusion area for workers there.

In determining the type of roads to be used, we decided that adequate roads and streets had to be provided as a matter of prime necessity, but adequate should mean no more than usable. Eventually, these roads were improved, but only after considerable use had shown that their improvement was absolutely essential. Considering the amount of heavy construction traffic and heavy equipment on the roads and streets, it was not feasible to surface the crushed-stone roads and streets with a bituminous-wearing surface during the main part of the construction program. As time went on, we found it desirable to hard-surface wherever it was not too expensive. The same treatment was given to the walks, curbs and gutters.

During the spring and summer of 1945, after the completion of much of the construction, it became feasible to institute a large-scale paving program. All of the important area roads and Oak Ridge streets were surfaced with hot-laid asphaltic concrete paving or with an asphaltic surface treatment, depending upon traffic conditions. Our most serious road problems were in connection with the access roads, for the area was practically devoid of anything approaching adequate ones. It was not possible for various reasons to make any real improvement in that direction until 1944.

APPENDIX VI

The entire electromagnetic plant was designed and constructed by Stone and Webster, who also co-operated in getting it into operation. Their personnel for development and construction were under the management of R. T. Branch, President; T. C. Williams and T. R. Thornberg, Project Managers; and F. C. Creedon, Resident Manager during the critical period from January 6, 1944, to January 31, 1945.

Tennessee Eastman was the operating contractor. Their key civilian personnel on the site included Dr. Frederick R. Conklin, Works Manager;

Dr. James G. McNally, Assistant Works Manager in Charge of Production from April, 1943, to September, 1945; and Mr. James Ellis, Assistant Works Manager in Charge of Engineering from March 1, 1943, until February 1, 1945, by which time his work was finished and he was transferred to other duties in the Tennessee Eastman Corporation in Kingsport. Off site there were two key men, without whom the work could not have been accomplished, Dr. A. K. Chapman, Executive Vice President of the Eastman Kodak Co. and Mr. James C. White, the President of Tennessee Eastman.

Many other organizations contributed to the success of this plant. The Tennessee Valley Authority, besides supplying power for the project, gave much time and effort in aiding in the selection of the site and later in the solution of various electrical supply and transmission problems. The U.S. Employment Service was of great assistance in the recruitment of personnel for the plant. The U. S. Office of Education helped organize the training classes. The U.S. Bureau of Mines helped to solve gas and ceramic problems. The U.S. Bureau of Standards co-operated on analytical problems. The Selective Service co-operated with the District at all times to keep trained personnel available.

For the gaseous diffusion plant, research, development, design, engineering, procurement and over-all job progress direction was provided by the Kellex Corporation, a special subsidiary of the M. W. Kellogg Company. Kellex was headed by P. C. Keith; A. L. Baker was Project Manager and J. H. Arnold was Director of Research and Development. The J. A. Jones Construction Company, Inc., with its own forces and some sixty-four subcontractors, did all the construction required by the facilities in the process area. Here the key man was Edwin L. Jones. The major subcontractors were the Midwest Piping and Supply Company, the Poe Piping and Heating Company, the L. K. Comstock Electric Company and the Bryant Electric Company. In co-operation with William A. Pope Company, the A. S. Schulman Electric Company and the Combustion Engineering Company, Jones also built the power plant, the main feature of which was the unusual requirement of a number of different frequencies.

Union Carbide was the operating contractor. Its key personnel on the site included G. T. Felbeck, in general charge both on and off site, C. E. Center, the superintendent, and H. D. Kinsey, the plant manager. Off-site, there were three key men without whom we could not have been successful: James A. Rafferty, a vice president of Carbide, Lyman A. Bliss, a vice president of a subsidiary, and L. M. Currie, who directed the SAM laboratory for Carbide.

The H. K. Ferguson Company was responsible for the design, engineering, construction, procurement and operation for the thermal diffusion plant.

APPENDIX VII

Written in April, 1943, to Dr. J. R. Oppenheimer:

DEAR ROBERT: This letter will serve to put in the record some of the things which we have discussed during the past week which have led to my decision not to accept a permanent connection with the Los Alamos project and to return to the Westinghouse Research Laboratories at East Pittsburgh.

First let me apologize for failure to emphasize the tentativeness of the situation in which I have been during the past month. This came about because I initially felt quite sure that I would decide to stay, and secondly because I thought that if the tentativeness were too much stressed it would interfere with my settling down and trying to be useful at once.

In trying to be clear about the reasons for the decision I suppose it boils down to this: With additional knowledge of detailed needs of the project I was unable to get a strong conviction that I am decidedly more useful to the war here than at Westinghouse. Since the change would entail considerable personal sacrifice, I do not feel justified in making it. I do not see how such a view could have been reached without my coming here to see the problem at first hand.

I am happy that you are generous enough to feel that I was of a little help during the first month. It will always be my hope to be able to help from a distance in any way that I can. There are many ways in which the technical resources with which Westinghouse could aid this project are as yet unexplored. Naturally, however, I will not take any initiative on this at home as being inconsistent with security policy. But if your people or others in related projects approach me with special needs I am sure that I will be in a better position to help because of the background I now possess.

There may be some point in making some general observations based on my brief experience. My own decision of course was weighted pretty heavily with personal factors which are not of general interest so I will skip them except insofar as they seem likely to be things that would also concern other people.

The thing which upsets me most is the extraordinarily close security policy. I do not feel qualified to question the wisdom of this since I am totally unaware of the extent of enemy espionage and sabotage activities. I only want to say that in my case I found that the extreme concern with security was morbidly depressing—especially the discussion about censoring mail and telephone calls, the possible militarization and complete isolation of the personnel from the outside world. I know that before long all such concerns would make me be so depressed as to be of little if any value. I think a great many of the other people are apt to be this way, otherwise I wouldn't mention it.

An aspect of this policy for which I am completely at a loss to find justification is the tendency to isolate this group intellectually from the key members of the other units of the whole project. While I had heard that there were to be some restrictions, I can say that I was so shocked that I could hardly believe my ears when General Groves undertook to reprove us, though he did so with exquisite tact and courtesy, for a discussion which you had concerning an important technical question with A. H. Compton. To me the absence from the conference of such men as A. H. Compton, E. O. Lawrence, and H. C. Urey was an unfortunate thing but up to that time in your office last Monday I had put it down simply to their being too busy with other matters.

I feel so strongly that this policy puts you in the position of trying to do an extremely difficult job with three hands tied behind your back that I cannot accept the view that such internal compartmentalization of the larger project is proper. My disturbance was complicated with the feeling that I might sooner or later unintentionally violate such rules through failure to comprehend them fully. On my way through Chicago coming out here I had a friendly chat with A. H. Compton about the project at his home which probably would be considered improper though if so I would say the scientific position of the project is hopeless.

To speak of something more on the positive side, I feel that the laboratory is extremely well-staffed on the basic physics side. You have several first-rate young experimentalists in Williams, Manley, and Wilson, who will do a splendid job in setting up the equipment and getting useful results from it. If to these can be added a couple of maturer experimentalists like Allison, and Bacher, in addition to McMillan, the success of this side of the project is assured. In the auxiliary fields like chemistry, with Kennedy and Segre, and metallurgy, with Cyril Smith, I do not think that you could have done better. The theoretical group is, of course, extremely brilliant. As you know the ordnance side is the weak spot and the one which will require some first-rate specialized mechanical engineering. This is one of the points at which I feel that Westinghouse might be of effective help on special problems.

The way the presence of such an excellent staff reacted on me was something like this: I found myself in a role analogous to that of a military man who would suddenly shift from the Air Force to submarines in the middle of a war. I saw that I would face a great task of learning a job while surrounded with people who understood it much better than I. At the same time, administratively, I would have to make decisions affecting their technical activities. Of course there are many minor matters that I could have handled as a stuffed shirt but I hope that that is not the best use of my abilities.

Now to get back again to less agreeable matters. I am worried about a situation which is not fully clear to me and perhaps is not as bad as my impression of it. I feel that an attempt is being made both by the Manhattan District and by the University of California to put too many restrictions on the activities of Dr. D. P. Mitchell. We have roomed together since he came up to the site and I know that he feels baffled and perplexed by some of the things confronting him. He is working with a high expenditure of nervous energy and I think that he should get more backing, otherwise an irreplaceable man may be lost. (Please let it be clear that these things are my own observations, put forth on my own responsibility, that he has in no way suggested that I take this up, that in fact, I am taking it up more as a laboratory problem than as Mitchell's personal problem.)

In the first place Dana Mitchell is an absolutely unique individual in America. He is a good physicist. In addition he has been buying equipment and supplies for experimental research projects for some 15 years, including recent experience in setting up such service for several war research projects. Situated as we are in remote isolation the supply problem is unusually difficult. It is moreover extraordinarily difficult owing to war shortages. A man in Mitchell's position needs to have complete authority, freedom of action, and responsibility within his field. A man of Mitchell's experience and competence should be given it unhesitatingly. The only criticism of Mitchell I have ever heard is that it is said that he oversimplified his records and accounting procedures. I do not know whether this is true, in any case it is the kind of matter on which there is bound to be strong difference of opinion.

What I would strongly recommend in this connection is that Mitchell be given complete control over all procurement, that it be absolutely definite that the Los Angeles office is under him, that he have full authority and responsibility in procurement matters, that the contractor do no more than set up accountants who record what he does, but that no person except yourself be in a position to question what he does.

Finally there is the matter of the working relations with the local military people. On the whole they have done remarkably well in getting the post started. But I feel that there is much room for improvement although some of the present fault lies on me for I did not carry on effectively a lot of detail which I would have done if not so preoccupied with my personal decision. The worst trouble seems to be a lack of close communications. I fully expect that this will remedy itself when Colonel Harman and his staff take up residence here and an effective town council is organized.

But there is also an unnecessary vagueness about many features of the town life with which the technical people are vitally concerned. The school situation is the most urgent. Many of the mothers are extremely anxious to know what is being done and how the schools are to be handled. These people come from good neighborhoods where there are good public schools. Many of them are more worried about the school here than any-thing else affecting their lives out of working hours. This matter gave me a great deal of personal concern and I know it is a factor that weighs heavily in Rabi's mind in his probable decision not to come. This situation was not helped with me by the way in which General Groves replied to my questions with a short plea for no "frills" when none had been asked for, together with what I felt were rather vague assurances that everything "necessary" would be provided. There is an awful lot of room for dis-agreement about the interpretation of the two words in quotes.

As I have said, I am afraid that I have been at fault in not better organizing our relations with Colonel Harman's staff. In consequence, his people have been pestered with many conflicting requests from different individuals of our group some of whom they did not know. On their side I believe they have not been entirely in the right since so many of his people have shown a tendency to be taciturn and uncommunicative so that our people find it difficult to learn what procedures they need to follow.

But these minor things are all of a kind that will work out in the next few months as people get acquainted. The school matter is much more critical. It will take decisive action very soon if a good school system is to be ready in the fall.

I hope this long tirade is of some help to you and that my association with the project adds up to something more positive than negative. With all best wishes for a complete and timely success in the solution of the primary problem of the project, and with the hope that I may be of some future help to you in it.

Sincerely,
(Signed) ED CONDON

(From testimony of Dr. Edward U. Condon before the Committee on Un-American Activities, U.S. House of Representatives)

APPENDIX VIII

(July 18, 1945)

MEMORANDUM FOR THE SECRETARY OF WAR

SUBJECT: The Test

    1. This is not a concise, formal military report, but an attempt to recite what I would have told you if you had been here on my return from New Mexico.

    2. At 0530, 16 July 1945, in a remote section of the Alamogordo Air Base, New Mexico, the first full-scale test was made of the implosion type atomic fission bomb. For the first time in history there was a nuclear explosion. And what an explosion! The bomb was not dropped from an airplane, but was exploded on a platform on top of a 100-foot-high steel tower.

    3. The test was successful beyond the most optimistic expectations of anyone. Based on the data which it has been possible to work up to date, I estimate the energy generated to be in excess of the equivalent of 15,000 to 20,000 tons of TNT, and this is a conservative estimate. Data based on measurements which we have not yet been able to reconcile would make the energy release several times the conservative figure. There were tremendous blast effects. For a brief period there was a lighting effect within a radius of 20 miles equal to several suns in midday; a huge ball of fire was formed which lasted for several seconds. This ball mushroomed and rose to a height of over 10,000 feet before it dimmed. The light from the explosion was seen clearly at Albuquerque, Santa Fe, Silver City, El Paso, and other points generally to about 180 miles away. The sound was heard to the same distance in a few instances, but generally to about 100 miles. Only a few windows were broken, although one was some 125 miles away. A massive cloud was formed which surged and billowed upward with tremendous power, reaching the substratosphere at an elevation of 41,000 feet, 36,000 feet above the ground, in about 5 minutes, breaking without interruption through a temperature inversion at 17,000 feet which most of the scientists thought would stop it. Two supplementary explosions occurred in the cloud shortly after the main explosion. The cloud contained several thousand tons of dust picked up from the ground and a considerable amount of iron in the gaseous form.

Our present thought is that this iron ignited when it mixed with the oxygen in the air to cause these supplementary explosions. Huge concentrations of highly radioactive materials resulted from the fission and were contained in this cloud.

4. A crater from which all vegetation had vanished, with a diameter of 1,200 feet and a slight slope toward the center, was formed. In the center was a shallow bowl 130 feet in diameter and 6 feet in depth. The material within the crater was deeply pulverized dirt. The material within the outer circle is greenish and can be distinctly seen from as much as 5 miles away. The steel from the tower was evaporated. Fifteen hundred feet away there was a 4-inch iron pipe 16 feet high set in concrete and strongly guyed. It disappeared completely.

5. One-half mile from the explosion there was a massive steel test cylinder weighing 220 tons. The base of the cylinder was solidly encased in concrete. Surrounding the cylinder was a strong steel tower 70 feet high, firmly anchored to concrete foundations. This tower is comparable to a steel building bay that would be found in a typical 15- or 20-story skyscraper or in warehouse construction. Forty tons of steel were used to fabricate the tower, which was 70 feet high, the height of a 6-story building. The cross bracing was much stronger than that normally used in ordinary steel construction. The absence of the solid walls of a building gave the blast a much less effective surface to push against. The blast tore the tower from its foundations, twisted it, ripped it apart and left it flat on the ground. The effects on the tower indicate that, at that distance, unshielded permanent steel and masonry buildings would have been destroyed. . . . I no longer consider the Pentagon a safe shelter from such a bomb. Enclosed are a sketch showing the tower before the explosion and a telephotograph showing what it looked like afterward. None of us had expected it to be damaged.

6. The cloud traveled to a great height first in the form of a ball, then mushroomed, then changed into a long trailing chimney-shaped column and finally was sent in several directions by the variable winds at the different elevations. It deposited its dust and radioactive materials over a wide area. It was followed and monitored by medical doctors and scientists with instruments to check its radioactive effects. While here and there the activity on the ground was fairly high, at no place did it reach a concentration which required evacuation of the population. Radioactive material in small quantities was located as much as 120 miles away. The measurements are being continued in order to have adequate data with which to protect the government's interests in case of future claims. For a few hours, I was none too comfortable about the situation.

7. For instance, as much as 200 miles away, observers were stationed to check on blast effects, property damage, radioactivity and reactions of the population. While complete reports have not yet been received, I know that no persons were injured nor was there any real property damage outside our government area. As soon as all the voluminous data can be checked and correlated, full technical studies will be possible.

8. Our long-range weather predictions had indicated that we could expect weather favorable for our tests beginning on the morning of the seventeenth and continuing for four days. This was almost a certainty if we were to believe our long-range forecasters. The prediction for the morning of the sixteenth was not so certain, but there was about a 90 per cent chance of the conditions being suitable. During the night there were thunderstorms with lightning flashes all over the area. The test had been originally set for 0400 hours and all the night through, because of the bad weather, there were urgings from many of the scientists to postpone the test. Such a delay might well have had crippling results due to mechanical difficulties in our complicated test setup. Fortunately, we disregarded the urgings. We held firm and waited the night through, hoping for suitable weather. We had to delay an hour and a half, to 0530, before we could fire. This was 30 minutes before sunrise.

9. Because of bad weather, our two B-29 observation airplanes were unable to take off as scheduled from Kirkland Field at Albuquerque and when they finally did get off, they found it impossible to get over the target because of the heavy clouds and the thunderstorms. Certain desired observations could not be made and while the people in the airplanes saw the explosion from a distance, they were not as close as they will be in action. We still have no reason to anticipate the loss of our plane in an actual operation, although we cannot guarantee safety.

10. Just before 1100 the news stories from all over the state started to flow into the Albuquerque Associated Press. I then directed the issuance by the Commanding Officer, Alamogordo Air Base, of a news release as shown on the enclosure. With the assistance of the Office of Censorship, we were able to limit the news stories to the approved release supplemented in the local papers by brief stories from the many eye-witnesses not connected with our project. One of these was a blind woman who saw the light.

11. Brigadier General Thomas F. Farrell was at the control shelter located 10,000 yards south of the point of explosion. His impressions are given below:

"The scene inside the shelter was dramatic beyond words. In and around the shelter were some twenty-odd people concerned with last-minute arrangements prior to firing the shot. Included were: Dr. Oppenheimer, the Director, who had borne the great scientific burden of developing the weapon from the raw materials made in Tennessee and Washington, and a dozen of his key assistants—Dr. Kistiakowsky, who had developed the highly special explosives; Dr. Bainbridge, who supervised all the detailed arrangements for the test; Dr. Hubbard, the weather expert, and several others. Besides these, there were a handful of soldiers, two or three Army officers and one Naval officer. The shelter was cluttered with a great variety of instruments and radios.

"For some hectic two hours preceding the blast, General Groves stayed with the Director, walking with him and steadying his tense excitement. Every time the Director would be about to explode because of some untoward happening, General Groves would take him off and walk with him in the rain, counseling with him and reassuring him that everything would be all right. At twenty minutes before zero hour, General Groves left for his station at the base camp, first, because it provided a better observation point, and second, because of our rule that he and I must not be together in situations where there is an element of danger, which existed at both points.

"Just after General Groves left, announcements began to be broadcast of the interval remaining before the blast. They were sent by radio to the other groups participating in and observing the test. As the time intervals grew smaller and changed from minutes to seconds, the tension increased by leaps and bounds. Everyone in that room knew the awful potentialities of the thing that they thought was about to happen. The scientists felt that their figuring must be right and that the bomb had to go off but there was in everyone's mind a strong measure of doubt. The feeling of many could be expressed by 'Lord, I believe; help Thou mine unbelief.' We were reaching into the unknown and we did not know what might come of it. It can be safely said that most of those present—Christian, Jew and atheist—were praying and praying harder than they had ever prayed before. If the shot were successful, it would be a justification of the several years of intensive effort of tens of thousands of people —statesmen, scientists, engineers, manufacturers, soldiers and many others in every walk of life.

"In that brief instant in the remote New Mexico desert the tremendous effort of the brains and brawn of all these people came suddenly and startlingly to the fullest fruition. Dr. Oppenheimer, on whom had rested a very heavy burden, grew tenser as the last seconds ticked off. He scarcely breathed. He held on to a post to steady himself. For the last few seconds, he stared directly ahead and then when the announcer shouted 'Now!' and there came this tremendous burst of light followed shortly thereafter by the deep growling roar of the explosion, his face relaxed into an expression of tremendous relief. Several of the observers standing back of the shelter to watch the lighting effects were knocked flat by the blast.

"The tension in the room let up and all started congratulating each other. Everyone sensed, 'This is it!' No matter what might happen now, all knew that the impossible scientific job had been done. Atomic fission would no longer be hidden in the cloisters of the theoretical physicists' dreams. It was almost full grown at birth. It was a new force to be used for good or for evil. There was a feeling in that shelter that those concerned with its nativity should dedicate their lives to the mission that it would always be used for good and never for evil.

"Dr. Kistiakowsky, the impulsive Russian [interpolation by Groves at this point: 'an American and Harvard professor for many years'], threw his arms around Dr. Oppenheimer and embraced him with shouts of glee. Others were equally enthusiastic. All the pent-up emotions were released in those few minutes and all seemed to sense immediately that the explosion had far exceeded the most optimistic expectations and wildest hopes of the scientists. All seemed to feel that they had been present at the birth of a new age—The Age of Atomic Energy—and felt their profound responsibility to help in guiding into right channels the tremendous forces which had been unlocked for the first time in history.

"As to the present war, there was a feeling that no matter what else might happen, we now had the means to insure its speedy conclusion and save thousands of American lives. As to the future, there had been brought into being something big and something new that would prove to be immeasurably more important than the discovery of electricity or any of the other great discoveries which have so affected our existence.

"The effects could well be called unprecedented, magnificent, beautiful, stupendous and terrifying. No man-made phenomenon of such tremendous power had ever occurred before. The lighting effects beggared description. The whole country was lighted by a searing light with the intensity many times that of the midday sun. It was golden, purple, violet, gray and blue. It lighted every peak, crevasse and ridge of the nearby mountain range with a clarity and beauty that cannot be described but must be seen to be imagined. It was that beauty the great poets dream about but describe most poorly and inadequately. Thirty seconds after the explosion came, first, the air blast pressing hard against the people and things, to be followed almost immediately by the strong, sustained, awesome roar which warned of doomsday and made us feel that we puny things were blasphemous to dare tamper with the forces heretofore reserved to The Almighty. Words are inadequate tools for the job of acquainting those not present with the physical, mental and psychological effects. It had to be witnessed to be realized."

(End of General Farrell's account)

12. My impressions of the night's high points follow:

After about an hour's sleep I got up at 0100 and from that time on until about five I was with Dr. Oppenheimer constantly. Naturally he was very nervous, although his mind was working at its usual extraordinary efficiency. I devoted my entire attention to shielding him from the excited and generally faulty advice of his assistants, who were more than disturbed by their excitement and the uncertain weather conditions. By 0330 we decided that we could probably fire at 0530. By 0400 the rain had stopped but the sky was heavily overcast. Our decision became firmer as time went on. During most of these hours the two of us journeyed from the control house out into the darkness to look at the stars and to assure each other that the one or two visible stars were becoming brighter. At 0510 I left Dr. Oppenheimer and returned to the main observation point which was 17,000 yards from the point of explosion. In accordance with our orders, I found all personnel not otherwise occupied massed on a bit of high ground.

At about two minutes of the scheduled firing time all persons lay face down with their feet pointing toward the explosion. As the remaining time was called from the loudspeaker from the 10,000-yard control station there was complete silence. Dr. Conant said he had never imagined seconds could be so long. Most of the individuals in accordance with orders shielded their eyes in one way or another. There was then this burst of light of a brilliance beyond any comparison. We all rolled over and looked through dark glasses at the ball of fire. About forty seconds later came the shock wave followed by the sound, neither of which seemed startling after our complete astonishment at the extraordinary lighting intensity. Dr. Conant reached over and we shook hands in mutual congratulations. Dr. Bush, who was on the other side of me, did likewise. The. feeling of the entire assembly was similar to that described by General Farrell, with even the uninitiated feeling profound awe. Drs. Conant and Bush and myself were struck by an even stronger feeling that the faith of those who had been responsible for the initiation and the carrying-on of this Herculean project had been justified. I personally thought of Blondin crossing Niagara Falls on his tightrope, only to me this tightrope had lasted for almost three years, and of my repeated, confident-appearing assurances that such a thing was possible and that we would do it.

13. A large group of observers were stationed at a point about twenty-seven miles north of the point of explosion.

14. While General Farrell was waiting about midnight for a commercial airplane to Washington at Albuquerque—120 miles away from the site—he overheard several airport employees discussing their reactions to the blast. One said that he was out on the parking apron; it was quite dark, then the whole southern sky was lighted as though by a bright sun; the light lasted several seconds. Another remarked that if a few exploding bombs could have such an effect, it must be terrible to have them drop on a city.

15. My liaison officer at the Alamogordo Air Base, sixty miles away, made the following report:

"There was a blinding flash of light that lighted the entire northwestern sky. In the center of the flash, there appeared to be a huge billow of smoke. The original flash lasted approximately ten to fifteen seconds. As the first died down, there arose in the approximate center of where the original flash had occurred an enormous ball of what appeared to be fire and closely resembled a rising sun that was three-fourths above a mountain. The ball of fire lasted approximately fifteen seconds, then died down and the sky resumed an almost normal appearance.

"Almost immediately, a third, but much smaller, flash and billow of smoke of a whitish-orange color appeared in the sky, again lighting the sky for approximately four seconds. At the time of the original flash, the field was lighted well enough so that a newspaper could easily have been read. The second and third flashes were of much lesser intensity.

"We were in a glass-enclosed control tower some seventy feet above the ground and felt no concussion or air compression. There was no noticeable earth tremor although reports overheard at the Field during the following twenty-four hours indicated that some believed that they had both heard the explosion and felt some earth tremor."

16. I have not written a separate report for General Marshall, as I feel you will want to show this to him. I have informed the necessary people here of our results. Lord Halifax after discussion with Mr. Harrison and myself stated that he was not sending a full report to his government at this time. I informed him that I was sending this to you and that you might wish to show it to the proper British representatives.

17. We are all fully conscious that our real goal is still before us. The battle test is what counts in the war with Japan.

18. May I express my deep personal appreciation for your congratulatory cable to us and for the support and confidence which I have received from you ever since I have had this work under my charge.

19. I know that Colonel Kyle will guard these papers with his customary extraordinary care.

                                                                    L. R. GROVES

APPENDIX IX

Two pertinent paragraphs from the opening statement by Secretary of War Robert P. Patterson to the Committee on Military Affairs, House of Representatives, on October 9, 1945:

The War Department has taken the initiative in proposing that it be divested of the great authority that goes with the control of atomic energy, because it recognizes that the problems we now face go far beyond the purely military sphere.

The atomic bomb is the most devastating weapon we know, but the means of releasing atomic energy which it employs may prove to be the greatest boon to mankind in the world's history. The wisest minds in our Nation will be required to administer this discovery for the benefit of all of us.

Opening paragraphs in statement by Major General Leslie R. Groves to the Committee on Military Affairs, House of Representatives, on October 9, 1945:

In coming before your committee today we are appealing for an opportunity to give you our existing powers. In the interest of the war effort, there was delivered into our care the responsibility for directing all activities relating to the release and use of atomic energy.

We have discharged that responsibility to the best of our ability. Thanks to the brilliant and selfless efforts of the thousands of scientists, engineers, industrialists, workers and Army and Navy officers associated with the project, and to the wise counsel of Secretary Stimson and his advisers and the members of the Military Policy Committee, our work achieved its purpose. It helped to shorten the war and to save the lives of American and Allied fighting men.

But the individual responsibility that was desirable in wartime should not be continued today. The hopes and fears of all mankind are so inextricably bound up with the future development of atomic energy, and the problems requiring immediate solution are so fundamental that control should be vested in the most representative and able body our democratic society is capable of organizing.

The bill you are considering today is intended to create such a body.

It would establish a commission of nine distinguished citizens, with a revolving membership to guard against political domination or the development within the commission of frozen attitudes that would act as a brake on experimentation and new ideas.

Within the limits of general policy, as defined by Congress, and of appropriations, as authorized by Congress, the commission would have broad power to conduct or supervise all research and manufacturing activities relating to the use of atomic energy for military or civilian purposes; to control the raw materials from which atomic energy may be derived and to provide for the security of information and property connected with the release of atomic energy. It is also the aim of this legislation that the commission capitalize on the initiative and ingenuity of American science and American industry by giving as much freedom and encouragement to private research and private enterprise in this field as it is possible to give consistent with the requirement of American security.

The success of the Manhattan District Project would have been impossible without the support it received from colleges and universities, from large and small industrial corporations, and from the skill of American labor.

APPENDIX X

Opening statement of Major General Leslie R. Groves to the Special Committee on Atomic Energy, United States Senate, on November 28, 1945:

It is essential, in the highest national interest, that further development in the field of atomic energy be pursued under controls which will preclude the utilization of atomic energy in a way which would imperil the national safety or endanger world peace. Future activity in this field is so important to the national welfare, and potentially to the enrichment of our living, that control should be exercised by a special commission independent of any existing Government agency with the sole duty of supervising and controlling the development of atomic energy. The commission should have complete authority over all activities in the field, subject only to the approval of Congress and the President. The commission should be composed of persons of recognized ability whose actions would be unquestionably in the public interest. Broad discretionary powers and adequate funds are essential to its success.

The War Department will always have a vital interest in the use of atomic energy for military purposes. In the field of practical administration and operation, the Army can furnish invaluable assistance. Civilian and military personnel who have acquired knowledge and experience on the project should continue to serve to the extent that their services are useful. The commission should be in complete control of policy and should exercise general direction and supervision of all activities.

Because of the current uncertainty, we are daily losing key people whose services should be retained. Until that uncertainty is resolved by the establishment of a national policy, we are not in a position to offer acceptable commitments to these key people. Prolonged delay will result in appreciable loss of the present efficiency of the vast combination of plants, scientific talent, and engineering skill.

We must recognize the clear distinction between domestic control and international control. The two can and should logically be separated. Domestic control is necessary no matter what international policy may be eventually worked out for the United States and the world. It is necessary to protect America's tremendous investment in atomic research and development and to insure that this development will go steadily forward.

Extracts from the statement by Secretary of War Robert P. Patterson to the Special Senate Committee on Atomic Energy, February 14, 1946:

Long before the bombs were dropped it was realized that there were unmeasured possibilities in the development of atomic energy for industrial and other peacetime purposes as well as for use as a war weapon. In May 1945 Secretary Stimson, with the approval of the President, formed a committee to consider the subject and to recommend legislation for the control and development of atomic energy. The committee consisted of Secretary Stimson; James F. Byrnes (prior to his appointment as Secretary of State); Will Clayton, Assistant Secretary of State; Ralph

Bard, Under Secretary of the Navy; George Harrison, president of the New York Life Insurance Co. and special assistant to the Secretary of War; Dr. Vannevar Bush, Chairman of the Office of Scientific Research and Development; Dr. Karl Compton, president of Massachusetts Institute of Technology; and Dr. James Conant, president of Harvard University.

I know of no committee appointed during wartime or thereafter that had higher talent than that committee. The committee had the assistance of a scientific panel composed of four of the leading scientists in the project, Dr. J. R. Oppenheimer, Dr. E. O. Lawrence, Dr. A. H. Compton, and Dr. Enrico Fermi. Legislation was drafted under direction of this committee and with the cooperation of the State Department, Interior Department, and Department of Justice, the bill being later introduced as the May-Johnson bill.

Senator AUSTIN. These names given in this paragraph of your statement really represent the authors of the May-Johnson bill?

Secretary PATTERSON. Yes, sir. The draftsmanship was done under their direction. I think the actual draftsmen were Captain Davis, who is sitting here, who had a hand in it, and Mr. Marbury, a Baltimore lawyer who was then in the employ of the War Department, and Under Secretary Royall, who was then a brigadier general.

I will not at this time go into the details of the May-Johnson bill. The objectives, fully stated in the bill, are to promote the national welfare, to secure the national defense, to safeguard world peace, and to foster the acquisition of further knowledge concerning atomic energy. Responsibility was transferred from the War Department to a new civilian agency, to be known as the Atomic Energy Commission, to take over and manage all source materials of atomic energy, all stock piles of materials, and all plants and property connected with development and use of atomic energy. Full ownership and control are vested in the United States. It is provided that the Commission interfere as little as possible with private research, and employ other governmental agencies, educational and research institutions, and private enterprise to the maximum extent. The Commission is given power to adopt the necessary security regulations to control the collection, publication, and transmission of information on release of atomic energy, as required by considerations of national defense or military security.

The House Military Affairs Committee held hearings on the bill. It was freely criticized as a measure drafted by the military and intended to perpetuate military control of atomic energy, although the true facts are that it was drafted by the committee of civilians named above, or under their supervision, and that it provides for transfer of responsibility from the

War Department to a civilian agency. The House committee, after adding several amendments including those further emphasizing freedom of research and investigation, reported the bill favorably.

I should add that the President, since introduction of the May-Johnson bill, has indicated that he is of the opinion that a number of changes should be made in it. The War Department will, of course, advocate such changes in discussion of any detailed legislation.

INDEX

Abelson, Philip, 23, 119
A-bomb, ix; assembly, 282-83; Corps of Engineers, 11; decision to use, 265; delay in using, 392; delivery (*see* 509th Group *under* Air Force *and* Air unit), deadline for, 124; designing, 60, 157-58; development (*see* Project Y); explosive force, 269; first, 308, 313-14; gun-type (*see* Implosion); haste and, 11; high air burst, 286; international implications of, 336; magnitude of project, 11; mechanical requirements, 159; names for, 260, 260n; operational plans, 307-14; power requirements of plants, 12, 12n; second, 308; take-off base (*see* Tinian); target date, 311-12; targets, 266-76; testing, 260, 261-62, 341 (*see* Alamogordo); U.S. entry into World War II, 9; Washington control over, 271, 272 (*see also* Manhattan Project, MED, *and* Project)
Accidents: gas diffusion process, 116; Oak Ridge, 110
Acheson, Dean, 411
Ackart, Everett G., 79
Advisory Committee on Uranium, 126
African Metals, 36
Air Force, U.S., 275-76; bomb and, 253-61; comprehension of bomb, 271; in England, 270-71; responsibilities, 254; Strategic Air Forces, 314; targets, 268; 3d Photo Reconnaissance Squadron, 316-17; 20th, 278, 283, 308, 311; 21st Bomber Command, 278, 283; 393d Heavy Bombardment Squadron, 259, 260; 509th

Composite Group, 259, 259n, 277, 278, 279, 284, 299, 308, 311, 314, 316-17, 341, delivery of bomb, 310, difficulty of, 354, staging, 283, strength, 259-60, training, 260-61, 284-86
Air unit, 253-62; commander, 258; navigational training, 261; staff for, 258-59
Akers, W. A., 117, 128-29, 130, 132
Alamogordo, 120, 269, 288-304, 326, 415; bomb (July 16, 1945), 40; choice of site, 289; countdown, 295-296; fallout, 298-99, 302; Farrell account, 435-38; Groves report, 303, 433-40; Jumbo, 288-89, 288n, 297; observation planes, 294-95; postponement, 294; press release, 301-2; radioactivity, 352; reaction to, 297-298, 303; results of, 305; security, 299; shock wave, 296; strength of explosion, 296; tension at, 293-94, 297; time of, 296; weather problem, 291-92; White House press release, 327
Alberta group, 283; aim, 353-54
Albuquerque, 65-66; Sandia Base, 374, 382
Allied Armies, 215
Allied Intelligence, 194-98
Allis, William, 193
Allis-Chalmers, 97, 103, 108, 114
Allison, Samuel, 418
Alpha separation process, 98
Alsos: advisory committee, 208; agents, 212; Calvert group, 215; commander, 192-93; on Continent, 208, 211; disbandment, 249; France,

445

210, 211-15; Germany, 210; Heidelberg headquarters, 242; interservice cooperation, 210; Italy, 191-98, 207, 208-9; Joliot-Curie interrogation, 213, 224; location of German scientists, 215-18, 220-21; London, 208, 215; Mediterranean section, 209, 215; officers, 236; Paris headquarters, 215; purpose, 191-92; Rome targets, 209-10; scientific members, 221-22; staffing, 210; second mission, 198ff.; Strasbourg, 230

Aluminum Company of America, 83

Amaldi, Dr., 209, 210

American Federation of Labor, 155

American industry: discretion, 47-48; involvement in Project, 44, 44n; patriotism, 48, 51; strength of, 83

American technology, x

Amherst College, 181, 182

Anderson, John, 132, 133-34, 136; British-French agreement, 224-27; postwar developments, 403-5; tripartite agreement, 70-73, 176

Anderson, Roland A., 420

André, Gaston, 218

Anglo-American intelligence liaison, 194-98

Annapolis, 160, 279

Antes, Donald E., 365

Arctic Circle, 178

Argonne Forest, 14, 52, 53, 68

Armed Forces Special Weapons Project, 376

Army, U.S., 7, 24; "about," 312; atomic program, 125, 194; counterintelligence, 139; du Pont and, 42-43; Fifth, 190, 193; G-2, 138, 139, 236; Alsos and, 207, 208, 210, 248, 249; atomic energy; 185, 186, 190, 191, 193; nuclear weapons position, 376; plutonium project, 45; security responsibility, 138

Army Corps of Engineers, 3, 4n, 10, 16-17, 18, 75, 162, 181; atomic bomb, 11; safety regulations, 110

Army District Engineer Offices, 73

Army Service Forces, 208

Army Services of Supply, xi, 3, 4n

Arnold, H. H., 188, 264, 278, 306; bomb control, 271, 272; Hiroshima, 315, 330; operational plan, 311; Project and, 253-54, 256-57; targets, 267, 268, 276

Arnold, J. H., 428

Ashbridge, W., 154n, 164

Ashworth, F. L., 261, 277, 278, 282n, 314, 343, 345, 353; Tinian choice, 278-79; weaponeer, 259, 284

"Atom Gives Up, The," 35

Atomic bomb (see A-bomb)

Atomic bomb project (see MED and Manhattan Project)

Atomic energy, ix, ixn; American intelligence, 185ff. (see also Alsos); American physicists, 5-6; by-products, 385-87; commissioners, 393, 395; diplomacy and, 338; European scientists, 5-6; French patents, 227; implications of, 336; interchange of information, 125ff., 406; international controls, 408, 412, 414; legislation, 389-95; military uses, 5; peaceful uses, 5, 386, 414; postwar developments, 390, 401-12; propaganda on, 391-92; research on, 5; voluntary international censorship, 6; War Department, 171

Atomic Energy Act, 137, 395

Atomic Energy Commission (AEC), 26, 164n, 376, 385, 387, 389-400, 406; establishment, 394

Atomic physics, ix, ixn

Atomic policy, making of, 23-25

Atomic weapons, types, 145

Attlee, Clement, 403, 404-6; international controls, 410

Auer-Gesellschaft, 197, 219n, 220, 221; Oranienburg, 230-31

Auger, Pierre, 211, 228-29

Axis Powers, 49, 140-41

B-29s, 254, 256, 268; altitude, 284; bomb delivery, 257-58, 261; modifications, 260, 284-85; Pacific Theatre, 257; take-off base, 277 (see also Hiroshima)

Babcock and Wilcox, 288

Bacher, Robert F., 164n

Badger & Sons, E. B., 15

Bagge, Erich, 214, 246, 248; reaction to Hiroshima, 335, 336

Bailey, Agent, 209
Bain, George W., 181-82, 183-84
Bainbridge, K. T., 164n, 289, 292n, 436
Baker, A. L., 428
Bankers Trust Co., N.Y., 177
Bard, Ralph A., 389n
Barker, C. D., 100
Barkley, Alben W., 363
Barnard, Chester, 411
Barnes, Philip, 343
Barnes, Thomas, 173
Bart Laboratories, 115
Baruch, Bernard, 412
Bateman, George C., 174n
Battelle Institute, 418
Battle of the Bulge, 222-23
Beahan, Kermit K., 259, 343, 345
Beatson, Agent, 211
Behler, Fred, 100-1
Belgian Congo: resources, 33, 34-35, 36, 37; uranium ore: richness, 179; tripartite agreement, 170-73, 176, 178
Belgium: German occupation, 218; government-in-exile, 170; invasion of, 34; tripartite agreement, 170-76; uranium ore, 33-34, 219, 236, 237-238 (see also Belgian Congo)
Bell, Daniel, 107, 108
Berg, Moe, 217
Berkeley lab, 17, 51, 60, 61, 62, 97, 98, 139, 149; demobilization, 377; espionage in, 138, 141; Japanese contacts, 187 (see also University of California)
Berlin: bombing of, 216; Dahlem, 231, 233, 240; heavy water, 188, 189; Institute of Physics, 239; Russian zone, 231; uranium plant, 197
Beta separation process, 98, 99
Bethe, Hans A., 154, 164n, 381
Betts, E. C., 173
Betts, T. J., 208
Bikini tests, 381, 383, 384-85, 384n; Joint Task Forces, 384
Bisingen-Hechingen area, 216-18
Bissell, Clayton L., 194, 207
Blackett, P. M. S., 248, 338
Blair, Robert C., 426n

Blandy, W. H. P., 384
Bliss, Lyman, A., 112, 428
Board of Economic Warfare, 35
Bohr, Niels, 197
Bomber missions, 261
Bonneville Dam, 74, 82, 89
Bothe, Walther, 213, 214, 231, 232, 233
Boulder Dam, 73
Bowen, H. G., 23
Bradbury, Norris, 378, 381, 382-84
Bradley, Omar, on Russians, 237
Branch, R. T., 427
Brazil, 184, 196
Bridges, Styles, 363
British Atomic Energy Office, 196
British Intelligence, 196, 197, 198, 217
British scientists, 7, 115, 338, 339; gaseous diffusion process, 117-19; interchange of information, 136-37; Manhattan Project, 143-44, 407 (see also Fuchs, Klaus)
British 21st Army Group, 219
British Supply Cabinet, 175
Brodie, Capt., 340
Brown, Edward J., 99
Brussels, 34, 197
Bull, H. R., 235
Bullock, J. C., 236
Bundy, Harvey, 20, 23, 31, 226, 264, 304; interchange of information, 131, 132, 135; Smyth Report, 350
Bush, Vannevar, xi, xii, 7, 8, 9, 10, 12, 16, 23, 45, 53, 60, 61, 70, 96, 160, 393; Alamogordo, 290, 294, 296, 303, 438; Alsos, 207; atomic intelligence activities, 190; Combined Policy Committee, 136n; Congressional briefing, 362, 363; decision making, 344; Groves and, 20-22; Interim Committee, 389n; letter to Roosevelt on U.S. responsibility, 58-59; Military Policy Committee, 24; nuclear program report (1942), 10; patents, 418-19; postwar developments, 403, 411; U.S.-British interchange of information, 125, 127, 128-36; -White House liaison, 125, 131
Buxton, G. E., 186

Byrnes, James F., 100, 304; Interim Committee, 389n, 390; postwar developments, 402-5, 411-12

California Institute of Technology, 44, 64

Calvert, Horace K., 140, 194-98, 208; in France, 211, 212, 213, 220; London mission, 215-18; Stassfurt mission, 237; Union Minière ore, 236

Canada, 36; Anglo-American agency, 170; Combined Development Trust, 174n; Combined Policy Committee, 136; National Research Council, 214; postwar development, 403-405; scientists in Manhattan Project, 407; Tube Alloys Project, 212, 225; uranium ore, 37, 178, 179; U.S. heavy water supply, 15n; U.S. interchange of information, 130

Cannon, Congressman, 364

Carnegie Institute (Washington), 119

Carpenter, Walter S., Jr., 47, 48, 49, 51, 54, 58

Carpenter, William, 331-32

Censorship, office of, 140, 146, 301, 325 (see also Press censorship)

Centerboard, 314

Centre Nationale de la Recherche Scientifique, 224, 225, 227

Chadwick, James, 136, 137, 144, 182, 229, 407; Smyth Report, 350-51

Chapman, A. K., 428

Chemical engineering, 43

Chemistry, in Project, 110-11

Cherwell, Lord, 130-31, 132, 304

Chicago lab, 14, 39, 40-42, 43, 50, 62, 70, 348; demobilization of, 377; du Pont liaison, 48, 79-80; European scientists, 44-45; friction, 46; patent claims, 45; peacetime project, 386-87; pile, 16; pilot reactor, 14-15; plutonium process, 68ff.; research, 17, 38; responsibility, 53; safety factor in design, 87 (see also University of Chicago)

Chipman, John, 418

Chrysler Corp., 114, 116

Church, Gilbert P., 73, 74

Churchill, Winston S., 30, 128, 248, 254; Groves' Alamogordo report, 304; Hyde Park aide-mémoire, 401-402; interchange of information, 130-33, 135; Manhattan Project, 408; optimism, 407; Quebec Agreement, 225; tripartite agreement, 174; use of bomb, 265

CIC: in Alsos, 193, 210, 211, 215; Alsos agents, 249; Stassfurt, mission, 237

Clayton, William L., 389n

"Clinton Engineer Works, The," 25, 26

Clinton Engineer Works, 80, 94, 364; electromagnetic plant, 95; medical program, 423; pile, 41, 78, 85, 88; semi-works, 68, 70, pile, 85, 88, primary purpose, 79, reason for building, 78, speed factor, 78 (see also Oak Ridge)

Collbohm, F. R., 268n

College of France, 212, 213

Colorado Plateau, uranium ore, 37, 178, 179

Colt, Sloan, 177

Columbia River, 75, 81; fish conservation, 82, 93

Columbia University (lab), 6, 7, 62, 164n; demobilization, 377; gas diffusion process, 113; SAM laboratory, 111, 112, 428; U-235 research, 51, 51n

Combined Chiefs of Staff, 205, 206; bomb control, 271; knowledge of bomb, 272

Combined Development Trust, 170-176, 238, 403-4; American funds for, 176-77; American trustees, 174n, 176-77; "Declaration of Trust," 173-74; German supplies, 242; territorial jurisdiction, 184

Combined Policy Committee, 133-34, 136-37, 174, 401, 403-5; establishment, 136

Committee of the Privy Council for Scientific and Industrial Research, 224

Communists, 138, 141, 142

Compton, Arthur H., 9, 14, 39, 41, 43-44, 48, 49, 53, 54, 80, 430; estimate on bomb production, 55-56; group (see Chicago lab); Nobel

Prize winner, 9, 62; physics of bomb development, 60; radioactive poisoning, 200; unexpendable, 62
Compton Effect, 9
Compton, Karl T., 389n
Conant, James B., xi, xii, 8, 10, 23, 43, 46, 52, 54, 55, 57, 60, 61, 63, 70, 96, 116, 393; Alamogordo, 290, 294, 296, 303, 438; British-French atomic agreement, 228; Combined Policy Committee, 136n; Interim Committee, 319n; letter to Oppenheimer on Los Alamos, 150-51; Los Alamos, 150, 151, 162; Military Policy Committee, 24; NRRC, chairman, 149; postwar developments, 411; radioactive poisoning, 200; scientific adviser to Groves, 44-45; Smyth Report, 348-52; U.S.-British interchange of information, 125-31
Condon, Edward U., 154-55, 154n, 156; resignation, 429-32
Congressional investigations, 70
Congressional Record, 148
Conklin, Frederick R., 428
Conrad, George B., 195, 208
Consodine, W. A., 173n, 177, 325, 326
Consolidated Mining and Smelting Company of Canada, 15
Cook County Forest Commission, 14-15
Cooper, Governor, 26-27
Cooling (pile) methods, 41
Cornell University, 154
Corps of Engineers (see Army Corps of Engineers)
Creedon, F. C., 102, 103, 427
Crenshaw, Thomas T., 426n
CSS, 198
Curie, Irène, 211
Curie, Marie, 211
Curie, L. M., 428
Cyclotrons, 41, 95n; Bothe and, 232; Japanese, 367-72; Joliot-Curie's, 213

Darwin, Charles, 338
Daniels, Farrington, 386-87
Davis, George C., 196
D-Day, 198, 200
Defense Plant Corp., 108

de Gaulle, Charles, 224
Degussa, 219n, 232
Deionizing plants, 81
Dempster, A. J., 95n
Dennis, W. R., 154, 154n
Dennison, D. M., 268n
Derry, J. H., 268, 320
DeSilva, Peer, 289, 307-8
Dickinson, W. T., 268n
Diebner, Kurt, 213, 239-40, 244, 246, 247; capture of, 242
Diebner, Frau, 247
Dill, John, 189; Combined Policy Committee, 136n
District, 80; organization, 100; security, 109 (see also Manhattan Project and MED)
District Engineer, 28, 63, 140, 181
Dix, Howard, 186
Doan, R. L., 418
Donovan, William, 186
Dow, David, 154n
"DSM," 17
Dudley, J. H., 64, 65
Dunning, John R., 111
du Pont, 70, 74, 101, 302, 331; in Alsos, 241; -Chicago liaison, 79-80; Compton on, 56; fee, 59; Hanford administrators, 91; heavy-water program, 15n; patent rights, 58; plutonium project, 42-44, 46-52, 52n, 54, 55-59, 68, 69; profit, 58-59; safety precautions, 87, 88; TNX, 79
Durkin, M. P., 101

Eastman Kodak Co., 96
Echols, Oliver P., 255
Eindhoven, uranium ore in, 219-20
Einstein, Albert, letter to Roosevelt, 7
Eisenhower, Dwight D., 173, 188; atomic policy, 380, 391; Chief of Staff, 369; Japanese cyclotrons, 369, 370; military in Project, 375-76; Operation Harborage, 235, 240; President, 395; radioactive warfare threat, 200-6; scientific intelligence mission, 208
Eldorado Mining Co., 178
Electromagnetic plant, 94, 95-97, 103-111, 127-28; hazards, 109-10; pri-

ority, 111; recruitment for, 99; support, 103 (see also Y-12)
Electromagnetic process, 17, 62; British efforts in, 128; chemistry, 110-11; Lawrence and, 348; pilot plant, 17, 18
Element 93, 333
Ellis, James, 428
Engel, Albert J., 361, 363, 364
England (see Great Britain)
Enola Gay, 317
Ent, U. G., 289, 299
Essau, Abraham, 213
ETOUSA (European Theater of Operation, U.S.A.), 195, 195n, 202
Europe: atomic energy projects, 195, 248; U.S. Occupation Forces, 375
European scientists, 56, 186-87, 197-198; Chicago group, 44-45; use of bomb against Japan, 266
Evans, R. M., 302
Exponential experiment, 53
Exponential pile, Berlin, 240

Fallout, 269, 286; Alamogordo, 298-299, 302; rain and, 291; radioactive, 291, 291n
Farm Hall, 333, 340
Farrell, Thomas F., 32, 267, 268, 283; Alamogordo, 291, 294, 298, 303, 435-38; Hiroshima, 315, 317, 319, 323, 327, 328, 329; Nagasaki bomb, 342, 344, 346; operational plan, 311, 312; transition period, 383
Fat Man bomb, 157, 163, 255-56, 260-62, 285; assembly, 282n, 343; explosive force, 269-70; flight testing, 256; postwar development, 381; production of, 309, 341, 342; testing date, 288 (see also Implosion bomb)
FBI: MED cooperation, 139, 140; security and, 138, 145
Federal Reserve Bank, 37
Feis, Herbert, 35
Felbeck, George T., 96, 112, 116, 428
Fercleve Corp., 120n
Ferebee, Thomas, 259, 317
Ferguson, H. K., Co., 120, 120n, 121, 122
Ferguson, Senator, 365-66

Fermi, Enrico, 6, 39, 197, 418; Alamogordo, 296-97; chain reaction, 70; experimental atomic pile, 54; test pile, 41
FFI, 212
Fields, K. E., 375, 395
Fine, Paul C., 349
Finletter, Thomas K., 35, 36
Fisher, William P., 268
Fisk, James, 193
Foreign-born, in Project, 141-42, (see also under nationality)
Fort Knox, 35
France: atomic energy position, 224; atomic energy project, 228-29; atomic patents, 224-25, 227; -British Atomic agreement, 224-25; fall of, 224; Project, 228; underground resistance movement, 213 (see also Alsos)
Franck, James, 39, 418
Free French Provisional Government (Algiers), 225
French Army, 240-41; 2nd Armored Division, 211; Moroccan troops, 241-42
French scientists: problem of, 224-29; uranium fission, 33-34
Friedel, H. L., 421
Frisch, O. R., 197
Fuchs, Klaus, 119, 143-45, 167, 407
Furnan, R. R., 194, 196, 209, 219, 305, 306
Fusion bomb, 158

G-2 (see under Army)
Gamma rays, tolerance dose, 87
Gary, Tom C., 52n, 79
Gaseous diffusion process, 111-19, 121; accidents, 116; all-American effort, 117; British and, 117-19, 127-128, 405, 407; Columbia success, 113; construction objectives, 117; gas corrosion, 114-15; plant, 428; postwar, 379-80; priority, 111; Scientific Safety Committee, 116; semiworks, 113; Union Carbide, 94, 95, 96, 100 (see also K-25 project)
Gattiker, David, 196; Stassfurt mission, 237
Gaudin, A. M., 183

Geiger counters, 200, 200n, 202; Alamogordo, 298
General Electric, 56, 97, 103, 411
General Motors Corp., 331
General Policy Group, 134; Sept., 1942, meeting, 126, 128
General Staff, U.S., 207; Japanese cyclotrons, 367, 369, 370; military in Project, 374
George, H. L., 306
George, Warren, 426n
Gerlach, Walther, 213, 231, 239, 244, 247; capture of, 242; reaction to Hiroshima, 335, 336
German Army Research, 213
German Chemical Society, 232
German National Research Council, 239, 245
German scientists, 6, 210, 215-16, 218; Alsos location of, 221-22; anti-Nazi, 244; capture of, 185, 239, 241, 242-248; deficiency of work, 244-45; detention of, 246-48; Farm Hall, 333; in France, 213; Heidelberg group, 231-33; Hiroshima, reaction to, 333-37; Joliot-Curie and, 213-215, 246; return to Germany, 338-340; search for, 197-98; in U.S. custody, 231; after V-J Day, 337-340; and war effort, 230, 245
Germany, 141, 154; Allied bombings, 246; Allied invasion, 199-200; Alsos in, 230-49; Ardennes offensive, 222-23; atomic energy program, 172, 185-97, 199-200, 210, 213-15, 222, 231-33, 239-40, 244-48, 335-337, Heidelberg group, 231-33; in Belgium, 34; DE, 231; division into four zones, 233; Fuchs, 143; General Staff, 231; Italian scientists and, 209-10; Jews, 200; Louvain Library, 367; Ministry of Education, 213, 245; Ministry of Trade and Finance, 219n; "Oak Ridge," 218; surrender, 265; uranium, 218-219, 219n, 220; U.S. bomb plan, 184 (see also Nazis)
Gertner, Wolfgang, 214, 232-33, 239
Giles, Barney M., 278; Hiroshima, 323, 328
Giordani, Dr., 209

Göring, Hermann, 245
Goudsmit, Samuel A., 207-8, 210, 213, 230, 334-35, 336; German scientists, 242
Grand Coulee Dam, 73, 74, 89
Graphite, 15n; pile, 128
Gray, Gordon, 63
Great Bear Lake, 178
Great Britain: atomic project code name (see Tube Alloys); Combined Development Trust, 174n; Combined Policy Committee, 136; contributions to Manhattan Project, 406-8; -French atomic agreement, 224-25, 228; gaseous diffusion process, 117-18; German bomb threat, 199; German scientists in, 247-48; intelligence liaison with U.S., 194-198; radium shipped to, 34; security measures, 143-44; Supply Council, 175; Travancore negotiations, 184; treason, feeling on, 144; tripartite agreement, 170-76; Tube Alloys Project, 196, 212; uranium ore, 33; -U.S. interchange, 114, 125-37, 227; -U.S. postwar collaboration, 401-8; voluntary international censorship, 6
Greenewalt, Crawford H., 52, 52n, 54; liaison, 79-80
Grinnell Corp., 122
Grogan, Les, 77
Groves, Leslie R., 13, 73; a-bomb, 415; Alamogordo, 288-304, 433-40; Alsos, 192-94, 207, 217, 219; Arnold, 253-58; atomic energy intelligence, 185-91; atomic policy, 380, 391, 395-96; bomb control, 271; brigadier general, 23; British-French agreement, 225-29; Bush, 20-22; on combat, 29-30; Combined Development Trust, 170-71, 174, 174n, 176-178; on command, 371-72; committee selection, 162-63; Conant, 162, 200; on Condon, Edward, 156; director of Project, 5, 23, 417-18; family, 21, 320-21; funds, 361; German scientists, 338-39; Hanford, 76-77; Hiroshima, 315, 317, 319-31; irreplaceability, 31-32; isotopes, 385-86; Japanese cyclotrons,

367-69, 371; Japanese invasion, 264; Joliot-Curie, 234; Laurence, William L., 325-26; Leahy, 271-72; legislation, 389, 391; LeMay, 283-284; Marshall, George C., 201, 201n, 230, 278, 324; May-Johnson Bill hearings, 391, 392; mistakes, 30, 258-59, 311-12; Nagasaki, 342, 343; Nimitz, 279; operational plans, 306-14; Oppenheimer, 61, 63, 65-66, 156, 160, 168, 377-78, 381-84, at Alamogordo, 288ff.; Parsons, 160; Patterson, 359, 371, 375-76; postwar developments, 403-5, 411; precautions, 269; press censorship on name, 147; procedures, 201, 201n, 320; Roosevelt, 227; safety regulations, 110; security rule, 140; Smyth Report, 348-51; statement to House Military Affairs Committee, 440-41; statement to Special Senate Committee on Atomic Energy, 441-42; Stimson, 273-76, 302-304; tact, 20; targets, 267, 273-76; third bomb, 352-53; transition period, 373-84; Truman, 266; uranium, search for, 181-84; von Neumann, 268; Washington headquarters, 27; Wilson, Roscoe C., 255-56 (see also Operation Harborage)
Groves-Anderson Memorandum, 405, 406
Guam, 275, 277, 278-79, 280, 311, 312; correspondents, 347; military operations, 280
Guarin, Paul L., 181-83
Guderian, Heinz, 231
Gueron, Prof., 227
Gun-type bomb, 157, 158, 163, 254, 260; testing, 288
Gunn, Ross, 23

Hahn, Otto, 215, 231, 240, 246, 337; capture of, 241; Hiroshima, reaction to, 333-35; Nobel Prize, 339-40
Haigerloch, 231, 233, 241; pile, 242
Half-life (radioactive isotope), 82
Halifax, Lord, 439
Hall, A. E. S., 73, 74
Ham, R. C., 209, 242

Hambro, Sir Charles, 174n, 177, 182; Stassfurt mission, 237, 238
Handy, T. T., 188, 188n, 309, 309n, 312-13; Hiroshima, 320-21; military in Project, 374-76; third bomb, 353
Hanford, 26, 29, 64, 68-77, 78-93, 302; deionizing plant, 81; design, 78, 80, 82-85, 88-89; developmental problems, 83; electric power, 89; industry involvement, 44n; labor trouble, 101-2; land procurement, 75-77; living accommodations, 89-90; medical program, 422-24; morale, 90-93; pile, 78; plutonium, 163, plant site, 70, 73, 74-77; press security, 147; procedure with subcontractors, 103; radiation, tolerance dose, 87-88; reactors, 41 (see also design); Red Cross, 92; religious facilities, 93; security, 145; Senators at, 365-66; separation process, 79; water-cooling, 81; women at, 90-93
Harmon, J. M., 154n
Harmon, M. F., 278
Harrington, Willis, 46, 47, 48, 49
Harrison, Eugene L., 235-36, 241
Harrison, George L., 174n, 177, 275, 302, 303; Hiroshima, 319, 321, 324; Interim Committee, 389n; Smyth Report, 350
Harteck, Dr., 232, 233, 246; capture of, 244
Harvard University, 54, 149, 156, 162n, 163, 164n
Hawkins, David, 154n
Hawley, Paul R., 202-5
H-bomb, 145, 158, 158n, 387
Health safeguards, 86, 420-24 (see also under process)
Heavy-water, 15, 15n, 17, 188-89; Canadian project, 15, 17, 18, 136; French, 224; Germany, 231, 242, 337; Italian scientists, 209-10; Norway, 231
Hechingen, 217, 221, 231, 233, 240, 242, 246; capture of, 241; in French zone, 233 (see Operation Harborage)

Heidelberg, Alsos in, 231-33, 239, 242
Heisenberg, Werner, 216-17, 231, 233, 239, 240, 242, 244, 246, 247, 339; capture of, 243-44; Hiroshima, reaction to, 333-37
Helium-cooling (see under Piles)
Hermanns, Ilse, 220
Hilberry, Norman, 39, 49, 81, 418
Hill, the, 152, 153 (see also Los Alamos)
Hill, Tom B., 277, 281
Himmler, Heinrich, 245
Hiroshima, 21, 28, 40, 259, 308, 316-332; air-sea-rescue plans, 315; bomb, 110, 157, 306, 309, assembly, 317, production, 379; casualties, 319, 324; crews, 319; devastation, 319; Farrell report, 323; German scientists' reaction, 333-37; log of flight, 318; military objective, 316; news ban on, 346-47; observer planes, 316, 326; photographs, 316-317; planes, number of, 316; Presidential release, 325; report to Marshall, 312; result of, 319; strike (bombing), 346; Smyth Report, 351; target, 272-73, 275; U-235 for bomb, 96; visual bombing, 316; weather, 315, 317
Hirschfelder, Joseph, 291
Hitler, Adolf, 200, 215, 222, 265, 266; New Order, 5; scientists, 197
Hitlerism, growth of, 6
Hodgson, John S., 426n
Hoover, J. Edgar, 138
Hopkins, Harry, 125, 130
Howard, N. R., 146
Howe, Mr., Combined Policy Committee, 136n
Hubbard, Dr., 436
Hughes, A. L., 154n, 164n
Hull, Cordell, 171
Hull, J. E., 234-35; Japanese cyclotrons, 370
Hungary, 158n
Hyde Park aide-mémoire, 401-2
Hydrogen bomb, 145, 158, 158n, 387

Ickes, Harold L., 289; U.S. atomic development, 389-90

Ihee, Dr., 220, 221
Imperial Chemicals, Ltd., 128
Imperial College of Science and Technology, 33
Imperial Trust for the Encouragement of Scientific and Industrial Research, 224
Implosion, 145, 147-48, 157; bomb, 254, 260, 269, 305; theory, 157-58
India, 32, 183, 196
Indianapolis, 306
Information exchange, U.S.-British, 114, 125-37, 227
Ingles, H. C., 321
Insurance, 57
Intelligence, 185-91; Project-London liaison, 194-98 (see also Alsos and Military intelligence)
Interim Committee, 327, 327n, 381, 389, 389n, 393
International Brotherhood of Electrical Workers, 99
International negotiations: Project, 27; on uranium, 184 (see also Combined Development Trust)
International Nickel Co., 407
International relations, xii; Roosevelt personal handling of, 35
Invasion (Western Europe), 194, 199, 207; Groves and, 206; intelligence opportunities from, 207ff.; radioactive poisons threat, 199, 201
Ismay, Gen., 206
Italian scientists, 190, 209-10
Italy, 141; Alsos in, 191-98, 207, 208-9; American intelligence activities, 190-98; armistice, July, 1943, 209; atomic project, 193; Navy, 193; scientific research, 209
Iwo Jima, 285, 316; emergency facilities, 281; Japanese on, 281

Jadwin, Gen., 162
James, Edwin L., 325-26
Jansen, Dr., 220-21
Japan, 124, 141; American air raids on, 316, 325; Army, 272, headquarters, 316; atomic atack on, xiii, 293; atomic program, 187; bombing, 260, 298; cyclotrons, destruction of, 367-72; decision to use bomb

against, 265-66; first bomb, 308, 313-14, date for, 304, 305; invasion, 263, 264; Leahy on, 272; Navy, 272; operations against, 309-11; Pacific Coast threat, 65; Potsdam ultimatum, 304; psychological warfare against, 346-47; second bomb, 308; surrender, 346, 353, 354; targets, 267, 268, 272-76; U.S. in bombing range, 286; U.S. Occupation Forces, 370 (*see also* Hiroshima *and* Nagasaki)
Japanese scientists, 187
Jepson, Morris, 317
Joachimsthal mines, 197, 210, 240
Johnson, John R., 193, 209
Joint Chiefs of Staff, xi, 206; bomb: control, 271, knowledge of, 272; enemy equipment, 368; German division, 233; Japanese invasion strategy, 263-64; targets, 287
Joliot-Curie, Frédéric, 6, 33-34, 211-215, 224-29; German scientists and, 246; Gertner and, 233; Groves and, 234; knowledge of U.S. atomic effort, 214, 224, 226-27; Sir John Anderson, 228
Jones, Edwin L., 428
Jones, J. A. Co., 112
Jones & Lamson, 162n
Jumbo, 288-89, 288n, 297

K-25 project, 111, 116-19; British scientists and, 117-19, 127-28, 405, 407; cleanliness control measures, 116-17; costs, 117n; plant, 121; power needs, 112; sabotage, 113; special conditioning area, 116; work force, 117n (*see also* Gaseous diffusion process)
Kaiser Wilhelm Institute (Berlin), 7, 213, 214, 215-16, 231, 245, 337
Katanga ores, 35 (*see* Union Minière *and* Uranium ore supply)
Kearton, C. F., 119
Keitel, Wilhelm, 213
Keith, P. C., 112, 114, 116, 428
Keller, K. T., 114-15
Kellex Corp., 111-12, 116, 428; British scientists and, 119
Kellogg Co., M. W., 111

Kennedy, J. W., 164n
Kettering, Charles, 331-32
Kilgore, Senator, 365
King, Ernest J., 121, 277, 280; power, 281
King, Mackenzie, 403, 410
"Kingman," 260
Kirkpatrick, E. E., 279-80, 281, 353
Kistiakowsky, George, 163, 164n, 295, 436, 437
Knoxville, Tenn., 14, 25, 69, 99; press censorship, 147
Kokura, 308, 316, 342-43, 345; Arsenal, 272, 316
Korshing, Dr., 246; reaction to Hiroshima, 335
Kowarski, Lew, 212, 214, 224-25, 227
Kuhn, Richard, 231, 232
Kyle, William, 234, 303; Smyth Report, 350
Kyoto, 316; Imperial University, 367; Stimson and, 352; target, 273-76
Kyushu, 263; invasion date, 264

"Laboratory for the Development of Substitute Materials" or "DSM," 13
Lane, James, 241
Lansdale, John Jr., 139, 140, 173n; Operation Harborage, 235; Stassfurt mission, 236, 237
Laurence, William L., 35-36, 325-27, 347
Lavender, R. A., 45, 418-20
Lawrence, Ernest O., 8-9, 51n, 61-62, 214, 348, 377, 430, 443; Alamogordo, 303; cyclotrons, 95n, 105; electromagnetic separation, 96
Leahy, William D., 401; disbelief in bomb, 271-72; *I Was There*, 401
Leclerc, Jean, 212
Lee, Frank G., 174n
Lee, Gen., 206
Leith, C. K., 174n
LeMay, Curtis, 283-84; bomb delivery, 260-61, 284; Hiroshima, 315, 317, 328, 329-30; Nagasaki bomb, 344; operational plans, 308, 311, 312
Letter of intent, 57, 58, 111

Lewis, W. K., 52n, 162n
Liaison, within Project, 167-68
Lilienthal, David E., 411
Linsay plant (Chicago), 183
Lipkin, William P., 361
Little Boy bomb, 260, 260n, 262, 341, 354; assembly, 282n, 343; explosive force, 269-70; Hiroshima, 317 (see also Thin Man bomb)
Llewellin, John, 136n
Lobito, Portuguese Angola, 34
Lockhart, Jack, 146, 301, 325
London: American Embassy, 196; Alsos in, 208, 215; Project liaison office, 194-98; tripartite agreement, 170-76
Los Alamos, 26, 29, 96, 149-69, 254, 261; aircraft safety studies, 286; Alberta division, 262; Army at, 154-55; Ballistics Group, 261; Bikini tests, 383; British at, 408; censorship of mail, 168-69; Certificate of Appreciation, 355; commanding officer, 154; coordinating council, 166-67; Delivery Group, 261; dependence on Project, 149; director, 154 (see Oppenheimer); engineering, 159; equipment procurement, 155-56; Explosives Division, 163-64; First Technical Service Detachment, 282-83; governing board, 164, 164n; liaison with Project, 167-68; living conditions, 153, 165-66; Navy at, 160-61; ordnance program, 159-60; peace talks, 353; personnel, 154-55, 289; postwar reorganization, 377-79, 381-83; press security, 147; problems, 164-66; recruitment, 149-50; research organization, 158-59; Review Committee, 162, 162n; rumors about, 151-52; scientists, 383, 430; second implosion bomb test program, 341; security, 160-62, 168-169, 307; site location, 66-67; Smyth Report, 349; speed, necessity of, 153; targets, 267, 268; technical series, 381; thermal diffusion plant, 379; Tinian, 282-83, 297; University, 381; uranium, 110, 124; Wen-
dover Field, 259 (see also Project Y)
Lothian, Marquess of, 126; aide-mémoire, 126, 127
Lovett, Robert A., 328
Lowe, Frank, 365
Lowlands, Alsos in, 215, 219
Luce, Clare Boothe, 392
Luedecke, A. R., 395
Lyons, George, 299, 384

MacArthur, Douglas, 263, 264, 286-287, 309; Japanese cyclotrons, 368-369, 370, 371
McCarthy, Col., 321, 322
McCarthy, Frank, 236
McCloy, John J., 411
McCone, John, 395
McCormack, John W., 362
McMahon, Brian, 393-94
McMahon Act, 406
McMillan, E. M., 151, 164n
McMorris, Charles H., 277
NcNally, James G., 428
McNarney, Joseph T., 207, 375
McNinch, J. H., 204, 205
Madigan, Jack, 360
Mahon, Congressman, 364
Makins, Roger, 405; Smyth Report, 350
Mallinckrodt, 16
Manhattan District, 174; intelligence activities, 185
Manhattan Engineer District (see MED)
Manhattan Project, ix, 36, 129, 236; Air Force support of, 257-58; Alsos in Germany, 230; American lives and, 264; British contributions to, 406-8; British Scientific Group, 136-37; Bush liaison with White House, 125; Churchill, 408; code names, 191; courtesy policy, 186; Groves in charge of, ix; headquarters, 27-28; Los Alamos liaison, 167; mission, 140; morale, 90-93; number in, 414; officer messengers, 202; reasons for telling story of, x; scientists, postwar, 410; security, 138-45; salaries, 150; success factors, 414-15; speed paramount,

141; use of bomb, 265; workers' lives, 89-90 (see MED and under names of plants)
Manley, J. H., 151, 283
Maris, Mrs. (see Steinmetz, Beatrice M.)
Marshall, George C., xii, 10, 20, 23, 24, 49, 70, 102, 135, 264; Alamogordo, 304; Alsos, 230; atomic energy intelligence, 185, 186, 189, 190; bomb: control, 271, delivery, 307-8; Congressional briefing, 362-363; Groves, 201, 201n, directive, 266-67, operational plan, 309-11; Hiroshima, 319-22, 324, 328, 330; policy on atomic energy, 391; at Potsdam, 309, 309n; power, 281; radioactive warfare threat, 201-2, 205-6; scientific intelligence, 207; Stassfurt mission, 239; targets, 268, 273-76; third bomb, 352-53 (see also Operation Harborage)
Marshall, James C., 10-17, 19, 21, 23, 25, 26, 27, 29, 74, 107
Martin, Joseph W., Jr., 362
Matthias, F. T., 72-73, 74
Maurer, Werner, 214
May-Johnson Bill, 391, 392-93, 394, 443-44
MED, xi, xii, 29, 54, 60, 70, 72, 160, 180, 208, 257; Alsos, 222; atomic intelligence, 185, 189; bases, 19; beginnings, 3-18, 19-32; Bikini tests, 384; Bush-White House liaison, 125; characteristic (example), 112; command channels, xiii; Congress, 359-66; "creeps," number of, 139n; engineering difficulties, 19-20; FBI cooperation, 139, 140; functioning, 21; funds, 385; German scientists, 213; Groves, director of, 19, confidence in, 371-72; intelligence, 193; major effort of, 95; medical section, 420-24; naming, 17; ore supply, 36; Pacific Coast installations, 290; patent work, 418-20; personnel limitations, 266-67; press security, 146, 147; priorities, 17, 19-20, 22-23; radioactive isotopes, 385-86; Scientific Intelligence Mission (see Alsos);

security, 138-39; Sengier, 177; silver, responsibility for, 108-9; Smyth Report, 348; targets, 268; thermal diffusion process, 119-21; transition period, 373-88 (see Manhattan Project)
Medical research, 420-24
Mehring & Hanson Co., 122
Meitner, Lise, 5
Merritt, Phillip L., 179
Metal Hydrides, 16
Metallurgical Laboratory, 9, 9n, 14, 39, 57; Greenewalt, 79; Hanford site, approval of, 76; Los Alamos liaison, 167-68 (see also Chicago lab)
Metallurgical Project, 418
Metals Reserve Corp., 34
Military Intelligence, 185-91; Alsos, 191-98, 207-23, 230-49; Project, 27; "ungraded," 216, 216n
Military Policy Committee, 23-25, 43, 97, 150, 159-60, 162, 200; Akers' proposals, 129; British-U.S. interchange, 135; Bush and, 125; demise, 380; in operational plan, 311; opposition to Oppenheimer, 62-63; report to Roosevelt (Dec. 1942), 129-30, 131; Smyth Report, 348; targets, 267; uranium ore supply, 170
Mills, Earle, 388
Minissini, Vice Admiral, 193
Mission orders, 27
MIT, 52n, 162n, 183, 418
Mitchell, Dana P., 164n, 431
Mitchell, Senator, 365
Monazite, sources of, 183-84
Monsanto Chemical Co., 411
Montgomery, Bernard L., 339
Moynihan, John F., 347-48
Murphree, E. V., 52n

Nagasaki, 40, 259, 308, 316, 341-46; bomb, 157; bombing, 319, 342-46, date of, 343; casualties, 346; destruction, 345, 346; importance of, 343; news dispatch, 327; photo planes, 342, 346; POW camp, 312-13; shock waves, 346; strike plane,

342, 343; target date, 341; weather, 344, 346

National Advisory Committee on X-Ray and Radium Protection, 87

National Defense Research Committee (see NDRC)

Naval Research Laboratory, 22, 23, 119, 120, 121

Navy, U.S., 6, 7, 24, 64; Bikini tests, 384; bombing action, 278, 279; Bureau of Ships, 388; intelligence, 185, 186, 190, 208; Los Alamos, 160-61; and Project, 22, 280; scientists, 193; thermal diffusion process, 119-21; uranium research, 23

Nazis, 197, 217, 221, 335; Norway, 188-89

NDRC, xi, 7, 8, 45, 127, 162n

Neddermeyer, S. H., 157-58

Nelson, Donald, 22-23

Netherlands East Indies, 184

New York, 14; Project suboffices, 29; uranium ore shipped to, 34

New York *Herald Tribune*, 369

New York Times, 325-26

Nichols, K. D., 12-16, 19, 20, 27, 30, 36, 49, 72, 107, 116, 117, 124, 177, 364, 373, 387, 395; District Engineer, 29; interchange of information, 135; Oppenheimer letter, 110; replaceability, 31-32

Niigata, 308, 342; target, 273

Nimitz, Chester W., 263, 264, 286-87, 309; power, 281; Tinian, 277-80

Nobel Prize, 9, 33, 39, 62; status of holders, 62

Nolan, Capt., 282n, 305, 306

Norstad, Lauris, 267, 267n, 268, 272; Hiroshima, 317; take-off base, 278

North Africa, invasion, 3

Norway, 187-89; heavy water, 231; underground, 216

Nuclear energy, ixn, 5, 387 (see Atomic energy)

Nuclear program, production methods, 10

Oak Ridge, 25-26, 29, 64, 68, 69, 94-124, 171, 424-27; accidents, 110; atomic engineering training, 387-388; complexity of problem, 123-24; Congressmen at, 363-65; electromagnetic plant, 94, 95-97, 103-11 (see also Y-12); Fuchs and, 144; gaseous diffusion process, 111-19; labor force, 99-100; leadership, 103; medical program, 423-24; morale, 102-3; operational plan, 124; postwar employment, 379; press security, 147; procedure with subcontractors, 103; radiation training, 385; security, 145; school system, 526; thermal diffusion plant, 94, 95, 121-23; union leaders, 100

Occupation Forces, U.S.: Europe, 375; Japan, 370

Office of the Chief of Engineers, 13, 74, 100

Office of Scientific Research and Development (see OSRD)

Okinawa, 276; Japanese on, 281

Old, Bruce S., 193

O'Leary, Mrs., 21, 28, 29, 302, 303

ONI (see Navy, intelligence)

Operation Crossroads (see Bikini tests)

Operation Harborage, 234-36, 237, 240

Operational plans: directive, 308-9; Groves' memorandum, 309-11; organization, 311

Operations Planning Division (OPD) of General Staff, 266, 267

Oppenheimer, J. Robert, 60, 61, 62, 65, 124, 149, 151, 153, 254; Alamogordo, 288-98, 436-38; Conant letter on Los Alamos, 150-51; Los Alamos: director, 154, 154n, 155, 156, 158, 159, 161, 164, 164n, 166, 168, postwar program, 381-84; loyalty, Groves on, 63; Nagasaki bomb, 343, 344; Nichols uranium estimate, 110; ordnance development, 163; postwar developments, 355, 411; Project Y, 63; resignation, 377; salary, 150; security risk, 63; Smyth Report, 349; targets, 267, 269-70; thermal diffusion process, 120

Oranienburg, 221; bombing of, 230-31

Osaka Imperial University, 367
Osenberg, Prof., 239
Oslo, 188
OSRD, xi, xii, 23, 24, 150, 208, 418; Alsos, 210, 222; atomic program, 194; establishment, 127; S-I Executive Committee, 111, 128; scientists, 190, 193; Smyth Report, 348; Uranium Committee, 7, 9
OSS, 208; in Alsos, 216, 217; atomic energy, 185, 186
Overlord, 206

Pacific Coast, radio announcement of Alamogordo test, 301, 302
Paris: atomic research, 233 (see also Joliot-Curie); Alsos in, 211-15; German occupation, 225; liberation of, 212
Parish, W. A., Jr., 299, 300, 301
Park, Richard, 73
Parker, T. B., 14
Parsons, Capt., 284, 294-95, 314, 317-18
Parsons, William B., 140
Parsons, William S., 160-61, 162, 164, 164n, 261, 268, 282n; Bikini test, 383; decision making, 343; Hiroshima, 315, 322, 323, 327; Nagasaki bomb, 344
Pash, Boris T., 192-93, 207-10; Brussels, 218; chief of German mission, 243; France, 211-12, 220; German scientific roundup, 239-43; Operation Harborage, 236; Stassfurt mission, 237; Strasbourg T-force, 221; uranium ore capture, 219
Patents, 418-20; claims, 45, 418-20; du Pont rights, 58; French, 224-25, 227
Patterson, Robert P., 99, 100, 265, 359-60; atomic policy, 380; Japanese cyclotrons, 367, 368, 370, 371; legislation, 390, 391; military in Project, 376; postwar developments, 402, 403, 405; statement to House Military Affairs Committee, 440; statement to Special Senate Committee on Atomic Energy, 442-44
Patterson-Brown plan, 99

Pearson, Lester, 403, 405
Peenemünde, 197
Pegram, George B., 6, 127
Peierls, R., 117, 119
Penney, William G., 268, 282n; decision making, 343
Pentagon, 4-5; construction, 72; Smyth Report, 351
Perrin, Francis, 211
Perrin, Michael, 196, 213, 228; Stassfurt mission, 237
Personnel Security Board, 63
Pétain, Henri, 246
Peterson, Arthur V., 54, 124, 202, 205, 206
Phelps Dodge Copper Products Co., 108
Philadelphia Navy Yard, 120, 121, 122, 385
Philippines, 65
Pile, 38; cooling methods, 41; control system, 88; design, 82, 85-86; Fermi message, 54; helium vs. water-cooled, 80; location, 52-54, 85; number needed, 71; radioactive isotopes, 385; shielding for, 83-84; test (Dec. 2, 1942), 54-55; water supply for, 84-85 (see also Reactor)
Pilot reactor plant, 14-15
Planck, Max, 216
Plutonium, 8; Alamogordo test, 120; Germany, 197; Los Alamos, 163; military possibilities, 8-9; obtaining, 10; process, 16, 56; production, 15n, 72, 305; transportation of, 341
Plutonium plant: health hazards, 50; priority, 14; site, 70-77 (see Hanford)
Plutonium project, 10, 38-59; amount needed per bomb, 39-40; Compton estimates, 55-56; construction, 42, 43, 44, 47; du Pont, 42-44, 46-52, 52n, 54, 55-59; engineering, 42, 43, 47; operation, 42, 43, 47; phenomenal achievement, 38; production phase, 43; quantity production, 40; theoretical basis, 39
Potsdam Conference, 248, 275, 302, 303-4; Truman at, 304; ultimatum, 292-93, 304, 311, 392

Press, American, cooperation of, 148
Press censorship, 146-48; banned word list, 146; principles of, 146
Press releases: on a-bombings, 346-348; Alamogordo, 327; Hiroshima, 325-31; Pacific Theater, 346-48
Price, Byron, 146, 301
Princeton University, 348
Project, the, xiii-xiv, 156; a-bomb design (*see* Project Y); Centerboard, 314; close of on Tinian, 353; Congress and, 359-66; construction program, 15; cooperation as success factor, 69; copper problem, 107-8; estimated cost, 97; decision making, 81, 343-44; early phases in U.S., 245; extravagance, 359-60; Fuchs mistake, 143; funds, 15-16, 360-66; Groves in charge, 5; hazards, 57, 58; information, handling of, 22; insurance, 57; internal organization, 28-29; Japanese cyclotrons, 369; letter of intent, 57, 58; members of, 332; military importance, 49, 50, 51; military-scientific friction, 45-46; military staffing, 373-76; naming, 13; new organizations involved in, 47; operational procedure, 308; personnel recruitment, 16-17; pilot reactor plant, 14-15; policy supervision, 23-25; power needs, 14; press: censorship, 146, releases, 325-31; priority, 16; production plants, 72; raw materials (*see* Uranium ore supply); regular officers and, 29-30; reorientation (1942), 10-11; replacements, 30; scientific staff, 374-75, demobilization of, 376; secrecy, 280; security, 47, 69, 80, 100-1, 140, 185, 332; site, 13-14, 16, 17-18, 19, 25-27; Smyth Report, 348; speed, necessity for, 98; supply contracts, 16; transactions, handling of, 37; transition period, 373-88; uncertainty, 72 (*see also* Air unit *and* MED)
Project Y, 60, 149-69; construction requirements, 155; Oppenheimer and, 61-66; site, 64-67 (*see* Los Alamos)

Pumpkin bombs, 285; missions over Japan, 353; planes for, 342
Purnell, W. R. E., xii, 23, 24, 160, 186, 277, 280; Nagasaki bomb, 342, 344; operational plan, 311; two bombs theory, 342

Quebec Agreement, 135-36, 143, 225, 227, 228; Article IV, 402; breach in, 225-29; postwar reappraisal, 402-5; use of bomb, 265

Rabi, Isidor, 432
"Race track," 98, 104, 105
Radiation, 269; exposure, 58; hazard, 87-88; training in, 385
Radioactive isotopes, 385-86
Radioactive poisons, Germans and, 199, 200
Radioactive warfare: Germans and, 207; possibility of, 199-205
Radioactivity, 286; hazard, 87-88; in plutonium formation, 85
Radium, 34, 178
Rafferty, James A., 112, 428
Rand, South Africa, 182-83
Rank, in academic world, 5
Raw Materials Board, 34-35
Rayburn, Sam, 362
Rea, Charles E., 423
Reactor, 38; design (*see under* Hanford); Clinton, 41; Hanford, 41; theory, 50
Read, Granville M., 79, 81, 91, 101
Reich Research Council, 213
Review Committee (Los Alamos), 162-63, 162n
Reybold, Eugene, 11, 13, 17, 19, 21
Richland, 89, 92
Rickover, Hyman G., 387-88
Rittner, T. H., 333, 340
Rjukan plant, 188-89, 337
Robins, T. M., 3, 11, 13, 16, 19, 21, 82
Roges GMBH, 219n
Rome, 192, 193; Alsos in, 208-10; U.S. capture of, 209
Roosevelt, Franklin D., xi, xii, 3, 6, 10, 16, 49, 59, 70, 97, 100, 102; British-French atomic agreement, 227; Bush, 125, 131, 134, 135, 136; Cabinet, 126, 128; Churchill, 407;

death, 364, 401; Einstein letter, 7; Executive Order creating OSRD, 127; Groves, 184; Hyde Park *aide-mémoire*, 401-2; international relations, 35; Navy and Project, 22; Quebec Agreement, 135-36, 225; Top Policy Group, 7-8; tripartite negotiations, 171, 172, 174; uranium negotiations, 184; U.S.-British interchange of information, 126, 128, 130-35; use of bomb, 265
Rose, E. L., 162n, 163
Royal Institute (London), 339
Russia, 132; Alsos German operations and, 242; espionage Berkeley lab, 138, 141; France and, 228, 234; -German relations, 143; German scientists, 244, 247, 248, 338-340; international control of atomic energy, 338; Sweden, 184; U.S. atomic development, 389-90, 409-412; -U.S. cooperation, 263
Russians: Fuchs and, 144-45; -Truman at Potsdam, 304

S-1 Committee, 12n, 15, 17, 36, 111, 128
S-1 Project, 171, 172 (*see* Manhattan Project)
Sabotage, 105; bomb test and, 293, 295; K-25 plant, 112-13
Sacks, Alexander, 6-7
Safety precautions, 86-88, 420-24
Sahara Desert, 34
Salton Sea, 259, 261
SAM laboratory, 111, 112, 428
Sandia Base, 374, 382
Santa Fe, 151, 152, 153, 166
Saturday Evening Post, 36
Schriver, Agent, 237, 238
Schumann, Erich, 213
Schurcliff, W. S., 349
Scientific intelligence missions (*see* Alsos)
Scientists: Alsos, 192; limitations on, 168; Los Alamos, 154-55; morality of a-bomb, 334; postwar, 410; Project, 374-75, demobilization, 376; security, 377 (*see also under* nationality)
Seaborg, Glenn T., 8, 418

Security, 138-45; aims, 141; Alamogordo, 299; basic problem, 140; bomb control, 271; breaks, 147-48 (*see also* Fuchs); British scientists, 143-44; drawings, 80; foreign-born employees, 142; Los Alamos, 160-162, 167, 168-69; major consideration, 141; Oak Ridge, 100-1; press (*see* Press censorship); scientists, 377; silver program, 109; slacking of, 353; Smyth Report, 349-50; test delay and, 293; Tinian mission, 280; Tinian-Los Alamos, 282-83; top government officials, 360; yardsticks, 142-43
Selective Service, 428
Selfridge, George, 181
Semi-works, 48, 48n, 50, 72; plutonium process, 68
Sengier, Edgar, 33-37, 170; Combined Development Trust, 175-76, 178; MED, 177, 178; Medal of Merit, 178
Separation plants: design, 78-79, 85; number needed, 71 (*see* Oak Ridge)
SHAEF, 202, 208, 237, 246; G-3, 235
Shasta Dam, 73
Shell Oil Co., 181
Sherman, Gen., 28
Shinkolobwe Mine, 33, 34, 176; unit production cost, 178
Sibert, E. L., 208, 237
Silver program, 107-9
Simon, F., 117
Skidmore, Owings and Merrill, 425
Smith, Clayton F., 14-15
Smith, Cyril, 430
Smith, F. J., 230-31
Smith, Walter Bedell, 202, 208; Alsos, 219; Operation Harborage, 235, 240; Stassfurt mission, 237
Smuts, Jan, 183
Smyth, Henry D., 348-50, 352
Smyth Report, 348-49
Snyder, Congressman, 364
Somervell, Brehon, xi, 3-4, 5, 13, 20, 23, 25, 417; Congressional investigations, 70
South Africa, uranium, 37, 182-83
Soviet Union (*see* Russia)

Spaak, Paul Henri, 176
Spaatz, Carl, 206, 231, 272, 346; Hiroshima, 323, 328; operational plans, 308, 311, 312-13
Spedding, F. H., 418
Speer, Albert, 231
Sproul, Robert G., 149
Stagg Field, 53, 54
Stalin, Josef, 336
Standard Oil Development Corp., 52n
Stark, H. R., 206
Stassfurt mission, 236-39
Staten Island, uranium storage, 34, 35, 36-37
Stearns, J. C., 268, 418
Steinmetz, Buena M., 90, 91-93
Stettinius, Edward, 35; uranium negotiations,' 184
Stevens, W. A., 289
Stimson, Henry L., xii, xiii, 3, 8, 20, 23, 24, 25, 30-31, 49, 70, 75, 102, 126, 174n, 264; atomic energy orders, 171; bomb control, 271; Bush, 125; Combined Policy Committee, 136n; Congressional briefing, 362-63; Eisenhower, letter to, 208; Groves (on Alamogordo), 302-4; Hiroshima, 319, 324; Hyde Park aide-mémoire, 402; information on bombing, 308; Interim Committee, 389, 389n; international controls, 408-9; Kyoto, 352; operational plans, 311, 313; at Potsdam, 303-4; power, 281; Presidential representative, 226; press release, 327; resignation, 359; Roosevelt, 227; Smyth Report, 350; targets, 273-76; U.S.-British interchange, 128, 131, 132, 133, 135; use of bomb, 265 (see Operation Harborage)
Stine, Charles M. A., 46, 47, 48, 49, 52, 55-56, 56n
Stone and Webster, 12, 13, 14, 15, 17, 42, 43, 96, 98, 178, 425, 427; Los Alamos, 151; Oak Ridge, 102
Stonehaven, Lord, 33
Strasbourg, Alsos in, 221-23, 230
Strassman, Fritz, 215, 240, 337
Strauss, Lewis, 395
Strikes, 100

Strong, G. V., 138, 139, 186; atomic intelligence activities, 188-91, 194, 195
Styer, W. D., xi, xii, 4, 4n, 5, 10, 12, 13, 16, 19, 20, 21, 23, 185-86, 380; Military Policy Committee, 25
Submarine, atomic-powered, 388
Super-bomb, 383
Sutherland, R. K., 370
Sweden, 188; Alsos in, 217; anti-Nazi scientists, 197; uranium ore, 184
Sweeney, Charles W., 343, 344-45, 346
Switzerland, 214; Alsos in, 217; anti-Nazi scientists, 197; intelligence from, 216, 217
Synthetic rubber program, 17
Szilard, Leo, 39, 41

Taber, Congressman, 364-65
Tailfingen, 231, 241
Tamm, E. A., 138
Targets: choosing, 266-76; cities selected, 272-76; committee, 268, 269, 270, 271, 272-73, 275; POW camps, 312; practice landings, 270; reservation of, 287; visual bombing, 270
Taylor, R. A., Jr., 307
Teller, Edward, 158n
Tennessee Eastman Co., 96-97, 98-99, 110, 427-28
Tennessee site, 24-25, 68
Tennessee Valley Authority, 14 (see TVA)
Thermal diffusion process, 119-23; isotope separation column, 121; S-50 security, 121; site location, 121-23
Thin Man bomb, 157, 163, 255, 309 (see also Gun-type bomb)
Thomas, Charles, 411
Thomas, Elmer, 363
Thorium, 183-84, 196; search for, 180-84
Thornberg, T. R., 427
Tibbets, Paul W., 258, 259, 260, 311, 314, 317, 318, 354; take-off base, 278
Tinian, 278-87, 316, 326; 509th

Group on, 283; bomb delivery to, 306-7, 308, 309; choosing, 278-79; closing, 353; during peace talks, 353; field crews, 282, 282n; Los Alamos group, 297; Nagasaki bomb, 341, 342, 343; security, 280, 283

Tizard, Henry, 33, 34, 126-27

Tokyo, 263; cyclotron news story, 369-70; Institute for Physical and Chemical Research, 367; MacArthur headquarters, 368, 369

Tolman, Richard, 44-45, 63, 116, 162n; Alamogordo, 303; decision making, 343; Smyth Report, 349-50

Top Policy Group, 7-8, 10

Travancore, India, 183-84

Traynor, H. S., 171-74, 176

Trinity test, 288, 354 (see also Alamogordo); site, 385

Truman, Harry S., xiii, 229; atomic legislation, 391, 394, 395; Cabinet, 390; Congressional visit to Oak Ridge, 364; Groves (Alamogordo report), 304; Hiroshima press release, 330-31; information on bombing, 308; international controls, 409-10; Japanese invasion, 264; postwar developments, 403-6; Potsdam ultimatum, 311; press release, 327; responsibility for use of bomb, 265-66; Senate Investigation Committee, 365; Smyth Report, 351; targets, 275

Truman-Attlee-King Declaration, 403-405, 410

Tube Alloys Project, 196, 212, 213, 225, 402; Canada, 225, 228; Washington, 228

Turner Construction Co., 425

TVA, 425, 428; power for, 112

Twining, Nathan, 323, 328

Tyler, G.R., 152-53, 154n, 161-62, 164

U-235, 8, 9, 39, 94, 95, 119, 167, 305; amount needed for bomb, 124; in Hiroshima bomb, 40, 305-8; production, 305; separating, 10

U-238, 8, 9, 94, 240

Union Carbide and Carbon Corp., 112, 113, 114, 121, 179, 428; uranium search, 180-84

Union Minière, 33, 35, 37, 170, 175, 178, 197; German purchases from, 218-19, 219n; ore in Europe, 218-220, 236-39; purchase agreement with, 176

Unions: Ferguson Co., 120n; security and, 100

United Nations: Article 102, 405; nuclear controls, 403, 406

United Press, 369-70

United States: Britain: interchange, 125-37, 144, 227, postwar collaboration, 401-8; Canadian interchange, 130; Combined Policy Committee (see Combined Policy Committee); committed to use of bomb, 265; German atomic energy program, 234ff.; German bomb threat, 199; negotiations for uranium, 184 (see also Combined Development Trust); patent laws, 45; postwar position, 275, 408-12; radium shipped to, 34; responsibility for hazards, 57, 58-59; risks during war, 199; scientists, 5-6; -Russia: atomic position and, 409-12, cooperation, 263; tripartite agreement, 170-76

University of California, 8, 9, 17, 60, 96, 149, 154, 154n, 164n; Los Alamos project, 155; Radiation Laboratory, 96, 99; scientific discoveries, 22; U-235 research, 51, 51n (see Berkeley lab)

University of Chicago, 9, 39, 53, 385; personnel policy, 382 (see Chicago lab)

University of Hamburg, 246

University of Illinois, 156

University of Munich, 242

University of Strasbourg, 221, 222

Uranium: American payments for, 176-77; atom, 5; deadline for delivery, 124; first shipment to Los Alamos, 110; Germany and, 196, 218-19, 219n, 242, 338; irradiated, 78, 79; search for, 180-84; separation, 12, 97-98 (see also under

name of process *and* Oak Ridge);
suppliers, 178-79, 197; tripartite
agreement, 170; value, 178
Uranium Committee, 7-9, 12, 348
(*see* S-1 Committee)
Uranium ore: Belgian stockpile, 197;
Congo supply, 37; shipping, 179-80;
source, 33; supply, 33-37 (*see*
Stassfurt mission)
Urey, Harold C., 51n, 62, 111, 127,
430; heavy water, 189; radioactive
poisoning, 200
U.S. Bureau of Standards, 7, 428
U.S. Congress, 107, 176; legislation,
385, 391; MED, 359-66; Project:
briefing on, 362-63, demobilization,
379
U.S. Employment Service, 155, 428
U.S. House of Representatives: legislation, 393, 394; Military Affairs
Committee, 391, Groves, 391-92,
440-41; May-Johnson Bill hearings,
392, Patterson, 440
U.S. Senate, 174; Groves testimony,
395-96; legislation, 393-94; Special
Committee on Atomic Energy, 361,
Groves, 441-42, Patterson, 442-44
U.S. State Department, 34-35, 126;
entry into atomic picture, 184; German division, 233, 234, 235; legislation, 390; postwar developments,
402, 403; Project, first informed of,
35; Russia, 409, 412; Sengier and,
35, 36
U.S. Treasury, 107-9, 177
U.S. War Department, 10, 27, 102;
atomic energy: diplomacy, 171,
policy, 391; Information Office,
347; Japanese cyclotrons, 367, 369,
370-71; observation at bombing,
308; press branch, 351; restrictions
on flight personnel, 313-14; scientists, 193; security, 138, 139

Vaughan, Harry, 365
V-E Day, 248, 249, 265
V-J Day, 59, 337, 354, 365, 373, 376,
382, 390
Vleck, J. H. Van, 162n
von Halban, Hans, 212, 214, 224-25,
226-27, 228

von Laue, Max, 231, 246, 339; capture, 241
von Neumann, John, 343; targets,
267, 268, 269-70
von Weizsäcker, C. F., 222, 240, 241,
242, 246; reaction to Hiroshima,
335, 336, 338
"V" weapon, 197

WACs, 151, 196
Wallace, Henry, 8, 24
War, impossibility of, 414
War Manpower Commission, 155
War Production Board, 16, 22, 89
Wardenburg, F.A.C., 241
Warren, Lindsay C., 58
Warren, Stafford, 298, 299, 384, 421
Washington, D.C., 14; Alamogordo
press release, 327; Capitol, 362;
control, 310, 313, 346-47; Hiro-
shima press release, 325-31; Project
headquarters, 27
Wegener, Al, 99
Welsh, Eric, 196
Wendover Field, 258, 259, 259n, 260,
299; code name, 260
West Point, 21, 30, 195, 374;
Academic Board, 374; Depository,
107, 108
West Stands pile (*see* Stagg Field)
Westinghouse, 16, 56, 97, 103; Re-
search Laboratories, 154
Whitaker, Martin D., 418
White, James C., 428
White, Wallace H., 363
White House, 100, 390; Alamogordo
press release, 327; Hiroshima an-
nouncement, 347; interchange con-
ference, 130-31; Project liaison, 125
Wick, Dr., 209-10
Wigner, Eugene, 39, 418
Williams, R. B., 299
Williams, Roger, 52n
Williams, T. C., 427
Wilson, E. B., 162n
Wilson, Roscoe C., 255, 256-57
Winant, John G., 170-73, 176, 196;
British-French atomic agreement,
225-27
Winne, Harry, 411

Wirtschaftliche Forschungs Gesellschaft (WIFO), 236, 237, 238
Wirtz, Dr., 241, 246, 340
World War I, 30, 58, 151
World War II: American military requirements, 11; German surrender,
  265; outbreak, 6, 34; U.S. entry
  into, 9

Y-12, 95; copper, 107; hazards, 109-
  110; lie detector, 107; plant, 98;
  problems, 103-11; security, 104,
  109; shutdowns, 106; silver, 107-9;
  value of materials, 107 (see also
  Electromagnetic plant)
Yalta Conference, 35, 124, 184, 233,
  263

ABOUT THE AUTHOR

Lieutenant General Leslie R. Groves, United States Army (Ret.), was born in 1896 in Albany, N.Y. He entered the University of Washington in 1913, transferring to the Massachusetts Institute of Technology in 1914. Two years later he left MIT to enter West Point under a Presidential appointment, graduating in 1918.

General Groves is also a graduate of the Army Engineer School (1918-21); Command and General Staff School (1935-36); and Army War College (1938-39).

After he was commissioned as a second lieutenant in November, 1918, he served on various assignments in the United States, Hawaii, Europe and Nicaragua. In 1939 he joined the War Department General Staff, then acted as Special Assistant to the Quartermaster General for the Army construction program. Later he was Deputy Chief of Construction under the Chief of Engineers.

General Groves headed the Manhattan Project almost from its inception in 1942 until 1947, when atomic energy affairs were turned over to the newly created civilian Atomic Energy Commission. He was in charge of all phases of it—scientific, technical and process development, construction, production, security and military intelligence of enemy activities, and planning for the use of the bomb. He was promoted to brigadier general in 1942, to major general in 1944, and to lieutenant general in 1948.

General Groves served, after the end of the Manhattan Project, as the Chief of the Armed Forces Special Weapons Project. He retired from active duty on February 29, 1948.

General Groves has received honorary degrees from the University of California, Hamilton, St. Ambrose, Lafayette, Williams, Hobart, Ripon, and Pennsylvania Military College. He holds the Distinguished Service Medal (U.S.); Legion of Merit (U.S.); Nicaraguan Presidential Medal of Merit; Comdr. Order of the Crown (Belgium); and Hon. Companion, Order of the Bath (U.K.).

General Groves was, until the fall of 1961, a vice president of the Remington Rand Division of the Sperry Rand Corporation. He lives in Darien, Connecticut.